Calculus and Statistics

Michael C. Gemignani

Dover Publications, Inc.
Mineola, New York

Copyright

Copyright © 1970, 1998 by Michael C. Gemignani
All rights reserved.

Bibliographical Note

This Dover edition, first published in 2006, is an unabridged republication of the work originally published in 1970 by Addison-Wesley Publishing Company, Inc., Reading, Massachusetts. The Table of Integrals which appeared on the inside front and back covers in the original edition have been moved to pages 355–358 in the Dover edition.

Library of Congress Cataloging-in-Publication Data

Gemignani, Michael C.
    Calculus and statistics / Michael C. Gemignani.
      p. cm.
    Includes index.
    Originally published: Reading, Mass. : Addison-Wesley Pub., 1970.
    ISBN-13: 978-0-486-44993-7 (pbk.)
    ISBN-10: 0-486-44993-9 (pbk.)
      1. Calculus. 2. Statistics. I. Title.

QA303.2.G46 2006
515—dc22

                                                                                                                   2005040168

Manufactured in the United States by LSC Communications
44993907   2019
www.doverpublications.com

To my parents

Preface

The "knowledge explosion," the interests of efficiency, and the improved comprehension that comes from seeing subject matter in a context in which it is applied, have all been reasons for the development of courses in which topics which used to be studied separately are studied in relation to one another. The present text is for just such a course, and for the reasons cited.

Calculus is becoming ever more necessary as a tool in modern economics, psychology, biology, and a host of other sciences in which it was not a widely used tool in the past. Many sciences, such as biology and psychology, have always used statistics as a tool, but more and more scientists and specialists in fields where statistics was not widely applied are finding a knowledge of statistics valuable or necessary in their work. The time seems right then for a combined calculus and statistics course for those who would like a working knowledge in both of these areas, yet who do not want to spend four or five semesters taking separate courses in each subject. There is, moreover, a natural relationship between calculus, probability, and statistics through which a study of one of these subjects can lend motivation and insight to a study of the others. This book is not merely a compendium of various facts relating to calculus, probability, and statistics, but rather a coordinated study in which the interrelationships between these fields are developed and exploited.

Since this text is designed for a two-semester course, we have necessarily omitted much of the material found in the standard courses in calculus and statistics. Our primary omissions have been of analytic geometry and the trigonometric functions. We have tried to include sufficient theory to enable the student to understand what he is using and why, thus avoiding a "cookbook" situation. We do not, however, recommend the use of this text in classes intended for mathematics majors. It was written primarily for those who need a working knowledge of differential and integral calculus and basic statistical methods.

We have assumed that the student is familiar with at least basic analytic geometry and can graph straight lines and simple curves. Other than this, the text is self-contained. Certain proofs early in the book use finite induction, but these proofs may be omitted without prejudice to later material.

ACKNOWLEDGMENTS

I am indebted to the Literary Executors of the late Sir Ronald Fisher, F.R.S., to Dr. Frank Yates, F.R.S., and to Oliver & Boyd Ltd., Edinburgh, for permission to reprint Table III (our Table 7 in the Appendix) from their book *Statistical Tables for Biological, Agricultural, and Medical Research;* to Iowa State University Press for permission to reprint the table of critical values of F (our Table 5) from *Statistical Methods*, by George W. Snedecor and William C. Cochran; and to Jack W. Dunlap for permission to reprint the table of squares and square roots (our Table 8) from *Handbook of Statistical Nomographs, Tables, and Formulas* by J. W. Dunlap and A. K. Kurtz (New York: World Book Company, 1932).

Northampton, Mass.   M.G.
January 1970

Contents

Chapter 1 The Basic Concepts of Function and Probability

1.1 Sets and functions 1
1.2 The notion of probability 6
1.3 The basic laws of probability 11
1.4 More basic facts about probabilities 15

Chapter 2 Some Specific Probabilities

2.1 Sampling without replacement but with regard to order . 22
2.2 Sampling without replacement or regard to order . . . 27
2.3 Sampling with replacement and with regard to order . . 31
2.4 Bayes' Theorem 34

Chapter 3 Random Variables. Graphs

3.1 Random variables. Admissible ranges 38
3.2 Graphs of equalities and inequalities 42
3.3 Properties of functions and graphs 48
3.4 Continuity 52
3.5 Summation notation 57
3.6 Probability distributions 61

Chapter 4 The Derivative

- 4.1 The limit of a function 70
- 4.2 The derivative of a function 77
- 4.3 Basic rules for finding a derivative 83
- 4.4 The Chain Rule. Implicit differentiation 88

Chapter 5 Applications of the Derivative

- 5.1 Maxima and minima 94
- 5.2 More about maxima and minima. Increasing and decreasing functions 99
- 5.3 Some theorems about continuous functions . . . 105
- 5.4 Higher derivatives and their applications 109
- 5.5 The density function of a continuous distribution. The mode 113

Chapter 6 Sequences and Series

- 6.1 Sequences and series 120
- 6.2 Tests for convergence of series 126
- 6.3 Power series 131
- 6.4 Taylor's series. Interval of convergence 137

Chapter 7 Integration

- 7.1 The definite integral. Area under a graph . . 142
- 7.2 The fundamental theorem of the calculus 149
- 7.3 Some basic integrals. The indefinite integral 154
- 7.4 Integration by parts and change of variable 159
- 7.5 Improper integrals. Tables of integrals 163
- 7.6 Numerical methods of integration 169

Chapter 8 The Integral and Continuous Variates

- 8.1 Measures of central tendency 174
- 8.2 Variation from the norm 179
- 8.3 Probability of extreme values. Moment-generating functions 185

Chapter 9 Some Basic Discrete Distributions

- 9.1 The rectangular and hypergeometric distributions . . . 190
- 9.2 The binomial distribution 194
- 9.3 Distributions involving the number of trials until success . 199
- 9.4 The Poisson distribution 204

Chapter 10 Other Important Distributions

- 10.1 The normal distribution 208
- 10.2 Student's t-distribution 216

10.3	More about the t-distribution	220
10.4	χ^2-distribution	225
10.5	Some other distributions	231

Chapter 11 Hypothesis Testing

11.1	Statistical inference	235
11.2	More about critical regions	239
11.3	Some remarks on the design of experiments	242
11.4	An example	245

Chapter 12 Functions of Several Variables

12.1	Multivariate functions	250
12.2	Partial differentiation	255
12.3	Multiple integration	263

Chapter 13 Regression and Correlation

13.1	Linear regression	271
13.2	Measures of correlation	275
13.3	The coefficient of correlation	279
13.4	The significance of r and r^2	282

Appendix

Table 1	Common logarithms	289
Table 2	Areas under the standard normal curve	291
Table 3	Critical values of chi square	294
Table 4	Critical values of t	295
Table 5	Critical values of F	296
Table 6	Exponential functions	300
Table 7	$z = \frac{1}{2}\ln\left(\frac{1+r}{1-r}\right)$	301
Table 8	Squares, square roots, and reciprocals from 1 to 1000	302
	Answers to the Exercises	309
	General Index	349
	Index of Symbols	353
	Table of Integrals	355

1 The Basic Concepts of Function and Probability

1.1 SETS AND FUNCTIONS

Definition 1. *A set is any well-defined collection of objects.* By "well-defined" we mean that we can tell what objects are in the collection and what objects are not in the collection. Any member of a set is called an **element**, or **point**, of the set.

Example 1. The collection of people who own a home within the city limits of Chicago is a set. Each person owning a home within the city limits of Chicago is an element of this set.

Example 2. If a group of students take an examination, then the collection of scores obtained by the students forms a set. Each individual score is an element of the set.

Example 3. A deck of data cards for use in a computer program forms a set. Each card is an element of the set.

Certain or all, of the elements in one set may be related in some way to certain, or all, of the elements of another set. This point is illustrated in the following examples.

Example 4. Let S be the set of all people. Then the phrase "is the parent of" relates each element of S, that is, each person, to those elements of S (persons) of which he is the parent. If x is a person who is not the parent of anyone, then "x is the parent of y" will not be satisfied for any person y.

Example 5. Let S be a set of students who took an examination and T be the set of scores obtained by the students. Then the phrase "has the score" assigns some element of T to each element of S.

Example 6. Let R be the set of real numbers. Then the rule $f(x) = 2x^3$ assigns to each real number x another real number $f(x)$ which is twice the cube of x.

Although the phrase "is the parent of" (Example 4) relates some people to no one at all, it also relates those who are the parents of several children to more than one person. In Example 6, however, not only does the rule $f(x) = 2x^3$ relate each real number x to some real number, but x is related to a *unique* real number $2x^3$. Given x, there is no choice as to what $f(x)$ is. When each element of one set is related to one and only one element of another set (which may also be the same as the first set), then we say that we have a *function* from the first set into the second set. More formally, we make the following definition:

Definition 2. *A rule, phrase, or relationship, which assigns to each element of a set S one and only one element of a set T is said to be a **function from S into T**.*

Thus, the rule in Example 6 is a function, while the phrase in Example 4 does not give a function.

Consider Example 5 again. The phrase "has the score" is a function from S into T since each student has one and only one score. If a student s has a score t, then we may, if we wish, represent this fact by means of the "ordered pair" (s, t).* More generally, we may represent the function "has the score" of Example 5 by all ordered pairs (s, t), where s is a student and s has the score t.

Definition 3. *If S and T are any two sets, then any object of the form (s, t), where s is an element of S and t is an element of T, is said to be an **ordered pair** with s as its **first coordinate** and t as its **second coordinate**.*

---

* On the other hand, we might also represent the fact by means of (t, s), or st; the number of ways is limited only by our imagination. However, (s, t) is the notation that mathematical history has passed on to us to be used to represent the pair consisting of s and t, where, for one reason or another, we wish to consider these elements in a particular order with s "coming before" t.

An ordered pair is a pair of elements, one from each of two sets, where the order in which the elements are given is important. An *ordered* pair is the opposite of an *unordered pair*, that is, a pair of elements, one from each of two sets, where the order is not taken into account. From the unordered pair containing, say 1 and 2, we can form two ordered pairs: (1, 2) and (2, 1).

If some rule relates each element of a set S to a unique element of a set T, then the rule gives rise to the set of all ordered pairs (s, t) such that s and t are elements of S and T, respectively, and t is related by the rule to s. Moreover, since no element of S is related to more than one element of T, but each element of S is related to some element of T, each element of S will appear as a first coordinate once and only once in the ordered pairs that the rule determines. The collection of ordered pairs determined by such a rule (function) is a subcollection of the set of all ordered pairs that can be formed with an element of S in the first coordinate and an element of T in the second coordinate. This inspires the following definition.

Definition 4. *Let S and T be any sets. Then the* **Cartesian product,** *or simply the* **product,** *of S with T is defined to be the set of all ordered pairs (s, t) such that s and t are elements of S and T, respectively. We denote the product of S with T by $S \times T$.*

If f is a function from a set S into a set T, then f determines a particular kind of subcollection of $S \times T$, specifically one in which each element of S appears as a first coordinate once and only once. On the other hand, if we begin with a subcollection of $S \times T$ having the property that each element of S appears as a first coordinate exactly once, then this subcollection itself determines a function, namely, the function which relates the element s of S to that element t of T such that (s, t) is the only point of the subcollection which has s as a first coordinate. Functions, therefore, can be considered either as rules relating the elements of one set S to elements of a set T, or as a subcollection of $S \times T$. In sum, we can say:

Characterization of functions in terms of ordered pairs. *A function f from a set S into a set T is a collection of elements of $S \times T$ such that each element of S appears as a first coordinate in f once and only once.**

We may also indicate that (s, t) is an element of f by writing $t = f(s)$.

---

* In set notation, no element of a set is ever repeated. For example, $\{a\}$ is never written $\{a, a\}$. In other words, we delete repetitious elements when we represent a set. We do this to eliminate the possibility of having some element of S appear twice in a function even though it is paired with the same second coordinate in both instances.

Example 7. Suppose S and T both consist of the elements 1, 2, and 3. Braces { } are customarily used to set off the elements of a set. Hence an equivalent form of the first sentence of this example is "Suppose $S = T = \{1, 2, 3\}$."

Then $S \times T$ contains all the elements in the array shown in Table 1.

Table 1

(1, 1)   (1, 2)   (1, 3)
(2, 1)   (2, 2)   (2, 3)
(3, 1)   (3, 2)   (3, 3)

Each of the following are functions from S into T. Each element of S appears as a first coordinate once and only once in each function. Observe that this is equivalent to saying that each function contains one and only one element from each row of Table 1.

$f_1 = \{(1, 1), (2, 2), (3, 3)\}$,
$f_2 = \{(1, 3), (2, 3), (3, 1)\}$,
$f_3 = \{(1, 1), (2, 1), (3, 1)\}$.

These are certainly not all the functions from S into T; in all, there are 27 such functions. Note that f_1 can also be characterized by the rule $f_1(s) = s$ for each element s of S. The function f_3 can be characterized by the rule $f_3(s) = 1$ for each element s of S. We may also specify f_2 by stating that

$$f_2(s) = \begin{cases} 3 & \text{if } s = 1 \text{ or } s = 2, \\ 1 & \text{if } s = 3. \end{cases}$$

Example 8. Let R be the set of real numbers. Then $R \times R$ is the set of all ordered pairs (x, y), where x and y are both real numbers. The reader should recognize $R \times R$ as the ordinary coordinate plane. If the subset (= subcollection) f of $R \times R$ is a function, then each real number x must appear as a first coordinate in f once and only once. If x_0 is a fixed real number, then the line whose equation is $x = x_0$ is a line parallel to the y-axis of the coordinate plane (Fig. 1). There is exactly one point $(x_0, f(x_0))$ of f whose first coordinate is x_0; hence there is exactly one point of f on the line $x = x_0$.

On the other hand, if f is a subset of $R \times R$ such that each line of the form $x = x_0$ meets f in exactly one point, then f is a function. If some line $x = x_0$ does not meet f, then x_0 does not appear as a first coordinate in f; and if some line $x = x_0$ meets f in more than one point, then f contains at least two points with first coordinate x_0.

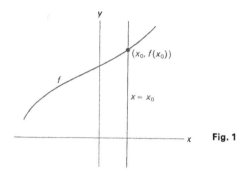

Fig. 1

Example 9. Let S and T be as in Example 7, and

$g_1 = \{(1, 2), (1, 3), (2, 3), (3, 1)\}$,
$g_2 = \{(1, 3), (3, 2)\}$.

Then g_1 and g_2 are not functions from S into T. In the case of g_1, the element 1 of S appears as a first coordinate twice; hence $g_1(1)$ is not clearly defined. On the other hand, 2 does not appear as a first coordinate in g_2; hence $g_2(2)$ is not defined at all.

The following terminology is standard; we include it for the sake of completeness.

Definition 5. *If f is a function from a set S into a set T, we call S the **domain** of f and T the **range** of f. The **image** of f is defined to be the set of all elements t of T such that $t = f(s)$ for some element s of S.*

Example 10. In Example 7, the domain and range are both $\{1, 2, 3\}$ for each of the functions f_1, f_2, and f_3. The image of f_1 is $\{1, 2, 3\}$, of f_2 is $\{1, 3\}$, and of f_3 is $\{1\}$. The image of f is merely the set of elements of the range which appear as second coordinates in f.

EXERCISES

1. We have denoted certain sets by listing their elements between braces, for example $f_1 = \{(1, 1), (2, 2), (3, 3)\}$. We might also have denoted a set by giving an arbitrary element of the set together with a condition that the element must satisfy to be in the set, all written in the following format: $\{x \mid \text{condition that } x \text{ must satisfy to be in the set}\}$. Thus $\{y \mid y \text{ is a house}\}$ is the set of all houses. Express verbally each of the following sets.

   a) $\{1, 45\}$ b) $\{a, b, 6, 7, 9\}$
   c) $\{\{1\}\}$ d) $\{1, \{1\}\}$
   e) $\{x \mid x \text{ is an animal}\}$ f) $\{w \mid w \text{ is a citizen of Canada}\}$
   g) $\{y \mid y \text{ is an Indian and } y \text{ lives in Iowa}\}$
   h) $\{z \mid z \text{ is an integer divisible by 2}\}$

2. A set S is said to be a *subset* of a set T if each element of S is an element of T. Thus $\{1, 2\}$ is a subset of $\{1, 2, 3\}$. A function f from a set S into a set T is a special kind of subset of $S \times T$. We use $S \subset T$ to denote that S is a subset of T. Which of the following statements are true? If a statement is true, prove it; if it is false, try to find an instance in which the statement should apply, but does not. In the following, S, T, and W are sets.
   a) If $S \subset T$ and $T \subset W$, then $S \subset W$.
   b) If $S \subset W$, then $W \subset S$.
   c) $S \subset S$
   d) If $S \subset T$, but T is not a subset of W, then S is not a subset of W.

3. Write out all the elements of the following products. Try to arrange the elements in an array similar to that given in Table 1.
   a) $\{1, 2\} \times \{3, 4\}$ b) $\{a, b, c\} \times \{6, 7\}$ c) $\{q\} \times \{l\}$
   d) $\{1, 2, 3, 4\} \times \{5, 6, 7, 8\}$ e) $\{A, B, C, D\} \times \{A, B, C, D\}$

4. Find four distinct functions from S into T, where S and T are given as in each of the following. It may help to use the arrays constructed in Exercise 3. Compute the range, domain, and image of each function found.
   a) $S = \{a, b, c\}$ and $T = \{6, 7\}$
   b) $S = \{1, 2, 3, 4\}$ and $T = \{5, 6, 7, 8\}$
   c) $S = T = \{A, B, C, D\}$

5. Let S be the set of living human beings. Which of the following phrases define a function from S into S? If a phrase fails to define a function, explain why it fails.
   a) Is the cousin of   b) Is the father of
   c) Is the same age as d) Has as mother

   If S were the set of all human beings who are living or have ever lived, would any of the phrases (a) through (d) define a function from S into S?

6. Which of the following define functions from the set R of real numbers into R? Indicate those functions whose image is the entire set of real numbers.
   a) $f(x) = x$ b) $g(y) = y + 2$ c) $h(w) = \pm 3w$
   d) $f(x) = x^{1/2}$ e) $g(u) = u^2$ f) $h(x) = \begin{cases} x & \text{if } x \geq 0, \\ -x & \text{if } x < 0 \end{cases}$

1.2 THE NOTION OF PROBABILITY

The general purpose of probability theory is to make more "mathematical" such statements as "very likely" and "not much chance." Given some event E, we wish to assign a number to E, the *probability* of E, which will measure in some suitable fashion the chance that E will occur. We would also like to know how to manipulate probabilities once they have been assigned. For example, given the probabilities of the events E and E', we would like to be able to derive the probability of "either E occurs, or E' occurs."

We formulate the following definitions to help make the discussion more precise.

Definition 6. *An* **experiment** *is a particular procedure to be performed, or a set of circumstances to be present simultaneously. The particular procedure to be performed, or the set of circumstances to be present, must be clearly defined.*

A **trial** *is one particular run of an experiment; that is, a trial is one actual performance of the procedure specified by an experiment, or a particular situation in which all the circumstances called for by the experiment are present.*

A **simple event** *is a possible outcome of a particular trial.*

A **sample space** *is the set of all simple events associated with an experiment.*

The following examples illustrate the concepts presented in Definition 6.

Example 11. Taking a particular coin, tossing it into the air, and letting it come to rest on the floor is an experiment. A particular toss of the coin in accordance with the directions is a trial. The simple events associated with the experiment would be "heads" and "tails"; thus the sample space is {heads, tails}.

If the experiment were to flip the coin twice, then heads could occur on both the first and second tosses, or heads could occur on the first toss and tails on the second toss, etc. If H and T represent heads and tails, respectively, then the sample space for this experiment is

$$S = \{HT, TH, HH, TT\}.$$

In this latter experiment the occurrence of exactly one head is equivalent to the occurrence of one of the simple events HT and TH. We can associate the event "one occurrence of heads" with the subset $\{HT, TH\}$ of the sample space S.

Definition 7. *An* **event** *is any subset of a sample space.*

Example 12. If the coin of Example 11 is flipped twice, then $\{HT\}$, $\{HH\}$, $\{HH, TT\}$, and S itself are all events. The event $\{HT, TH, HH\}$ can be thought of as the "nonoccurrence of TT," or as "the occurrence of at least one head."

Two events A and B may be *mutually exclusive.* Informally, this means that A and B cannot occur together. In terms of a sample space, it means that A and B share no simple events in common; that is, their intersection is empty.

Two events may also be independent, that is, the occurrence or nonoccurrence of one of the events in no way affects the occurrence or nonoccurrence of the other event. More will be said about mutually exclusive and independent events later.

We may also "build other events from two given events A and B." In particular, we make the following definition.

Definition 8. *Let A and B be two events (subsets of a sample space S). Then $A \cap B$, the **intersection** of A and B, is defined to be*

$$\{s \mid s \text{ is an element of } S \text{ in both } A \text{ and } B\}.$$

$A \cap B$ is the event "A and B." The event $A \cup B$, the **union** of A and B, is defined to be

$$\{s \mid s \text{ is an element of } S \text{ in either } A \text{ or } B\};$$

$A \cup B$ is the event "either A or B" (Fig. 2). The event $\overline{A}$, the **complement** of A, is defined to be

$$\{s \mid s \text{ is an element of } S, \text{ but not an element of } A\};$$

$\overline{A}$ is the event "not-A."

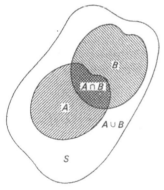

Fig. 2

Example 13. In Example 12, $\{HT, TH\} \cup \{HH\} = \{HT, TH, HH\}$. This may be interpreted to mean, "If either one head and one tail occurs, or two heads occur, then two tails do not occur." Note that $\{HT, TH, HH\}$ is the complement of $\{TT\}$. The events $\{HH, TT\}$ and $\{HT\}$ have no elements in common; therefore their intersection is the *empty set*, the set which contains no elements.

Sample spaces furnish us with a precise way to formulate events and enable us to use the machinery of set theory in the development of the theory of probability. Nevertheless, events can still be thought of "informally," that is, by merely considering what we wish to consider without explicit reference to a sample space. For example, "rain" can be thought of as an event without reference to a sample space. Once a suitable sample space has been constructed, however, then "rain" will be the subset of all elements of the sample space which involve rain.

Example 14. Let A and B be the events of wind and rain, respectively. Then $A \cap B$ is the event of wind and rain together, while $A \cup B$ is the event of either wind or rain. $\overline{A}$ is the event of not-wind. Note that no sample space has been mentioned explicitly.

Definition 9. *Two events (subsets of the sample space S) A and B are said to be **equivalent** if $A = B$, that is, if A and B contain the same elements of S.*

Less formally, we call two events *equivalent* if each occurs whenever the other occurs.

Example 15. Let A, B, and C be events. Then $A \cap (B \cup C)$ is the event of A occurring together with either B or C. In terms of set theory, it is the set of elements (of the sample space) which are both in A and either B or C (Fig. 3). The event $(A \cap B) \cup (A \cap C)$ is the event of A occurring together with B, or A occurring together with C. Set-theoretically, this is the set of elements which are either both in A and B or both in A and C; this again is the set represented in Fig. 3. Thus, the events $A \cap (B \cup C)$ and $(A \cap B) \cup (A \cap C)$ are equivalent.

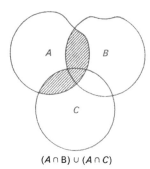

$(A \cap B) \cup (A \cap C)$ **Fig. 3**

Given two events A and B, we can define an important event associated with these events which is not, however, a subset of the sample space of A and B.

Definition 10. *Given events A and B, the event "A given B," or "A if B" is denoted by $A \mid B$. We may call $A \mid B$, the event A **conditioned by** B.*

Example 16. If A and B are rain and wind, respectively, then $A \mid B$ is *rain if there is wind*, while $B \mid A$ is *wind if there is rain*. Note that in $A \mid B$, wind is a *given* event; we may read $A \mid B$ as A *given* B, or A *if* B.

EXERCISES

1. Let A, B, and C be the events of rain, wind, and lightning, respectively. Write out in words each of the following events.

   a) $\bar{B}$ b) $B \cap C$ c) $\bar{A} \cap C$ d) $A \cup C$
   e) $\bar{A} \cup \bar{C}$ f) $A \cap (B \cup \bar{C})$ g) $(A \cup \bar{C}) \cap B$ h) $A \mid \bar{C}$
   i) $\bar{B} \mid (A \cap C)$ j) $A \cap B \cap \bar{C}$

2. Letting A, B, and C be as in Exercise 1, write symbolically each of the following events.

   a) Rain if it is windy and there is lightning
   b) Rain without lightning
   c) Lightning with neither wind nor rain
   d) Lightning if there is neither wind nor rain
   e) Lightning and rain without wind

3. In each of the following an experiment is described. Devise a sample space to represent the outcomes of each experiment.

   a) A coin is flipped three times.
   b) A die is tossed twice.
   c) Two dice are tossed together.
   d) One slip is drawn from an urn which contains ten slips numbered 1 through 10.
   e) Two slips are drawn one at a time from an urn containing ten slips numbered 1 through 10.

4. Let A, B, and C be events (in a sample space S). Prove that the two events in each of the following pairs of events are equivalent. Do this by showing that the sets contain the same elements; a diagram such as that in Fig. 3 may be used for the proof. Write out the meaning of each of the events in words; if done properly, it should be clear from the verbal description of the events that they are equivalent.

   a) $(A \cap B) \cap C$; $A \cap (B \cap C)$ b) A; $\bar{\bar{A}}$
   c) $(B \cup C) \cap A$; $(B \cap A) \cup (C \cap A)$ d) $\bar{A} \cup \bar{B}$; $\overline{A \cap B}$

5. Formulate three distinct experiments that might be performed with an ordinary deck of cards. For each experiment, formulate a sample space.

6. We have not yet defined what is meant by the probability of an event; however, the reader should already have some intuitive ideas about the properties of probabilities. If A is any event, we let $P(A)$ denote the probability of A. In each of the following, A and B are events. Decide which of the following statements are true and which are false. Justify each of your assertions.

   a) $P(A \cup B)$ is at least as large as $P(A)$.
   b) $P(A \cap B)$ is no larger than $P(B)$.
   c) $P(A \cup B) = P(A) + P(B)$.
   d) $P(A \cap B)$ is less than or equal to $P(A)P(B)$.

e) If A and B are mutually exclusive, then $P(A \cap B)$ will have the smallest value permitted for a probability.
f) If A is twice as probable as B, then $P(A) = 2P(B)$.

1.3 THE BASIC LAWS OF PROBABILITY

We would like to define the *probability* $P(A)$ of an event A in such a manner that if the experiment associated with A is performed n times, then A should occur about $nP(A)$ times. For example, if the probability of obtaining a head on any toss of a certain coin is $\frac{1}{2}$, and the coin is flipped 1000 times, then heads should occur about 500 times. It follows then that if we wish to find the probability of an event A, we could perform the appropriate experiment n times. If A occurs m times out of the n trials, then we may estimate $P(A)$ to be m/n.

Example 17. A balanced coin is flipped in an unbiased manner 1000 times and 495 heads are obtained. Then we can estimate $P(\text{heads})$ to be .495.

Example 18. The weatherman finds that 80 out of 100 days on which the temperature in Chicago was above 75° saw rain. The probability of rain in Chicago on a day when the temperature is above 75° can therefore be estimated as .8.

Definition 11. *If some particular experiment is performed n times and the event A occurs on m of the n trials, then m/n is called the **relative frequency** of A. We shall denote the relative frequency of A by $R(A)$.*

We have said that $R(A)$ could be used to estimate $P(A)$. However, this "relative frequency" approach to probability has certain limitations.

First, in order to obtain a reasonably good estimate of a probability, a fairly large number of trials would have to be performed. As an extreme example, if we flip a coin only once, then we obtain either heads or tails. Using just one flip, we would then estimate the probability of heads to be either 0 or 1, depending on whether heads or tails had come up. About the only thing that one flip might tell us is that whatever was flipped has a nonzero probability.

Although a probability estimate based on a large number of trials should be fairly accurate, there is always the possibility that the unusual will happen. For example, even though the true probability of flipping heads with a fair coin is $\frac{1}{2}$, it is not impossible to flip 1000 consecutive tails. Thus, in computing an estimate of the probability using the relative frequency approach, we can never be really sure that the unusual has not occurred and the estimate is far off the mark. What does increase with the number of trials is the probability that our estimate will "approach" the true probability.

In certain instances, it might be hard to get enough information to make an intelligent estimate of the probability. The following example illustrates this point.

Example 19. A veterinarian wishes to know the probability that a certain species of fox will develop rabies. The veterinarian may, however, be able to study only a very limited number of foxes. Moreover, the fact that the foxes are observed, particularly if the observation removes them from their natural habitat, may grossly distort the veterinarian's conclusions. For example, all sick foxes may hide in their burrows and hence cannot be observed, or those captured may not contract the disease because they have been removed from its source.

Despite its limitations, we shall accept the relative frequency approach to probability as having some validity. We shall use it to motivate the axioms for a theory of probability.

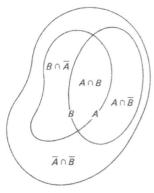

Fig. 4

Let us assume now that A and B are two events associated with some experiment. Then the outcome of any particular trial of the experiment must fall into one and only one of the following categories: $A \cap B$, $\bar{A} \cap B$, $A \cap \bar{B}$, and $\bar{A} \cap \bar{B}$ (Fig. 4). That is, A occurs with or without B, or B occurs without A, or neither A nor B occurs. Suppose n trials are performed. We summarize the results as shown in Table 2. We shall use this table to derive important relationships between relative frequencies.

Table 2

Type	$A \cap B$	$\bar{A} \cap B$	$A \cap \bar{B}$	$\bar{A} \cap \bar{B}$	Total trials
Number of type	p	q	r	s	n

We first compute $R(A)$. The event A is equivalent to $(A \cap B) \cup (A \cap \bar{B})$. Since $A \cap B$ occurs p times and $A \cap \bar{B}$ occurs r times, A occurs $p + r$ times. This means that

(1) $\quad R(A) = (p + r)/n.$

Similarly,

(2) $\quad R(\bar{A}) = (q + s)/n.$

We therefore have $R(A) + R(\bar{A}) = (p + q + r + s)/n = n/n = 1$. Hence we have proved

(3) $\quad R(A) + R(\bar{A}) = 1.$

Since p, q, r, s, and n are all nonnegative integers, we may also say

(4) $\quad R(A)$ is a real number between 0 (if $p + r = 0$) and 1 (if $p + r = n$).

Now consider $A \cup B$. Since $A \cup B$ is equivalent to

$(A \cap B) \cup (A \cap \bar{B}) \cup (\bar{A} \cap B),$

$A \cup B$ occurs $p + q + r$ times. We therefore have

(5) $\quad R(A \cup B) = (p + q + r)/n.$

As we have seen, $R(A)$ is $(p + r)/n$, while $R(B)$ is $(p + q)/n$. Thus, we can say

(6) $\quad R(A) + R(B) = (p + r)/n + (p + q)/n = (p + q + r)/n + p/n.$

But since $p/n = R(A \cap B)$, using (5), we find that (6) becomes

(7) $\quad R(A) + R(B) = R(A \cup B) + R(A \cap B),$

or

(8) $\quad R(A \cup B) = R(A) + R(B) - R(A \cap B).$

Finally, consider $R(A \cap B)$. We have $R(A \cap B) = p/n$. Dividing $R(A \cap B)$ by $R(A)$, we obtain

(9) $\quad \dfrac{R(A \cap B)}{R(A)} = \dfrac{p/n}{(p+r)/n} = p/(p+r).$

Now $p + r$ is the number of trials in which A occurs, and p is the number of trials in which both A and B occur. Therefore $p/(p + r)$ is the relative frequency of B in those trials in which A occurs; that is, $p/(p + r)$ is nothing other than $R(B \mid A)$. Equation (9) thus becomes

(10) $\quad \dfrac{R(A \cap B)}{R(A)} = R(B \mid A),$

which is equivalent to

(11) $R(A \cap B) = R(A)R(B \mid A)$.

If we assume that for a large number of trials, $R(A)$ is a "good" estimate of $P(A)$, then the probabilities should satisfy the same relationships that relative frequencies satisfy. In particular, motivated by the relationships among relative frequencies derived in this section, we make the following *assumptions* about probabilities.

Assumption 1. *If A is any event, then $P(A)$, the probability of A, is a real number which lies between 0 and 1, inclusive.* $P(A) = 0$ *if A is the empty set (an impossible event) and $P(A) = 1$ if A is the entire sample space (a certain event).*

Assumption 2. *If A and B are events, then*

$$P(A \cup B) = P(A) + P(B) - P(A \cap B).$$

Assumption 3. *For any events A and B,*

$$P(A \cap B) = P(A)P(B \mid A) = P(B)P(A \mid B)$$

if $P(A)$ and $P(B)$ are not 0. If either $P(A)$ or $P(B)$ is 0, then $P(A \cap B) = 0$.

In the next section we shall derive some basic consequences of these assumptions.

EXERCISES

1. Table 3 represents the results of 1000 trials of an experiment.

Table 3

Type	$A \cap B$	$\overline{A} \cap B$	$A \cap \overline{B}$	$\overline{A} \cap \overline{B}$	Total
Number of type	400	150	300	150	1000

  a) Find $R(A)$, $R(B)$, $R(\overline{A})$, and $R(\overline{B})$.
  b) Find $R(\overline{A} \cup B)$ directly from Table 3. Compute $R(\overline{A} \cup B)$, using (8).
  c) Find $R(A \mid B)$ directly from Table 3. Compute $R(A \mid B)$, using (10).
  d) Confirm each of the following statements.
   i) $R(A \cup B)$ is at least as large as $R(A)$.
   ii) $R(A \cap B)$ is no larger than $R(B)$.
   iii) A and B are not mutually exclusive.
   iv) $R(A \mid B)$ is greater than $R(B \mid A)$.
  e) If A and B are the events of rain and lightning, respectively, what interpretation may be given statement (iv) of (d)?

2. There are instances in which the only way in which the probability of an event A can be estimated is by using its relative frequency. Cite three such instances.

3. Explain what a weatherman means when he says that there is an 80% chance of precipitation. Was the "80% chance" arrived at through using a relative frequency?

4. Table 4 represents the results of 850 trials of a certain experiment. A, B, and C are events.

Table 4

Type	$A \cap B \cap C$	$A \cap \bar{B} \cap C$	$A \cap B \cap \bar{C}$	$A \cap \bar{B} \cap \bar{C}$
Number of type	50	100	50	200
Type	$\bar{A} \cap B \cap C$	$\bar{A} \cap \bar{B} \cap C$	$\bar{A} \cap B \cap \bar{C}$	$\bar{A} \cap \bar{B} \cap \bar{C}$
Number of type	150	100	150	50

Prove each of the following:

a) $R(A \cup B \cup C) = R(A) + R(B) + R(C) - R(A \cap B) - R(B \cap C) - R(A \cap C) + R(A \cap B \cap C)$

b) $R(A \cap B \cap C) = R(A)R(B \mid A)R(C \mid A \cap B)$

1.4 MORE BASIC FACTS ABOUT PROBABILITIES

If A is an event in a sample space S, then $A \cup \bar{A} = S$, while $A \cap \bar{A}$ is the empty set. From Assumptions 1 and 2, we find that

$$P(A \cup \bar{A}) = P(S) = 1 = P(A) + P(\bar{A}) - P(A \cap \bar{A}) = P(A) + P(\bar{A})$$

since $P(A \cap \bar{A}) = P(\text{empty set}) = 0$. Thus, we have proved:

Proposition 1. *If A is any event, then $P(\bar{A}) = 1 - P(A)$.*

If A and B are mutually exclusive events, that is, if $A \cap B$ is the empty set, then $P(A \cap B) = 0$. Consequently,

$$P(A \cup B) = P(A) + P(B).$$

We generalize this result in the following proposition.

Proposition 2. *Suppose $A_1, A_2, \ldots, A_n$ are finitely many events such that the intersection of any two of these events is the empty set (no two of these events can occur together). Then*

$$P(A_1 \cup A_2 \cup \cdots \cup A_n) = P(A_1) + P(A_2) + \cdots + P(A_n).$$

That is, the probability that at least one (and hence exactly one) of the n events will occur is the sum of their individual probabilities.

Proof: We shall use finite induction. We have already seen that the proposition is true for $n = 2$; that is,

$$P(A_1 \cup A_2) = P(A_1) + P(A_2) \quad \text{if} \quad P(A_1 \cap A_2) = 0.$$

We now show that if the proposition is true for n events, it must also be true for $n + 1$ events. Assume that Proposition 2 holds for n events and let

$$A_1, A_2, \ldots, A_n, A_{n+1}$$

be $n + 1$ events such that no two of these events can occur together. Set

$$B = A_1 \cup A_2 \cup \cdots \cup A_n.$$

Then $B \cap A_{n+1}$ is the empty set. For if $B \cap A_{n+1}$ contained some element of S, then A_{n+1} and some A_i, $i = 1, \ldots, n$, would have to contain some common element of S, which by assumption is impossible. Therefore

$$P(B \cap A_{n+1}) = 0.$$

But

$$A_1 \cup A_2 \cup \cdots \cup A_n \cup A_{n+1}$$

is equivalent to $B \cup A_{n+1}$. Therefore

$$
\begin{aligned}
(12) \quad P(A_1 \cup A_2 \cup \cdots \cup A_n \cup A_{n+1}) &= P(B \cup A_{n+1}) \\
&= P(B) + P(A_{n+1}) \\
&= P(A_1 \cup A_2 \cup \cdots \cup A_n) \\
&\quad + P(A_{n+1}).
\end{aligned}
$$

However, $A_1, A_2, \ldots, A_n$ are n events any two of which have an empty intersection; therefore, since Proposition 2 is assumed to hold for n, we have

$$(13) \quad P(A_1 \cup A_2 \cup \cdots \cup A_n) = P(A_1) + P(A_2) + \cdots + P(A_n).$$

Combining (12) and (13), we have

$$P(A_1 \cup A_2 \cup \cdots \cup A_n \cup A_{n+1}) = P(A_1) + P(A_2) + \cdots + P(A_n) + P(A_{n+1}),$$

which is what we wanted to prove.

The following corollary of Proposition 2 enables us to compute a number of probabilities.

Corollary: If $A_1, A_2, \ldots, A_n$ are n events (*in a sample space S*) *such that the intersection of any two of these events is empty and*

$$A_1 \cup A_2 \cup \cdots \cup A_n = S$$

and all of these events have the same probability, that is,

$P(A_1) = P(A_2) = \cdots = P(A_n),$

then the probability of any one of the events is $1/n$.

Proof: $P(A_1 \cup A_2 \cup \cdots \cup A_n) = P(S) = 1$. Let $x = P(A_1) = P(A_2) = \cdots = P(A_n)$.

By Proposition 2, we have

$P(A_1 \cup A_2 \cup \cdots \cup A_n) = 1 = P(A_1) + P(A_2) + \cdots + P(A_n) = nx.$

Therefore $x = 1/n$.

Example 20. A fair coin is flipped; we may assume that heads and tails have the same probability. The sample space here is $\{H, T\}$. Then H and T satisfy the conditions of the corollary above. Hence

$P(H) = P(T) = \frac{1}{2}.$

Example 21. The sample space associated with the roll of a die is

$\{1, 2, 3, 4, 5, 6\}.$

If the trial is unbiased, then any face of the die is as likely to occur as any other face, that is, $P(1) = P(2) = \cdots = P(6)$. Applying the corollary to Proposition 2, we find $P(1) = P(2) = \cdots = P(6) = \frac{1}{6}$.

Example 22. The sample space S associated with the deal of a five-card poker hand is the set of all five-card poker hands; let M be the number of elements in this sample space. It can be shown using the corollary to Proposition 2 that the probability of any one poker hand (assuming that all are equally likely) is $1/M$. The event A of obtaining a poker hand which contains only spades is associated with the subset A of S of poker hands which contains only spades; suppose there are K elements in A. We identify each poker hand which contains only spades by an integer from 1 to K. The probability, then, of dealing a hand which contains only spades is $P(A) = P(1 \cup 2 \cup \cdots \cup K)$, which by Proposition 2 is

$P(1) + P(2) + \cdots + P(K) = K(1/M) = K/M.$

In the next section we shall deal with techniques of evaluating K and M.

Example 23. Suppose A, B, and C are events such that $P(A) = \frac{1}{2}$, $P(B) = \frac{1}{2}$, and $P(C) = \frac{1}{4}$. Then A, B, and C could not be pairwise mutually exclusive. For if these events were pairwise mutually exclusive, then $P(A \cup B \cup C)$ would be $P(A) + P(B) + P(C) = \frac{5}{4}$, which is greater than 1. But no number greater than 1 can be a probability. It is quite possible that two of the events have an empty intersection (and thus $A \cap B \cap C$ is empty); for example, this would be the case if A were "rain," B were $\overline{A}$, and C were "gale force winds."

We now generalize Assumption 2 to the case of three events.

Proposition 3. Let A, B, and C be any events. Then
$$P(A \cup B \cup C) = P(A) + P(B) + P(C) - P(A \cap B) - P(B \cap C)$$
$$- P(A \cap C) + P(A \cap B \cap C).$$

[Compare this with Exercise 4(a) of Section 1.3.]

Proof: $A \cup B \cup C$ is equivalent to $(A \cup B) \cup C$. Consequently, by Assumption 2, we have

(14) $\quad P(A \cup B \cup C) = P((A \cup B) \cup C)$
$\qquad\qquad\qquad = P(A \cup B) + P(C) - P((A \cup B) \cap C)$
$\qquad\qquad\qquad = P(A) + P(B) + P(C) - P(A \cap B)$
$\qquad\qquad\qquad\quad - P((A \cup B) \cap C).$

The event $(A \cup B) \cap C$ is equivalent to $(A \cap C) \cup (B \cap C)$ [Exercise 4(c), Section 1.2]; hence

$$P((A \cup B) \cap C) = P((A \cap C) \cup (B \cap C)),$$

which by Assumption 2 equals

$$P(A \cap C) + P(B \cap C) - P((A \cap C) \cap (B \cap C)).$$

Since $(A \cap C) \cap (B \cap C)$ is equivalent to $A \cap B \cap C$, we have

$$P((A \cup B) \cap C) = P(A \cap C) + P(B \cap C) - P(A \cap B \cap C).$$

Substituting for $P((A \cup B) \cap C)$ in (14), we obtain

$$P(A \cup B \cup C) = P(A) + P(B) + P(C) - P(A \cap B)$$
$$- P(A \cap C) - P(B \cap C) + P(A \cap B \cap C),$$

which is what we wished to prove.

We now generalize Assumption 3.

Proposition 4. *If A, B, and C are events in a sample space S, then*

$$P(A \cap B \cap C) = P(A)P(B \mid A)P(C \mid A \cap B).$$

Proof: Since $A \cap B \cap C$ is equivalent to $((A \cap B) \cap C)$, we have

$$P(A \cap B \cap C) = P((A \cap B) \cap C) = P(A \cap B)P(C \mid A \cap B)$$
$$= P(A)P(B \mid A)P(C \mid A \cap B).$$

Example 24. An urn contains ten red, five white, and seven blue balls. Someone reaches into the urn and draws a ball at random. Since there are 22 balls in all and no one of these 22 seems more likely to be drawn than any of the others, the probability that some one particular ball will be drawn is $\frac{1}{22}$. Since 10 of the 22 balls are red, the probability of drawing a red ball is $\frac{10}{22}$.

If three balls are drawn from the urn one at a time and set aside as they are drawn, what is the probability of drawing one red, one white, and one blue ball, in that order? Instead of constructing a sample space explicitly to solve the problem, we shall use Proposition 4. Let A, B, and C be the events of drawing red on the first draw, white on the second draw, and blue on the third draw, respectively. We want to find $P(A \cap B \cap C)$. By Proposition 4,

$$P(A \cap B \cap C) = P(A)P(B \mid A)P(C \mid A \cap B).$$

Now $P(A)$, as we saw above, is $\frac{10}{22}$. $P(B \mid A)$ is the probability that a white ball will be drawn on the second draw on the assumption that a red ball has been drawn on the first draw. If a red ball was selected on the first draw, then we are left with 9 red, 5 white, and 7 blue balls from which to choose on the second draw. Therefore $P(B \mid A) = \frac{5}{21}$. Similarly, $P(C \mid A \cap B) = \frac{7}{20}$. Applying Proposition 4,

$$P(A \cap B \cap C) = (\tfrac{10}{22})(\tfrac{5}{21})(\tfrac{7}{20}).$$

Example 25. Suppose A, B, and C are events such that the occurrence or nonoccurrence of any one or more of the events in no way affects the occurrence or nonoccurrence of the other events. Then $P(B \mid A) = P(B)$; that is, the probability of "B if A" is the same as the probability of B, since whether or not A occurs does not affect whether or not B occurs. Similarly, $P(C \mid A \cap B) = P(C)$. In this instance, Proposition 4 becomes $P(A \cap B \cap C) = P(A)P(B)P(C)$.

Definition 12. *If* $A_1, A_2, \ldots, A_n$ *are events (in a sample space S) such that*

$$P(A_1 \cap A_2 \cap \cdots \cap A_n) = P(A_1)P(A_2) \cdots P(A_n),$$

*then the events are said to be **independent**. If the events are not independent, then they are said to be **dependent**.*

Definition 12 essentially says that the events $A_1, \ldots, A_n$ are independent if the occurrence of one or more of the events does not influence the occurrence of any of the other events.

Example 26. The events A, B, and C of Example 24 are dependent. For, as we shall show,

$$P(A \cap B \cap C) \neq P(A)P(B)P(C).$$

1.4   More Basic Facts about Probabilities

From Proposition 4 we have

$$P(A \cap B \cap C) = P(A)P(B \mid A)P(C \mid A \cap B).$$

We now show that $P(B \mid A) \neq P(B)$. It can be shown in like manner that $P(C \mid A \cap B) \neq P(C)$; direct computation then establishes that

$$P(A \cap B \cap C) = P(A)P(B \mid A)P(C \mid A \cap B) \neq P(A)P(B)P(C).$$

To show that $P(B \mid A) \neq P(B)$, and hence that the occurrence of A influences the occurrence of B, we let E and F be the events of drawing a white and a blue ball on the first draw, respectively. Then B (drawing a white ball on the second draw) is equivalent to

$$(A \cap B) \cup (E \cap B) \cup (F \cap B).$$

Since $A \cap B$, $E \cap B$, and $F \cap B$ are pairwise mutually exclusive, by Proposition 2 and Assumption 3, we have

(15) $\quad P(B) = P(A \cap B) + P(E \cap B) + P(F \cap B)$
$\quad\quad\quad = P(A)P(B \mid A) + P(E)P(B \mid E) + P(F)P(B \mid F)$
$\quad\quad\quad = (\frac{10}{22})(\frac{5}{21}) + (\frac{5}{22})(\frac{4}{21}) + (\frac{7}{22})(\frac{5}{21}),$

which does not equal $P(B \mid A) = \frac{5}{21}$.

Example 27. Suppose Jack, Sam, and Harry each flip a fair coin. Let A, B, and C be the events that Jack, Sam, and Harry each obtain heads. We may reasonably assume that A, B, and C are independent events, that is, what one person flips will not affect what any other person flips. If $P(A) = P(B) = P(C)$, then $P(A \cap B \cap C) = (\frac{1}{2})^3 = \frac{1}{8}$.

EXERCISES

1. Prove that $P(A \cup B \cup C) = P(A \cap B \cap C)$ if and only if the events A, B, and C are equivalent.

2. Consider Table 5.
   a) Find $P(A)$, $P(B)$, and $P(C)$. Are A, B, and C independent?
   b) Find $P(A \cap B)$, $P(A \cap \overline{B})$, $P(\overline{A} \cap B)$, and $P(\overline{A} \cap \overline{B})$. Are A and B independent?
   c) Can $P(A \mid B)$ be computed from the information given in Table 5? If yes, compute $P(A \mid B)$. If not, decide what further information is necessary before $P(A \mid B)$ can be computed.
   d) If $P(A) = \frac{1}{2}$, $P(B) = \frac{1}{3}$, and $P(C) = \frac{1}{6}$, what would the values in Table 5 have to be in order for A, B, and C to be independent?

3. Find each of the probabilities asked for in the following. In the expression for each probability, you may encounter one (but no more) quantity which

Table 5

$X \cap Y \cap Z$	$A \cap B \cap C$	$A \cap \bar{B} \cap C$	$A \cap B \cap \bar{C}$	$A \cap \bar{B} \cap \bar{C}$
$P(X \cap Y \cap Z)$	$\frac{1}{8}$	$\frac{1}{16}$	$\frac{1}{4}$	$\frac{3}{16}$
$X \cap Y \cap Z$	$\bar{A} \cap B \cap C$	$\bar{A} \cap B \cap \bar{C}$	$\bar{A} \cap \bar{B} \cap C$	$\bar{A} \cap \bar{B} \cap \bar{C}$
$P(X \cap Y \cap Z)$	$\frac{1}{8}$	$\frac{1}{16}$	$\frac{1}{8}$	$\frac{1}{16}$

you cannot yet compute numerically. In the next chapter we shall show how to compute quantities of this kind.

a) Five people A, B, C, D, and E are in a room. Two leave. What is the probability that one of those who left is A?

b) What is the probability that a bridge hand (13 cards) contains only spades?

c) An urn contains three blue and four red balls. Two of the balls are drawn at random from the urn. What is the probability that one of the balls is red and the other is blue?

d) A committee of five is chosen from a group of 20 people. What is the probability that the four oldest people in the group will be on the committee?

4. Suppose $A_1, A_2, \ldots, A_n$ are events in a sample space S. Prove that

$$P(A_1 \cap A_2 \cap \cdots \cap A_n)$$
$$= P(A_1)P(A_2 \mid A_1)P(A_3 \mid A_1 \cap A_2) \cdots P(A_n \mid A_1 \cap A_2 \cap \cdots \cap A_{n-1}).$$

2 Some Specific Probabilities

2.1 SAMPLING WITHOUT REPLACEMENT BUT WITH REGARD TO ORDER

It often happens that people try to draw conclusions about a set of objects by examining a subcollection of that set. For example, suppose four parts are chosen at random from a shipment of 100 airplane parts and examined for defects. To be certain that there are no defective parts, we would have to examine each part in the shipment and find it free from defects. But probability theory can tell us what chance there is that there are no defects in the four randomly chosen parts if there are in fact k defective parts in the shipment.

The problem of the airplane parts, as well as many other problems, are related to the following type of problem: Given a specified collection of objects from which a specified random sample is to be taken, what is the probability that the random sample will be of a certain type? As to the airplane parts, this question might become: Given 100 airplane parts, what is the probability that m randomly chosen parts will be free from defects if there are k defective parts in the shipment?

We have used the word "random" to describe a manner of selection. By *random* we mean that the selection is made according to chance and not according to any rule or procedure that would tend to influence the outcome. For example, consider a collection of 10 red and 12 blue balls. Someone is to pick five balls from these 22 balls, and we want to find the probability that all five selections will be red. To obtain a random selection, the balls may be put in an urn and the person who is to do the drawing may be blindfolded. In this way we minimize the effect of any natural preference the selector may have for either red or blue. If the selection is made by someone who can see the balls as he is making a selection and he prefers red over blue, then the selection will not be random; consequently, we would not be able to apply probability theory to solve the problem.

In this chapter we are going to develop various devices which will help us computationally when we are dealing with certain types of random samples. The first situation we shall consider is illustrated in the following example.

Example 1. Five slips in an urn are numbered 1 through 5. Someone reaches into the urn and draws out three slips one at a time, putting each slip aside as it is drawn.

If M is the total number of ways in which we can draw three slips from the urn without replacing any slip after it has been drawn and taking the order of selection into account,* then the probability of any one particular sequence of three draws is $1/M$ (since any such sequence is as likely as any other sequence and the sequences are pairwise mutually exclusive). Finding the probability of a sequence therefore reduces to finding the value of M.

The general type of computation we shall investigate in this section is the following: A collection of n objects is given from which m objects are to be selected. Once an object has been selected, it will be put aside (that is, not returned to the collection with the consequent possibility of its being drawn again). We are to take the order of selection into account so that even if the same m objects happen to be chosen on two draws, if the same m objects are drawn in different orders, we consider the draws to be distinct. With these qualifications, in how many ways can the m objects be selected?

After answering this question (which is a problem in bookkeeping rather than probability theory), we shall apply the answer to the computation of probabilities.

The following proposition about sets is needed.

---

* That is, two draws will be distinct even if the same three slips are drawn, provided that the slips are drawn in different orders.

Proposition 1. *If S and T are sets which contain m and n elements, respectively, then $S \times T$ (Definition 4, Chapter 1) contains mn elements.*

Proof: We can assume $S = \{1, 2, \ldots, m\}$ and $T = \{1, 2, \ldots, n\}$. Then the elements of $S \times T$ are presented in the following array:

$$
\begin{array}{cccc}
(1,1) & (1,2) & (1,3) & \ldots & (1,m) \\
(2,1) & (2,2) & (2,3) & \ldots & (2,m) \\
\vdots & & & & \vdots \\
(n,1) & (n,2) & (n,3) & \ldots & (n,m) \\
\end{array}
$$
$$S \times T$$

The array contains m columns, and each column contains n elements; hence there are mn elements in all.

This proposition has the following corollary.

Corollary: *If $S_1, S_2, \ldots, S_t$ are sets containing $n_1, n_2, \ldots, n_t$ elements, respectively, then $S_1 \times S_2 \times \cdots \times S_t$ defined to be the set of ordered t-tuples $(s_1, s_2, \ldots, s_t)$, where s_i is an element of S_i, $i = 1, 2, \ldots, t$, contains $n_1 n_2 \cdots n_t$ elements.*

The proof of this corollary, an exercise in finite induction, is left to the reader.

Example 2. If S and T are sets of five and six elements, respectively, then $S \times T$ contains 30 elements. If $S = \{1, 2, 3, 4, 5\}$ and $T = \{1, 2, 3, 4, 5, 6\}$, then the 30 elements of $S \times T$ are presented in the following array:

$$
\begin{array}{cccccc}
(1,1) & (1,2) & (1,3) & (1,4) & (1,5) & (1,6) \\
(2,1) & (2,2) & (2,3) & (2,4) & (2,5) & (2,6) \\
(3,1) & (3,2) & (3,3) & (3,4) & (3,5) & (3,6) \\
(4,1) & (4,2) & (4,3) & (4,4) & (4,5) & (4,6) \\
(5,1) & (5,2) & (5,3) & (5,4) & (5,5) & (5,6) \\
\end{array}
$$
$$S \times T$$

Example 3. If $S_1, S_2, \ldots, S_7$ are seven sets of two elements each, then $S_1 \times S_2 \times \cdots \times S_7$ contains $2^7 = 128$ elements.

A particular draw of m objects (without replacement but with regard to order of selection) from a collection of n objects can be represented by an ordered m-tuple

(1) $(a_1, a_2, \ldots, a_m)$,

where the ith coordinate is the ith object drawn. However, since an object is not replaced after it is drawn, the same object cannot be drawn twice; hence all coordinates in (1) must be different. Each draw of m objects without replacement but with regard to order can be represented by an ordered m-tuple such as (1) and none of the coordinates will be repeated. Conversely, any ordered m-tuple with coordinates chosen from the set from which the m objects are drawn and without any repetition of coordinates represents a specific draw of m objects without replacements but with regard to order. There are therefore precisely as many ordered m-tuples without any coordinates repeated which can be formed from the elements of the set from which the selection of m objects is being made as there are draws of m objects from the set of n objects without replacement but with regard to order. The problem then of finding the number of draws of the specified type is equivalent to finding the number of m-tuples of the specified type. We now compute the number of these m-tuples.

There are n ways to fill the first coordinate of (1), that is, n ways to make the first draw, since there are n objects from which to choose. The set S_2 of possibilities for the second coordinate will contain only $n-1$ elements, since one of the original n elements will have been used to fill the first coordinate and coordinates may not be repeated. The set S_3 of possibilities for the third coordinate will contain only $n-2$ elements, since two of the original n elements will have been removed to fill the first and second coordinates. We continue in the same manner and eventually find that after the first $m-1$ coordinates have been assigned there are only $n-(m-1) = n-m+1$ possibilities for the mth coordinate.*

Applying the corollary to Proposition 1, we arrive at:

Proposition 2. *If S is a set of n elements, then the number of ways in which we can select m objects from S without replacing any objects selected and taking the order of selection into account is equal to the number of ordered m-tuples that we can form from the elements of S without repeating any coordinates. This number is the number of elements of the set $S_1 \times S_2 \times S_3 \times \cdots \times S_m$, where S_i is the set of possibilities for the ith coordinate, assuming the first $i-1$ coordinates have been filled. S_i contains $n-i+1$ elements, $i = 2, \ldots, m$, and S contains n elements. Therefore the number of draws of the specified type is*

(2) $\quad {}_nP_m = n(n-1)(n-2)\cdots(n-m+1).$

Definition 1. *A particular selection of m objects from n objects without replacement but with regard to the order of selection is said to be a **permutation of n objects taken m at a time**. The number ${}_nP_m$ of Proposition 2 is the number of permutations of n things taken m at a time.*

---

* If a draw of m elements is to make sense, of course m can be at most as large as n.

Example 4. Consider the five slips in the urn of Example 1. The total number of ways in which three slips can be drawn from the urn without replacement but with regard to order is $_5P_3 = 5 \cdot 4 \cdot 3 = 60$. The probability that slips 1, 2, and 3 will be drawn in that order is $\frac{1}{60}$.

Example 5. Four people A, B, C, and D are in a room. In how many ways can these four people file out of the room one at a time? Although this problem is not stated as a selection problem, it is equivalent to one. The answer to this question is the same as the answer to the question, "In how many ways can four people be chosen one at a time without replacement, but with regard to the order of selection?" The answer is $_4P_4 = 4 \cdot 3 \cdot 2 \cdot 1 = 24$.

What is the probability that the people will leave in the order $ABCD$ or $DCBA$? Since $ABCD$ and $DCBA$ are mutually exclusive events, we have $P(ABCD \cup DCBA) = P(ABCD) + P(DCBA) = \frac{1}{24} + \frac{1}{24} = \frac{1}{12}$.

EXERCISES

1. Numerically evaluate each of the following:
   a) $_1P_1$ b) $_4P_2$ c) $_5P_2$ d) $_{10}P_3$
   e) $_{78}P_0$ (In how many ways can we select 0 objects from 78 objects?)
   f) $_nP_1$, where n is any positive integer
   g) $_nP_0$, where n is any positive integer

2. For any nonnegative integer, n, we define n *factorial* to be 1 if $n = 0$, and the product of all positive integers less than or equal to n if n is positive. Thus 5 factorial would be $5 \cdot 4 \cdot 3 \cdot 2 \cdot 1 = 120$. We usually denote n factorial by $n!$.
   a) Evaluate $4!, 6!, 0!2!$ b) Prove that $_nP_m = n!/(n-m)!$

3. Answer each of the following. If the complete numerical evaluation of a result would be too lengthy, the answer may be left in any form that clearly indicates how a final numerical answer would be arrived at.
   a) In how many ways can each of five children select a different number from 1 to 10?
   b) In how many ways can a person draw five slips from an urn which contains ten slips without replacement, but with regard to the order of selection?
   c) Suppose five people A, B, C, D, and E are in a room. In how many ways can these people file out of the room if A is to go first and B last?
   d) How many committees of five people can be selected from a group of 20 people if the order of selection determines a person's rank on the committee?
   e) Groups A and B each contain eight people. The people in the two groups are to march two by two into a room, with a person from one group paired with a person from the other group. In how many ways can this be accomplished?

4. Compute each of the probabilities asked for in the following. It may not be necessary to use permutations to arrive at an answer.

a) Five people A, B, C, D, and E file out of a room one at a time. What is the probability that A leaves first? that A leaves last? that A leaves first and E leaves last?
b) Each of five children A, B, C, D, and E chooses a different number from 1 to 10. What is the probability that B chooses 5? that 3 is one of the numbers chosen? Another person selects a number from 1 to 10 at random (which may be a number picked by one of the children). What is the probability that it is one of the numbers picked by the children? that it is the number that A picked? that it is a number not picked by any of the children? that it is either 3 or 4?
c) Six people are chosen at random from a group of 15 people to fill six distinct offices. We shall denote the people from among whom the selection was made by 1 through 15. What is the probability that 1 is selected? that 4, 5, 6, 7, 14, and 15 are selected? Take care in answering this last question. The answer is not $1/{}_{15}P_6$. Why?

2.2 SAMPLING WITHOUT REPLACEMENT OR REGARD TO ORDER

In Section 2.1 we learned to compute the number of ways in which m objects can be selected from n objects, provided (1) no object is replaced in the pool of objects after it has been selected, and (2) the order of selection is taken into account, that is, two selections of the same m objects will be considered to be different if the m objects are selected in different orders. In this section we shall learn to compute the number of ways that m objects can be selected from n objects, provided (1) no object is replaced in the pool of objects after it has been selected, and (2) the order of selection is not taken into account, that is, all we shall concern ourselves with is the m objects that are selected. The following examples give specific instances of the general problem we shall be studying.

Example 6. In how many ways can we select five cards from a standard deck of 52 cards, taking only the cards selected into account (and not the order of selection)? In other words, how many five-card poker hands are there?

Example 7. In how many ways can a committee of six people be selected from among 20 people if the order of selection does not matter?

Example 8. What is the probability of obtaining a bridge hand which contains 12 spades?

We know that the number of ways of selecting m objects from a set of n objects without replacement, but with regard to order, is ${}_nP_m$. Suppose f is some particular permutation of n objects taken m at a time. We shall say that a permutation g of the n objects taken m at a time is *equivalent* to f if f and g each select precisely the same m objects.

Let S be a set of n objects and C be any choice of m objects of S without replacement or regard to order. Then we can associate C with the set of all permutations of n objects (the n elements of S) taken m at a time which select the same m objects that C does. In other words, if f is a permutation which selects the same m objects from S that C does, then we can associate C with f and all permutations equivalent to f. We now compute the number of permutations which are equivalent to f.

Suppose, for convenience, that f selects the objects $1, 2, \ldots, m$. Then any permutation g which is equivalent to f selects the objects $1, 2, \ldots, m$ and differs from f at most in the order of selection. There will therefore be the same number of permutations equivalent to f as there are ways of ordering the objects $1, 2, \ldots, m$. For each permutation equivalent to f determines an ordering of $1, 2, \ldots, m$ (namely, the order in which it selects these objects), and, conversely, any ordering of $1, 2, \ldots, m$ determines a unique permutation equivalent to f (namely, the permutation which selects $1, 2, \ldots, m$ in the order given by the ordering). There are, however, $_mP_m$ ways of ordering $1, 2, \ldots, m$; hence there are $_mP_m$ permutations equivalent to f. Therefore if C is a selection of $1, 2, \ldots, m$ objects without regard to order of selection, then C corresponds to $_mP_m$ permutations.

There are $_nP_m$ permutations of n objects taken m at a time and each unordered selection of m objects corresponds to $_mP_m$ of these permutations. Moreover, any unordered selection of m objects corresponds only to those $_mP_m$ permutations which select the same m objects that it does. Consequently, the $_nP_m$ permutations can be broken up into nonoverlapping groups, each containing $_mP_m$ permutations, in such a way that each of these groups corresponds to a unique unordered selection of m objects and each unordered selection of m objects corresponds to one of these groups. If, however, there are $_nP_m$ permutations broken up into nonoverlapping groups of $_mP_m$ permutations each, then there are $_nP_m/_mP_m$ groups.* Hence there are

$$(3) \qquad C(n, m) = {_nP_m}/{_mP_m} = \frac{n(n-1)\cdots(n-m+1)}{m(m-1)\cdots 2 \cdot 1}$$

unordered selections of m objects.

Definition 2. *A particular selection of m objects from a collection of n objects without replacement or regard to order of selection is called a* **combination of n objects taken m at a time.** *$C(n, m)$ is the number of combinations of n objects taken m at a time. [In some texts, $\binom{n}{m}$ is used instead of $C(n, m)$.]*

---

* This is analogous to the following: Twenty people are used to form committees of five members each such that each person is on one and only one committee. We must therefore have four committees.

Using the factorial notation of Exercise 2 of the preceding section, we find

(4) $C(n, m) = \dfrac{n!}{(n-m)!m!}.$

Example 9. The answers to the problems posed in Examples 6 and 7 are $C(52, 5)$ and $C(20, 6)$, respectively. Since the numerical evaluation of $C(n, m)$ and $_nP_m$ generally requires a strong constitution but does not involve any mathematics beyond that of elementary school, we shall, as a rule, leave such quantities unevaluated. The reader should be certain, however, that he knows how to evaluate the expressions if required to do so.

Example 10. In this example we solve the problem posed in Example 8. A bridge hand is one selection of thirteen cards (without replacement or regard to order) from a deck of 52 cards.* Therefore the total number of bridge hands is $C(52, 13)$. Thus the probability of any particular bridge hand is $1/C(52, 13)$ by the corollary to Proposition 2 of Chapter 1. If M is the number of bridge hands which contain 12 spades, then the probability of getting a bridge hand which contains 12 spades is $M/C(52, 13)$. The problem now is to evaluate M.

There are 13 spades and 39 other cards in the deck. Imagine the spades placed in one pile and the 39 remaining cards placed in another pile. To get a hand which contains 12 spades, we must select twelve cards from the spade pile and one card from the non-spade pile. The total number of ways of selecting 12 spades from the aggregate of 13 spades is $C(13, 12)$ and the number of ways of selecting the other card is $C(39, 1)$. Therefore the total number of ways in which we can get a hand with exactly 12 spades is $C(13, 12)C(39, 1)$ (Proposition 1 can be used to justify this multiplication). Consequently, the probability asked for in Example 8 is $C(13, 12)C(39, 1)/C(52, 13)$.

Example 11. Twelve men and ten women are at a meeting. A committee of five is chosen at random from those present. What is the probability that the committee consists entirely of men? There are 22 people from which five are to be selected (presumably without regard to the order of selection). Therefore the total number of committees that might be formed is $C(22, 5)$. There are $C(12, 5)$ ways to choose five men and $C(10, 0) = 1$ way to choose none of the women. Consequently, the number of committees of five which consist entirely of men is

$C(12, 5)C(10, 0) = C(12, 5).$

---

* That is, a bridge hand is a combination of 52 things taken 13 at a time.

Therefore the probability of choosing a committee which consists entirely of men is $C(12, 5)/C(22, 5)$.

EXERCISES

1. Numerically evaluate each of the following:
   a) $C(5, 4)$ b) $C(7, 4)$ c) $C(10, 7)$
   d) $C(n, n)$, where n is any positive integer
   e) $C(n, 1)$, where n is any positive integer
   f) $C(100, 98)$

2. Note that $C(n, 1) = C(n, n - 1)$. Using (4) prove:
   a) If $n = 2m$, then $C(n, k) = C(n, n - k)$, $k = 1, 2, \ldots, m$.
   b) If $n = 2m + 1$, does the equality in (a) remain valid?
   c) Determine the value of k for which $C(n, k)$ has its largest value.

3. Answer each of the following:
   a) In how many ways can a committee of five be selected from 100 candidates?
   b) How many seven-card poker hands are there?
   c) In a certain gambling game there are 32 numbers on which one can bet. Someone wishes to bet on exactly three of these numbers. In how many ways can this be done?
   d) Someone wishes to play five slot machines in a row of 20 machines. In how many ways can this be done?
   e) An urn contains 13 red and 14 white balls. How many ways are there of selecting nine balls (without replacement) such that exactly five of these are red?
   f) A man is to select 10 bills from a stack of 100 one-dollar bills, three bills from a stack of 25 five-dollar bills, and five bills from a stack of 10 ten-dollar bills. In how many ways can this be done?

4. Compute each of the probabilities asked for in the following.
   a) A committee of three is to be chosen at random from a gathering of 10 people; we designate the people as $1, 2, \ldots, 10$. What is the probability that the committee selected will have no. 1 as a member? that it will have no. 5 as a member? that it will consist of nos. 6, 7, and 8? that no. 4 will not be selected? that either no. 4 or no. 5 will be a member?
   b) An urn contains eight red and seven blue balls. One ball is drawn from the urn. What is the probability that the ball is blue? Two balls are drawn from the urn. What is the probability that both are red? that one is red and one is blue? that one is black? Five balls are drawn from the urn. What is the probability that exactly two of these are blue? that at least two, that is, two or more, are blue? that less than two are blue?
   c) An urn contains seven blue, five red, and six white balls. One ball is drawn from the urn. What is the probability that it is white? Two balls are drawn from the urn. What is the probability that one is red and the other is white? Seven balls are drawn from the urn. What is the probability that two are blue, three are red, and two are white? that all seven are red?

2.3 SAMPLING WITH REPLACEMENT AND WITH REGARD TO ORDER

Suppose there are three red and four blue balls in an urn, and a selection of two balls is made at random from the urn. What is the probability that one ball is blue and the other is red? If we assume that the draws are made without replacement, then the answer is $C(3, 1)C(4, 1)/C(7, 2)$. We can, however, also solve this problem without the use of combinations as follows: In order to draw one red and one blue ball in two draws, we must have

(red on the first draw $\cap$ blue on the second draw)
$\cup$ (blue on the first draw $\cap$ red on the second draw).

We shall denote red by R, blue by B, and use a subscript to indicate the draw; for example, R_1 will indicate a red ball on the first draw. Therefore drawing one red and one blue ball is equivalent to

$(R_1 \cap B_2) \cup (B_1 \cap R_2)$.

Using Assumption 2 of Section 1.3, we find

(4) $\quad P\big((R_1 \cap B_2) \cup (B_1 \cap R_2)\big) = P(R_1 \cap B_2) + P(B_1 \cap R_2)$
$\quad\quad\quad - P\big((R_1 \cap B_2) \cap (B_1 \cap R_2)\big).$

Noting that $\big((R_1 \cap B_2) \cap (B_1 \cap R_2)\big)$ is impossible (and hence has probability 0), and using Assumption 3 of Section 1.3, we find that (4) reduces to

(5) $\quad P\big((R_1 \cap B_2) \cup (B_1 \cap R_2)\big)$
$\quad\quad\quad = P(R_1)P(B_2 \mid R_1) + P(B_1)P(R_2 \mid B_1).$

Since we originally have three red and four blue balls, $P(R_1) = \frac{3}{7}$ and $P(B_1) = \frac{4}{7}$. After one ball has been removed, only six balls remain; which six depends on what color ball was taken away. Thus, if a blue ball was withdrawn, then there would be three red and three blue balls left. Consequently, $P(R_2 \mid B_1) = \frac{3}{6}$. Similarly, $P(B_2 \mid R_1) = \frac{4}{6}$. We now have all the information necessary for the numerical evaluation of (5).

Suppose now the problem were changed to read: There are three red and four blue balls in an urn, and a selection of two balls is made at random from the urn, *with the first ball being replaced before the second ball is chosen*. What is the probability that one of the balls is red and the other is blue? This problem can be solved using the same method as above where there was no replacement except that the values for $P(R_2 \mid B_1)$ and $P(B_2 \mid R_1)$ will be different. Since the first draw is replaced, what is drawn first has no effect on the probability of what will be drawn second. Consequently, $P(R_2 \mid B_1)$ and $P(B_2 \mid R_1)$ are the same as if we were selecting one ball from the urn without any previous draws. Thus,

$P(R_2 \mid B_1) = \frac{3}{7}$ and $\quad P(B_2 \mid R_1) = \frac{4}{7}.$

Many situations encountered in the application of probability theory are really instances of the following general situation: A selection of m objects is to be made from a collection of n objects. But after an object is drawn, it is returned to the collection before the next draw is made. We need to know the number of ways in which one can select the m objects, taking the order of selection into account.

Examples 12 and 13 illustrate this type of problem.

Example 12. An urn contains 12 red and 14 blue balls. A ball is drawn at random from the urn, its color is recorded, and the ball is returned to the urn. This process is repeated five times. How many different strings of five recordings are possible?

Example 13. A coin is flipped 15 consecutive times. How many different strings of heads and tails are possible? Although this is not explicitly a "choosing" problem, it is equivalent to the following: A selection is made from the set {head, tail}, the selection is recorded, and then returned to the set. This process is repeated 15 times. How many different strings of heads and tails are possible?

As in Section 2.1 we can represent a particular ordered selection of m objects as an ordered m-tuple

(6) $\quad (a_1, a_2, \ldots, a_m),$

where the coordinates of the m-tuple are elements of the set from which the selection is being made. Since we are now working with replacement, we can have repetition of coordinates; moreover, it is also possible to have m greater than n as, for example, in Example 13. Let S be the set of n objects from which the draws are made. Then the m-tuples (with possible repetition of coordinates) are nothing but the elements of

$S \times S \times \cdots \times$ (m times) $\times S$.

By the corollary to Proposition 1, there are n^m m-tuples; hence there are n^m ordered selections of m objects from S with replacement.

Example 14. The answer to the question posed in Example 12 is 2^5. Note that we are only recording the color of each draw; hence on each draw we have only two possibilities, red or blue. The answer to the problem in Example 13 is 2^{15}.

Example 15. In how many ways can we fill five distinct positions from among 20 candidates if any person may hold more than one position? We want the number of selections of five people that can be made from an aggregate of 20 people with replacement (since any person remains eligible to be chosen even if previously selected) and with regard to order (since the positions to be filled are distinct). The answer is 20^5.

Definition 3. *There is no generally accepted term for an ordered selection of m objects from n objects with replacement. We shall call such a selection an* **accumulation of *n* objects taken *m* at a time.** *We shall denote the number of accumulations of n objects taken m at a time by $A(n, m)$. Thus we have*

(7) $\quad A(n, m) = n^m$.

We illustrate the use of $A(n, m)$ in finding probabilities in the following examples.

Example 16. A fair coin is flipped fairly four times. What is the probability that all four flips are heads? The total number of (ordered) strings of heads and tails that can be obtained from the four flips is 2^4. Since the coin and flips are fair, all the strings can be assumed to have the same probability. Hence the probability of one particular string, and in particular the one consisting of four heads, is $(\frac{1}{2})^4$.

Example 17. A fair die is rolled fairly seven times. What is the probability that exactly three of these rolls will be 4's? Here we are making an ordered selection with replacement seven times from a set of six objects (the faces of the die). The total number of possible strings of seven rolls is therefore 6^7. Since the die and each roll are fair, all the strings have the same probability; hence each string has probability $(\frac{1}{6})^7$. If M is the number of strings in which exactly three 4's are found, then the probability asked for is $M/6^7$. We now compute M.

Each string of seven rolls can be represented by an ordered 7-tuple. If a 7-tuple is to represent a string containing exactly three 4's, then three of the coordinates of the 7-tuple will have to be 4. The number of ways in which three coordinates can be chosen without replacement or regard to order (since each of the three coordinates will be the same, namely, 4) is $C(7, 3)$. There are 5 choices for each of the remaining 4 coordinates. Therefore $M = C(7, 3)5^4$. Consequently, the probability asked for is $C(7, 3)5^4/6^7$.

EXERCISES

1. Compute numerically each of the following:
   a) $A(3, 4)$ b) $A(4, 3)$ c) $A(2, 10)$
   d) $A(n, 1)$, where n is any positive integer
   e) $A(n, 0)$, where n is any positive integer
   f) $A(n, n)$, where n is any positive integer

2. Order $A(n, m)$, $C(n, m)$, and $_nP_m$ according to size. Justify your answer not only by means of the formulas used for computing these quantities, but also by using the underlying notion that each of these symbols represents, that is, you should be able to justify your answer without appealing to the formula for each expression.

3. Answer each of the following:
   a) An urn contains three red and four blue balls. A selection is made from the urn, the color of the selection is recorded, and the ball is returned to the urn. The process is repeated six times. In how many ways can the (ordered) string of six draws turn out? Note that only the color is being recorded, hence we are drawing from a set of two objects rather than seven. Do all the strings have the same probability? Explain carefully.
   b) A balanced die is rolled five times. How many strings of numbers can this give?
   c) In how many ways can two offices be filled from 30 candidates if any candidate can hold both offices?
   d) In how many ways can each of two people be assigned 15 out of 30 jobs if both people are never to be assigned the same job?
4. Find each of the probabilities asked for in the following:
   a) A fair coin is flipped fairly five times. What is the probability that all five flips are heads? that two of the five flips are heads? that two or more of the five flips are heads?
   b) A fair die is rolled fairly six times. What is the probability of getting a 6 on every roll? of getting exactly four 6's? of getting no 6's? of getting three 4's and three 5's?
   c) Ten jobs are to be assigned randomly to three people with no two people getting the same job. Denote the people by A, B, and C, and the jobs by $1, 2, \ldots, 10$. What is the probability that A gets job 1? that B gets all 10 jobs? that C gets jobs 4 and 5? that C is not assigned any of the jobs?

2.4 BAYES' THEOREM

Suppose that event A can only occur in conjunction with one and only one of events $B_1, B_2, \ldots, B_n$. If A occurs, then it follows that some B_i will also have occurred since A can occur only in conjunction with some B_i. If A has occurred, what is the probability that it is B_1 that has occurred in conjunction with A; that is, what is $P(B_1 \mid A)$?

From Assumption 3 of Section 1.3, we have

(8) $P(A \cap B_1) = P(A)P(B_1 \mid A) = P(B_1)P(A \mid B_1).$

Consequently, it follows that

(9) $P(B_1 \mid A) = \dfrac{P(B_1)P(A \mid B_1)}{P(A)}.$

Now since A must occur in conjunction with one of the B_i's, event A is equivalent to

(10) $(A \cap B_1) \cup (A \cap B_2) \cup \cdots \cup (A \cap B_n).$

Since, by hypothesis, A can occur in conjunction with only one of the

B_i's, the events in (10) are mutually exclusive in pairs. For if, say, $A \cap B_1$ and $A \cap B_2$ occurred together, then A would occur together with both B_1 and B_2. Since A and (10) are equivalent, Proposition 2 of Chapter 1 gives us

(11) $P(A) = P(A \cap B_1) + P(A \cap B_2) + \cdots + P(A \cap B_n).$

This, in turn, gives

(12) $P(A) = P(B_1)P(A \mid B_1) + P(B_2)P(A \mid B_2)$
    $+ \cdots + P(B_n)P(A \mid B_n).$

Using (12) to substitute for $P(A)$ in (9), we obtain

(13) $P(B_1 \mid A)$
$$= \frac{P(B_1)P(A \mid B_1)}{P(B_1)P(A \mid B_1) + P(B_2)P(A \mid B_2) + \cdots + P(B_n)P(A \mid B_n)}.$$

Equation (13) is known as *Bayes' Theorem*.

Despite its rather formidable appearance, the evaluation of (13) is rather straightforward in most instances. The following examples illustrate the use of Bayes' Theorem.

Example 18. Urn 1 contains three red and four blue balls, while Urn 2 contains eight red and eight blue balls. Three balls are drawn at random (without replacement) from one of the urns and all are found to be red. What is the probability that the selection was made from Urn 1? The event that actually occurred was the selection of three red balls; hence we shall let this be event A. Since the three balls must have been selected from either Urn 1 or Urn 2, we shall let B_1 and B_2 be selections from Urn 1 and Urn 2, respectively. Then the probability we are looking for is $P(B_1 \mid A)$.

Since there are two urns and there is no reason to suppose that one was more likely to be chosen than the other, we set

$P(B_1) = P(B_2) = \frac{1}{2}.$

In this problem, (13) has the form

(14) $P(B_1 \mid A) = \dfrac{P(B_1)P(A \mid B_1)}{P(B_1)P(A \mid B_1) + P(B_2)P(A \mid B_2)}.$

We already have $P(B_1)$ and $P(B_2)$, hence all that remains to be done is to find $P(A \mid B_1)$ and $P(A \mid B_2)$.

Now $P(A \mid B_1)$ is just the probability of drawing three red balls from the first urn. Using the methods of Section 2.2, we find that

$P(A \mid B_1) = C(3, 3)/C(7, 3).$

Similarly, since $P(A \mid B_2)$ is the probability of drawing three red balls from Urn 2, we have $P(A \mid B_2) = C(8, 3)/C(16, 3)$. With the appropriate substitutions (14) becomes

(15) $\quad P(B_1 \mid A) = \dfrac{(\frac{1}{2})(C(3,3)/C(7,3))}{(\frac{1}{2})(C(3,3)/C(7,3)) + (\frac{1}{2})(C(8,3)/C(16,3))}.$

Example 19. It is known that one die out of a given set of six dice rolls only 4's, although the other dice in the set are unbiased (that is, fair). It is not known, however, which die is biased. One of the dice is chosen at random and rolled three times, and each time a 4 appears. What is the probability that the die chosen is the biased one? The event that actually occurred is rolling three straight 4's; hence this shall be event A. This roll was accomplished either with a fair die, or a biased die; therefore we let B_1 be the event of using the biased die and B_2 be $\bar{B}_1$. The probability sought is $P(B_1 \mid A)$; Bayes' Theorem for this problem again has the form (14).

Since there are six dice, only one of which is biased, the probability of choosing the biased die, that is, $P(B_1)$, is $\frac{1}{6}$; therefore

$$P(B_2) = 1 - P(B_1) = \tfrac{5}{6}.$$

Now $A \mid B_1$ is the event of rolling three 4's with the biased die. Since the biased die always rolls 4's, $A \mid B_1$ is certain; hence $P(A \mid B_1) = 1$. On the other hand, the probability of rolling three consecutive 4's with a fair die, that is, $P(A \mid B_2)$, is $(\tfrac{1}{6})^3$. Making the appropriate substitutions in (14), we obtain

(16) $\quad P(B_1 \mid A) = \dfrac{(\tfrac{1}{6})(1)}{(\tfrac{1}{6})(1) + (\tfrac{5}{6})(\tfrac{1}{6})^3}.$

There are problems that seem to lend themselves to solutions via Bayes' Theorem, although in reality the use of Bayes' Theorem is not legitimate. The following example illustrates this point.

Example 20. In a certain part of the Antartic snow is always accompanied by high winds or freezing temperatures. Let A, B_1, and B_2 be the events of snow, high winds, and freezing temperatures, respectively. We shall assume that

$P(B_1) = .7$, $P(B_2) = .98$, $P(A \mid B_1) = .1$, and $P(A \mid B_2) = .1$.

Although all the terms necessary to evaluate (14) are given, we cannot use (14) to find $P(B_1 \mid A)$. For $A \cap B_1$ and $A \cap B_2$ are not mutually exclusive; therefore (11), a necessary step in the derivation of Bayes' Theorem, is invalid in this situation.

EXERCISES

1. Find the probabilities asked for in each of the following. In some of the problems below, Bayes' Theorem may not be applicable. In such instances,

try to find the desired probability by other methods. If the problem cannot be solved from the information given, indicate what other information is needed before it can be solved.

a) Urn 1 contains four red and four blue balls, while Urn 2 contains six red, seven blue, and six white balls. Four balls are selected at random without replacement from one of the urns. What is the probability that the selection was made from Urn 1 if all the balls drawn are blue? What is the probability that the selection was made from Urn 2 if two of the balls are red and two are blue? What is the probability that the selection was made from Urn 2 if one of the balls drawn is white?

b) It is known that one of three dice rolls 6's twice as often as an unbiased die, but the other two dice are fair. It is not known which of the three dice is biased. One of the dice is selected at random and rolled five times. What is the probability that the die picked is the biased one if all five rolls are 6's? What is the probability that the die chosen is the biased one if none of the five rolls is a 6?

c) Let A, B_1, and B_2 be as in Example 20. What is $P(B_1 \mid A)$? what is $P(A)$?

d) Every housewife uses one of three brands of cleanser. Brand X controls 30%, Brand Y, 20%, and Brand Z, 50%, of the market. Three housewives meet at a party. What is the probability that all three use the same brand? If all three use the same brand, what is the probability that it is Brand Y?

2. What simplification can be made in (13) if $B_1, B_2, \ldots, B_n$ have the same probability, for example, as in Example 18.

3. An event A always occurs in conjunction with one of the events B_1, B_2, and B_3; also, B_1, B_2, and B_3 are mutually exclusive (that is, $B_1 \cap B_2 \cap B_3$ is the empty set) though not necessarily in pairs. Assume $P(B_1 \mid A) = \frac{1}{4}$, $P(B_2 \mid A) = \frac{1}{4}$, and $P(B_3 \mid A) = \frac{1}{2}$.

a) Was it necessary that $P(B_1 \mid A) + P(B_2 \mid A) + P(B_3 \mid A) = 1$? Explain carefully.

b) Assume that B_1, B_2, and B_3 are equiprobable (that is, have the same probability). What information, if any, can we gather about $P(A \mid B_1)$?

c) Suppose B_1, B_2, and B_3 are equiprobable and mutually exclusive in pairs. What information, if any, can we gather about $P(A \mid B_1)$?

3 Random Variables. Graphs

3.1 RANDOM VARIABLES. ADMISSIBLE RANGES

At times certain characteristics of a point of a sample space S are of great interest. Consider the following examples.

Example 1. A clinic tests the efficacy of a new treatment for a certain disease by using the treatment on 50 patients. After the treatment, a patient is classified as either improved (I) or *not-improved* (N). The sample space for the experiment consists of all possible strings of 50 observations; we know from Section 2.3 that there are 2^{50} such strings. The doctors conducting the test, however, are not so much interested in what particular string is actually observed, but in the number of I's in the string, since it is this number which is the best measure of how effective the treatment really is. Let S be the sample space of strings of observations. For each element s of S, define $X(s)$ to be the number of I's in s; X is a function from S into the set of real numbers.

Example 2. The temperature at noon in Miami is recorded each day for a whole year. The sample space S is all ordered 365-tuples of real numbers (assuming that any real number can serve as a temperature). We now define several functions from S into the set of real numbers:

a) $X(s)$ = temperature on the 10th day of the year for each s in S;
b) $Y(s)$ = average year-round temperature for each s in S;
c) $W(s)$ = average temperature in July for each s in S;
d) $Z(s)$ = numbers of days with a temperature above 70° for each s in S.

Definition 1. *A function X from a sample space S into the set of real numbers is said to be a* **random variable,** *or* **variate.** *The image of a random variable X, that is, $\{X(s) \mid s \text{ in } S\}$, is called the* **admissible range** *of X. Any element of the admissible range of X is said to be an* **admissible value** *of X.*

We shall use the terms *random variable* and *variate* interchangeably. A random variable will be denoted by a capital letter, for example, X, while specific values of the variate will be denoted by lower-case letters indexed, if necessary, to keep them distinct, for example $x_1, x_2, \ldots, x_n$.

Example 3. The admissible range of the random variable X of Example 1 is $\{0, 1, 2, \ldots, 50\}$. The admissible range of the variate X in Example 2 is the entire set of real numbers, while the variate Z of that same example has $\{0, 1, 2, \ldots, 365\}$ as its admissible range.

Just as we can consider an event "informally," that is, without explicit reference to a sample space, so a random variable can be considered informally as a numerical variable whose values are generated by a random selection, or process dependent on chance. The next example defines a random variable without explicit reference to a sample space.

Example 4. Let X be the number of particles emitted by a radioactive source during one hour. The admissible range of X is the entire set of nonnegative integers.

The admissible range of X in Example 4 is an infinite set of integers, whereas the admissible range of Z in Example 2 is a finite set of integers. Both X and Z are examples of *discrete* variates. The variates Y and W of Example 2 can take on any real number as their value; these are examples of *continuous variates*.

The following definition is preliminary to making the notions of a discrete and a continuous variate more precise.

Definition 2. *Suppose a and b are real numbers such that $a < b$. Then the* **open interval** *determined by a and b is defined to be $\{x \mid a < x < b\}$. The* **closed interval** *determined by a and b is defined to be $\{x \mid a \leq x \leq b\}$. We use (a, b) and $[a, b]$ to denote the open and closed intervals, respectively,* determined by a and b. We also let

$$[a, b) = \{x \mid a \leq x < b\}, \qquad (a, b] = \{x \mid a < x \leq b\},$$
$$(a, \infty) = \{x \mid a < x\}, \qquad (-\infty, a) = \{x \mid x < a\},$$

$[a, b)$ and $(a, b]$ are called **half-open intervals**, while (a, ∞) and $(-\infty, a)$ are **open half-lines**. The general rule is that parentheses are used to indicate the **exclusion** of an "endpoint," while brackets are used to indicate the **inclusion** of an "endpoint." It is left to the reader to find the definition and notation for a **closed half-line**.

Definition 3. A random variable X is said to be **continuous** if the admissible range of X is the entire set of real numbers, an open or closed half-line, an open or closed interval, or a half-open interval.

Definition 4. A random variable X is said to be **discrete** if the admissible range of X can be put in one-one correspondence with a set of integers.

Example 5. If the admissible range of a variate X is a set of integers, then X is discrete. It is in fact these discrete variates which occur most frequently and are of the greatest importance.

Example 6. If the admissible range of a variate X is a finite set of real numbers, say $\{x_1, \ldots, x_n\}$, then X is a discrete variate.

Example 7. The lifespan of a randomly selected person is a random variable X which is continuous if we assume that a length of time can be measured to an arbitrary degree of accuracy. The admissible range of x can be considered to be $(0, \infty)$.

The "theoretical" admissible range may be larger than the "practical" admissible range. Even though the longest any person may be known to have lived is, say, 175 years, it is still possible to take $(0, \infty)$ as the admissible range of X in this example because (1) there is nothing that intrinsically limits a lifetime to less than 175 years, (2) one must allow for the possibility of the unusual, and (3) it may be easier to fit a *distribution* to X if X has admissible range $(0, \infty)$ rather than $(0, a)$ for some positive number a. This last point will be discussed later in this book.

If X is a random variable defined on a sample space S and f is any function from the set R of real numbers into R, then the random variable $f(X)$ is defined by $\bigl(f(X)\bigr)(s) = f\bigl(X(s)\bigr)$ for each s in S. For example, if Z is the random variable defined as in Example 2 and g is defined by $g(x) = x(\frac{100}{365})$, then

$$g(Z) = Z(\tfrac{100}{365})$$

is the percentage of days in a year with a temperature above 70°.

We may also define algebraic operations with random variables. For example, if X and Y are random variables defined on a sample space S, then the random variable $X + Y$ is defined by

$$(X + Y)(s) = X(s) + Y(s)$$

for each s in S, and XY is defined by

$$XY(s) = X(s)Y(s)$$

for each s in S. We can also define such random variables as $3X$, $X^2 - Y^2$, etc. In Example 2, the random variable $W - Y$ would be the difference between the average temperature in July and the average year-round temperature.

EXERCISES

1. For which of the following admissible ranges would a random variable X be discrete? continuous? If a variate is neither discrete nor continuous, indicate which properties are lacking that would make it discrete or continuous.
    a) $\{1, 2, 3\}$
    b) $(-1, 1)$
    c) $\{1, 1.1, 1.11, 1.111, 1.1111, \ldots\}$
    d) $\{x \mid x \text{ is in } (-1, 1) \text{ or } x = 3\}$
    e) $\{w \mid w > 0\}$
    f) $\{r \mid 0 \leq r < 8 \text{ or } 10 < r < 11\}$

2. Confirm that any variate whose admissible range is a finite set of real numbers is discrete.

3. Suppose X and Y are random variables (on the sample space S) with admissible ranges A and B, respectively.
    a) What is the largest possible admissible range of $X + Y$? If X and Y are discrete, prove that $X + Y$ is discrete. If X and Y are continuous, is $X + Y$ necessarily continuous? Explain.
    b) What is the largest possible admissible range of XY? If X and Y are discrete, is XY discrete? If X and Y are continuous, is XY necessarily continuous? Explain.
    c) Suppose $A = \{3, 4, 5\}$ and $B = \{\frac{1}{2}, 2, \frac{5}{4}\}$. Find the largest possible admissible ranges of $X + Y$ and XY.
    d) Suppose $A = (1, 2)$ and $B = (-1, 1)$. Find the largest possible admissible ranges of $X + Y$ and XY.
    e) Let $g(x) = x^2 + 3x$. What is $g(X)$? What is the admissible range of $g(X)$?

4. Explain what each of the following random variables represent.
    a) $50 - X$, where X is as given in Example 1
    b) $2X$, where X is as given in Example 1
    c) $Y - X$, where X and Y are as given in Example 2

5. Give five examples of discrete random variables that we encounter often. What is the admissible range of each of these variates? Give five examples of continuous random variables that we encounter often. What is the admissible range of each of these variates?

3.2 GRAPHS OF EQUALITIES AND INEQUALITIES

Before proceeding further in our discussion of random variables, we shall investigate certain aspects of functions.

As we saw in Section 1.1, a function with domain S and range T can be defined as a special kind of subset of $S \times T$. We can graphically portray a function f from S into T by representing $S \times T$ in a suitable manner and indicating which elements of $S \times T$ are in f.

Example 8. Let S and T be as in Example 7 of Chapter 1. Then the functions f_1, f_2, and f_3 from S into T given in that example can be portrayed graphically as shown below. Note that we are doing nothing more than indicating which elements of $S \times T$ are in these functions.

$\underline{(1,1)}$	$(1,2)$	$(1,3)$		$\underline{(1,1)}$	$(1,2)$	$\underline{(1,3)}$
$(2,1)$	$\underline{(2,2)}$	$(2,3)$		$(2,1)$	$(2,2)$	$\underline{(2,3)}$
$(3,1)$	$(3,2)$	$\underline{(3,3)}$		$\underline{(3,1)}$	$(3,2)$	$(3,3)$
	f_1				f_2	

		$\underline{(1,1)}$	$(1,2)$	$(1,3)$
		$\underline{(2,1)}$	$(2,2)$	$(2,3)$
		$\underline{(3,1)}$	$(3,2)$	$(3,3)$
			f_3	

Definition 5. *If f is a function from a set S into a set T, then the subset of $S \times T$ consisting of all pairs $(s, f(s))$ for all elements s of S (that is, the set of all points of the function, where the function is considered as a collection of ordered pairs) is called the **graph** of f.*

Indeed if a function is defined by means of ordered pairs, then a function and its graph are identical. However, it often happens that a visual representation of the function in the context of $S \times T$ is of help in our work with the function.

Example 9. Let f be the function defined by $f(x) = x^2$ from the set R of real numbers into R. Then f, and the graph of f, will consist of all ordered pairs of real numbers of the form $(x, f(x)) = (x, x^2)$. The set f presented as a subset of $R \times R$, the usual coordinate plane of analytic geometry, is shown in Fig. 1. From this "picture" of f, we are able to gather certain information about f. For example, we see that $f(x)$ increases as x increases if x is positive, but $f(x)$ decreases as x increases if x is negative. We also see that $f(x)$ is never less than 0. The graph of f, or more technically, a picture of the graph, is, like most pictures, an aid to the intuition, imagination, and memory. One usually needs as much information about a func-

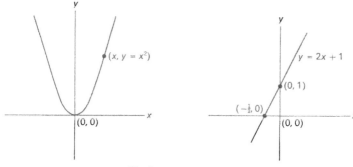

Fig. 1   Fig. 2

tion to draw an accurate graph of it as the graph itself might convey; but once the graph is obtained, it is a handy reminder of the properties of the function.

There is not much point in our considering methods of graphing a function now, since later on we shall have tools that will make our work much simpler.

A function f from a set S into a set T is one important subset of $S \times T$. There are, however, other important subsets of $S \times T$ that may arise in connection with our work. Many of these subsets can also be portrayed visually as subsets of $S \times T$.

Example 10. Let $S = T = \{1, 2, 3\}$. Then the graph of

$\{(s, t) \mid s$ is in S, t is in T, and $x < y\}$

is given below. If it is understood that we are dealing with the sets S and T, we can say that the graph given is that of $x < y$.

(1, 1) (1, 2) (1, 3)
(2, 1) (2, 2) (2, 3)
(3, 1) (3, 2) (3, 3)

Graph of $\{(x, y) \mid x < y\}$ or of $x < y$

Example 11. Let R be the set of real numbers. Then the graph of the straight line in $R \times R$ whose equation is $y = 2x + 1$ is as shown in Fig. 2. [This straight line consists of all points of the form $(x, 2x + 1)$. The graph is that of the function f defined by $f(x) = 2x + 1$.] We may, however, be interested in the points (x, y) of $R \times R$ which satisfy the inequality $y < 2x + 1$. This set of points does not form a function, since

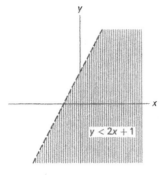

Fig. 3

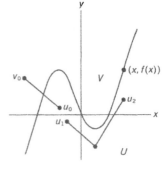

Fig. 4

one value of x can appear as a first coordinate many times. We can, nevertheless, portray $\{(x, y) \mid y < 2x + 1\}$. A graph of this set is shown in Fig. 3.

The following fact was used in arriving at Fig. 3 and is of great importance in graphing many inequalities in the coordinate plane.

Proposition 1. *Suppose f is a function from the set R of real numbers into R such that the graph of f separates $R \times R$ into two sets U and V (Fig. 4) having the following properties:*

1) *Any two points of U can be joined by a path of joined line segments no one of which meets the graph of f.*
2) *Any two points of V can be joined by a path of joined line segments no one of which meets the graph of f.*
3) *If a point of U is joined to a point of V by means of a path of joined line segments, then at least one of the segments meets the graph of f.* [*Recall that the graph of f consists of all points of the form (x, y), where $y = f(x)$.*] *Then all of the points (x, y) of U will satisfy one of the inequalities $y < f(x)$ or $y > f(x)$, and all of the points of V will satisfy the other inequality.*

Proposition 1 is an informal statement of a fairly profound mathematical fact. Because of its informality, it may be inaccurate for certain "pathological" functions, but it is certainly accurate in the case of "well-behaved" functions, and we shall assume that the reader will rarely, if ever, encounter pathological examples. Although Proposition 1 appears to be quite a mouthful, even "informally" stated, its application is usually rather straightforward. We illustrate its application in the next examples.

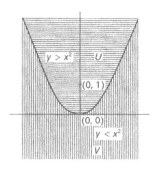
Fig. 5

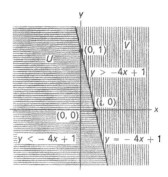
Fig. 6

Example 12. Consider the graph of f as in Example 9. This graph separates the plane into two sections labeled U and V in Fig. 5. If we take any two points of U, we can join these points by a sequence of joined line segments all of which lie in U and hence do not meet the graph of f. (Although a formal proof of this last assertion would involve a fair amount of work, the assertion should be geometrically clear. An argument by intuition is fraught with danger but an attempt at too much rigor is not without its own perils. Perhaps one of the strongest reasons for the huge number of calculus books on the market is that no one has found, or ever will find, the right mixture of intuition and formality to please everyone.) Similarly, any two points of V can be joined by a sequence of joined line segments all of which lie in V. But if we try to join a point of U with a point of V by means of a straight line segment, or any other sequence of joined line segments, we must cross the graph of f at some point. The hypotheses of Proposition 1 are therefore satisfied. Testing the point $(0, 1)$ of U, we find that it satisfies $y > x^2$ (since $1 > 0$). Therefore every point of U satisfies $y > x^2$. Every point of V then satisfies the inequality $y < x^2$.

Example 13. The graph of $y = -4x + 1$ is given in Fig. 6. We see that this graph separates the coordinate plane into two regions U and V. Given two points of U, the straight line segment joining these two points lies entirely in U. Similarly, the straight line segment joining any two points of V lies entirely in V. But if we try to join a point of U with a point of V by means of any path consisting of joined line segments, then that path will meet the graph. Testing the point $(0, 0)$ of U, we find that it satisfies the inequality $y < -4x + 1$ (since $0 < -4 \cdot 0 + 1$); hence each point of U satisfies $y < -4x + 1$, and each point of V satisfies $y > -4x + 1$.

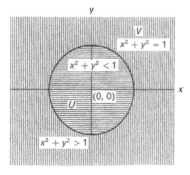
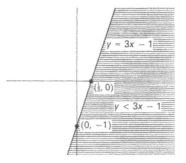

Fig. 7   Fig. 8

We see from the following example that Proposition 1 can work in the case of a graph of an equality which separates the plane in accordance with the hypotheses of Proposition 1 even if the equality does not define a function.

Example 14. Consider the graph of $x^2 + y^2 = 1$. This graph is a circle with center $(0, 0)$ and radius 1 (Fig. 7), which separates the plane into two regions U and V satisfying the hypotheses of Proposition 1. But $x^2 + y^2 = 1$ does not define a function. For, given a particular real number x, there is either more than one y or no y at all such that $x^2 + y^2 = 1$. For example, if $x = 0$, then y can be 1 or -1. If $x^2 + y^2 = 1$ were to define a function from some subset T of the real numbers R into R, then any real number could appear as a first coordinate in

$$\{(x, y) \mid x^2 + y^2 = 1\}$$

at most once.

Testing the point $(0, 0)$ in U, we find that it satisfies $x^2 + y^2 < 1$. In fact, every point of U satisfies this inequality, and every point of V satisfies $x^2 + y^2 > 1$.

We can represent the simultaneous solutions of several inequalities by taking the area common to the solutions of the individual inequalities. This is illustrated in the following example.

Example 15. We shall find all the points (x, y) of the coordinate plane such that $y < 3x - 1$ and $y > -x$. We first graph the equalities $y = 3x - 1$ and $y = -x$. Each of these graphs divides the plane in accordance with Proposition 1. Hence it follows that all points satisfying

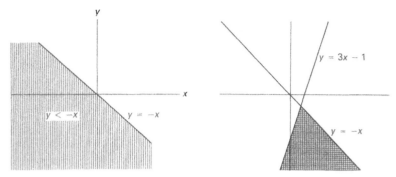

Fig. 9   Fig. 10

$y < 3x - 1$ form the set shown in Fig. 8, while the points satisfying $y > -x$ are shown in Fig. 9. All points which satisfy *both* $y < 3x - 1$ and $y > -x$ will have to be in both the area presented in Fig. 8 and the area shown in Fig. 9; that is, these points will have to be in the overlap, or *intersection*, of the two areas. This intersection is portrayed in Fig. 10. It is this area, then, that represents the simultaneous solutions of the two inequalities.

In order to graphically represent the simultaneous solutions of several inequalities, one must find the area representing the solution of each of the individual inequalities and then take the area common to all these solutions. It may be that there is no area common to all individual solutions. This would mean that there is no simultaneous solution to all the given inequalities.

EXERCISES

1. Sketch a graph of each of the following functions from R, the set of real numbers, into R.

   a) $f(x) = x$ b) $g(x) = 2x - 1$ c) $h(x) = -x + 1$
   d) $f(x) = 2x^2$ e) $f(x) = -x^3 + 3$ f) $h(x) = x^2 + 4x + 3$
   g) $h(x) = 3$ h) $g(x) = -x^2 + x$

   i) $f(y) = y^2$. Here y is the variable, and the x-coordinate will be the function value. One can simplify the graphing by interchanging the x- and y-axes.

2. Graph those points of the coordinate plane which satisfy each of the following inequalities. The best procedure for doing this is to graph first the equality associated with each inequality. Then test a point on each "side" of the graph

of the equality to find the region of the coordinate plane which represents the set of solutions to the given inequality.

a) $y < x$ b) $y > 3x + 1$ c) $y \leq x^2$ d) $y > x^3 + 1$
e) $y \leq 2x + 3$ f) $y \geq -x - 4$ g) $x < y^2$ h) $2x < y + 1$
i) $y + 3 > x^3 - 7$ j) $-y \geq x + 2$

3. Indicate the region of the coordinate plane, if there is one, which represents the set of simultaneous solutions to the inequalities in each of the following.

a) $y < x$,   b) $2y > x$,
    $y < 2x + 1$ $y \leq x^2$
c) $3y < 2x + 1$,   d) $y < 3x + 2$,
    $y > -x + 3$, $y > 2x + 6$,
    $y < x^2$ $y < -x - 3$,
                     $y > -3x + 4$

4. Find the areas of the plane which represent the set of solutions to each of the following inequalities. In doing this exercise, you may assume that if one of the points of any of the regions into which the graph of the associated equality separates the plane satisfies the given inequality, then every point of the region satisfies the given inequality.

a) $x^2 + y^2 \leq 4$ b) $x^2/4 + y^2/9 < 1$ c) $(x - 3)^2 + (y - 8)^2 > 9$
d) $y^2 < x^2$ e) $y^2 \geq x^3$

3.3 PROPERTIES OF FUNCTIONS AND GRAPHS

Henceforth we shall restrict our attention to functions from some set of real numbers into the set of real numbers. We shall always use R to denote the set of real numbers. If the domain of a function is not explicitly given, it will be assumed to be the largest subset of R which is consistent with the rule defining the function. For example, the function f defined by

$$f(x) = 3x + 1$$

makes sense for any real number x. The function defined by

$$g(x) = 1/x$$

makes sense for any real number x other than 0, while the function defined by

$$h(x) = x^{1/2}$$

makes sense only if x is nonnegative.

Two functions from R into R may appear quite dissimilar, yet be similar in many respects. Although certain differences and similarities between two functions can be read from the graphs of the functions, it is often necessary to know of these differences and similarities in order to be able to graph the functions. In this section we shall describe properties of functions

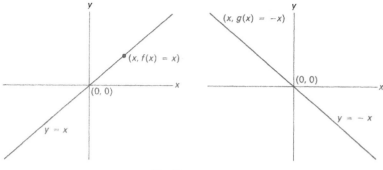

Fig. 11 Fig. 12

and their graphs. Most of this discussion will be informal; later in the text we shall develop techniques for working with certain of the properties introduced in this section.

Example 16. Consider the functions defined by $f(x) = x$ and $g(x) = -x$. The graphs of these functions are given in Figs. 11 and 12. Although the graphs look somewhat different, the two functions actually exhibit many more similarities than differences.

One difference between f and g is that as x increases, $f(x)$ also increases; but as x increases, $g(x)$ decreases. The function f has a *rising* graph, while g has a *falling* graph. (The reader may recognize the graph of f as a straight line with positive slope, while the graph of g is a straight line with negative slope.)

We now list some of the similarities between f and g:

1) The functions f and g never assume the same value for two distinct values of x. Put another way, if $x \neq x'$, then $f(x) \neq f(x')$ and $g(x) \neq g(x')$. Any line parallel to the x-axis in the coordinate plane has an equation of the form $y = a$, where a is a constant. Since a can appear as $f(x)$ for at most one x, any line whose equation is $y = a$ can meet the graph of f in at most one point, namely, in $(b, f(b) = a)$, if there is a b such that $f(b) = a$. If the line $y = a$ intersected the graph of f more than once, say in (b, a) and (b', a), then we would have $b \neq b'$, but $f(b) = f(b') = a$. Similar considerations hold for the function g.

2) The functions f and g take on every real number as a value at least once [and hence precisely once from (1)]. For, if r is any real number, then $f(r) = r$ and $g(-r) = r$. Graphically, any line whose equation has the form $y = a$, where a is any real number, must intersect the graphs of f and g at least once.

3) Neither the function f nor the function g assumes a largest or smallest value. In the case of f, for example, there is no real number a or b such that $f(a) \geq f(x)$ for any x in R, and $f(x) \leq f(b)$ for any x in R.

4) The graphs of f and g are "smooth"; that is, they have no ragged edges, breaks, or other apparent pathologies.

The following definitions make some of the notions presented in Example 16 more precise.

Definition 6. *Let f be a function from a set T of real numbers into S, a set of real numbers.*

*We say that f is **one-one** if different points of T yield different values of f; that is, if x and x' are distinct points of T, then $f(x) \neq f(x')$.*

*We say that f is **onto** S if given any element y of S, there is x in T such that $f(x) = y$.*

*We say that f is **increasing on** T if whenever x and x' are elements of T such that $x \leq x'$, then $f(x) \leq f(x')$. We say that f is **decreasing on** T if whenever x and x' are elements of T such that $x \leq x'$, then $f(x) \geq f(x')$.*

*We say that f assumes the **maximum** M on T if there is x in T such that $f(x) = M$ and if for any x' in T, $f(x') \leq M$. We say that f assumes the **minimum** m on T if there is x in T such that $f(x) = m$ and if for any x' in T, $f(x') \geq m$. If $f(x)$ is a maximum (minimum) for f on T, then f is said to have a **maximum (minimum) at** x.*

*If W is a subset of T, then f can be considered as a function from W into S. For $f(x)$ is defined for each element of T, and each element of W is also an element of T. The meanings then of f **is one-one on** W, f **is increasing (decreasing) on** W, or f has a **maximum (minimum) on** W are defined just as they would be if we considered f as a function from W into S.*

Example 17. We reconsider Example 16 in the light of Definition 6. Both f and g are functions from R into R. Both f and g are one-one [(1) of Example 16] and onto [(2) of Example 16]. Both f and g have no maximum or minimum on R [(3) of Example 16]. However, f and g each have both a maximum and minimum on any closed interval $[a, b]$. We prove this fact for f and leave the proof for g as an exercise. If $[a, b]$ is any closed interval, then f is increasing on $[a, b]$. This follows from the fact that f is increasing on all of R. For, since $f(x) = x$, the statement that $f(x)$ increases as x increases is merely a tautology saying that as x increases, x increases. But then $f(a)$ is the least value, that is the minimum, that f assumes on $[a, b]$ and $f(b)$ is a maximum for f on $[a, b]$.

Example 18. Consider the function defined by $f(x) = x^2$. The graph of this function is pictured in Fig. 1. The function f is not one-one since it has the same value, 4, for both -2 and 2. (Note that some lines parallel to the x-axis cut the graph of f more than once.) Neither is f onto; for example, -1 never occurs as a value of f. (Note that some lines parallel to the x-axis do not meet the graph of f at any point.) The function is decreasing on $(-\infty, 0]$ and increasing on $[0, \infty)$. The function f has a minimum at 0, since x^2 is never negative for any real number x. If $[a, b]$ is any closed interval which does not contain 0 except possibly as an endpoint, then f has a maximum on $[a, b]$ at one of the endpoints of the interval, and a minimum on $[a, b]$ at the other endpoint. For example, for $[4, 5]$, $f(4)$ is the minimum and $f(5)$ the maximum of f on $[4, 5]$.

We note that the graph of f appears to be smooth and without apparent pathologies.

In the next section we shall see some functions which do have some pathological aspects.

EXERCISES

1. Each of the following is to define a function from the open interval $(-1, 1)$ into R. Graph each of these functions. Determine which of the functions are one-one and which are onto. Determine for which subsets of $(-1, 1)$ each function is increasing and for which subsets of $(-1, 1)$ each function is decreasing. Find any maximum or minimum that each function might have on $(-1, 1)$.

   a) $f(x) = 3x$ b) $f(x) = 2x + 2$ c) $f(x) = x^3$
   d) $f(x) = x^3 + x^2$ e) $f(x) = |x|$ f) $f(x) = |x| + x$
   g) $h(x) = \begin{cases} -2 & \text{if } x \leq 0, \\ 2 & \text{if } x > 0 \end{cases}$ h) $f(x) = 1/(x - 1)$.

2. Each of the following defines a function from R into R. Determine which of these functions are one-one and which are onto. Whenever a function is one-one *and* onto, derive an explicit rule for computing the x one must use to get a function value y. For example, if $f(x) = 2x$, then if we want $f(x) = y$, x must be $y/2$. Sketch the graph of each function.

   a) $f(x) = x$ b) $f(x) = 5x$ c) $f(x) = x^3$
   d) $f(x) = x^2 + 2x$ e) $g(x) = |x| + x$ f) $f(x) = \begin{cases} x^2 & \text{if } x > 0, \\ x & \text{if } x \leq 0 \end{cases}$
   g) $g(x) = \begin{cases} x & \text{if } x \text{ is in } (-1, 1), \\ 0 & \text{if } x \text{ is not in } (-1, 1) \end{cases}$

3. If g and h are two functions from R into R, then $g + h$, the *sum* of g and h, is a function from R into R defined by $(g + h)(x) = g(x) + h(x)$. How would the difference, $g - h$, of g and h be defined?

   a) Explain how one can find the graph of $g + h$ from the graphs of g and h. That is, how do we "add" graphs?

b) Graph $g(x) = x^2$ and $h(x) = x$. Use these graphs to obtain the graph of $f(x) = x^2 + x$.
c) Suppose g and h are both increasing on $[a, b]$. Prove that $g + h$ is increasing on $[a, b]$.
d) Suppose h, g, and $h - g$ are increasing on $[a, b]$. Is it reasonable to say that h is *increasing faster* than g? Explain.
e) Suppose g, h, and $h - g$ are decreasing on $[a, b]$. Is $h + g$ decreasing on $[a, b]$? Is h decreasing more slowly than g?

3.4 CONTINUITY

Example 19. Let the function f be defined by

$$f(x) = \begin{cases} 2 & \text{if } x < 0, \\ x & \text{if } x \geq 0. \end{cases}$$

The graph of f is given in Fig. 13. At $(0, 0)$ there is a "break" in the graph.

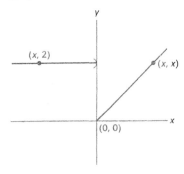

Fig. 13

As x approaches 0 from the positive direction, $f(x)$ also approaches 0. In other words, if x is positive and "close to" 0, $f(x)$ will also be "close to" 0. We intuitively feel that if the graph of f is to avoid breaking at $(0, 0)$, then, as x approaches 0 from the negative direction, $f(x)$ must also approach 0. In other words, if x is close to 0, regardless of whether x is positive or negative, $f(x)$ should also be close to 0. But according to the manner in which f has been defined, if x has any negative value whatsoever, $f(x) = 2$; hence $f(x)$ never gets closer to 0 than 2 for any negative value of x. This is what causes the break in the graph of f.

If f is any function from R into R, then if the graph of f is to avoid breaking at $(a, f(a))$, we must have $f(x)$ "close to" $f(a)$ whenever x is "close to" a. We need a more precise notion of "closeness" before we can formulate more rigorously that property which a function must have to avoid breaks in its graph.

It is customary to use the absolute value of the difference of two real numbers to measure their "nearness to," or distance from each other.* If a and b are two real numbers, then $|a - b|$ measures the length of the line segment joining them. The smaller $|a - b|$, the "nearer" a and b are to each other.

We now rephrase the notion "$f(x)$ is close to $f(a)$ whenever x is close to a" in terms of the absolute value. The absolute value of any real number, and hence the distance between any two real numbers, is always nonnegative. In fact, $|a - b| = 0$ if and only if $a = b$. Any positive real number p can act as a measure of nearness to any real number b; that is, choosing a positive number p arbitrarily, we shall get a real number x that will be p-near b if $|b - x| < p$. The phrase "$f(x)$ is close to $f(a)$" is too qualitative to serve any useful mathematical purpose, but given any positive number p, it does make sense mathematically to speak of $f(x)$ being p-near to $f(a)$, that is, $|f(x) - f(a)| < p$. What we want, then, is for $|f(x) - f(a)|$ to be less than p if x is "close to" a. Again, because of its vagueness, "close to" must be replaced with a more precise notion involving the absolute value. Specifically, $|f(x) - f(a)|$ is to be less than p if $|x - a| < q$ for some positive real number q. We therefore make the following definition.

Definition 7. *Suppose f is a function from a set T of real numbers into R. Then f is said to be* **continuous at a point** *a of T if, given any positive number p, there is a positive number q such that whenever x is a point of T and $|x - a| < q$, then $|f(x) - f(a)| < p$. The function f is said to be* **continuous** *if it is continuous at every point of T. (See Fig. 14.)*

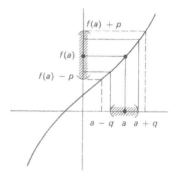

Fig. 14

---

* The reader is assumed to be familiar with the basic properties of absolute value. If b is any real number, then the absolute value of b, denoted by $|b|$, is defined to be

$$|b| = \begin{cases} b & \text{if } b \geq 0, \\ -b & \text{if } b < 0. \end{cases}$$

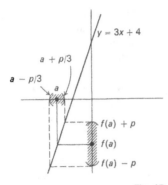

Fig. 15   Fig. 16

The following proposition whose proof is a simple exercise in the use of the absolute value is of some help in the application of Definition 7.

Proposition 2. *If a is any real number and p is any positive real number, then*

a) $\{x \mid |x - a| < p\} = (a - p, a + p)$,
b) $\{x \mid |x - a| \leq p\} = [a - p, a + p]$.

Example 20. The function f of Example 19 is not continuous at 0. For, if p is any positive number less than $2 (= |2 - 0|)$, then for *any* positive number q, there will be some x, for example, $-q/2$, which will have the property that $|x - 0| < q$, but $|f(x) - f(0)| = 2 > p$.

Example 21. The function defined by $f(x) = 3x + 4$ is continuous. For let p be any positive real number. Set $q = p/3$ (see Fig. 15). Then, if $|x - a| < q$, we have

$$|f(x) - f(a)| = |(3x + 4) - (3a + 4)| = |3(x - a)| = 3|x - a| < 3q$$
$$= 3(p/3) = p.$$

Since a can be any real number, f is continuous at any real number; hence f is continuous.

As a rule, a direct use of Definition 7 to show that a function is continuous is rather tedious. The q of Definition 7 that will work for a continuous function f depends not only on the function but on a and p; the relationship between f, q, a, and p can be quite involved. One classic rule of thumb frequently used to determine whether a function is continuous is the following: A function f is continuous if we can draw the graph of f without lifting the pencil from the paper. A more formal, although still somewhat imprecise, method of determining continuity is given without proof in the following proposition.

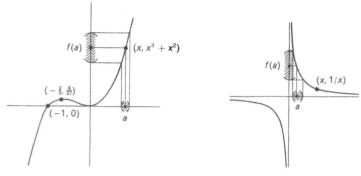

Fig. 17   Fig. 18

Proposition 3. *Let f be a function from T into R. For each real number a and positive real number p, set $U(a, p)$ equal to the set of all x in T such that*

$$f(a) - p < f(x) < f(a) + p.$$

That is, $U(a, p)$ consists of all x in T such that $f(x)$ is p-near $f(a)$. Then f is continuous if and only if each $U(a, p)$ consists of the intersection with T of a collection of open intervals or open half-lines, or contains no elements of all. (If $T = R$, this condition reduces to "$U(a, p)$ consists of a collection of open intervals, open half-lines, or is empty.") See Fig. 16.

Example 22. Consider the function f from R into R defined by

$$f(x) = x^3 + x^2.$$

The graph of f is given in Fig. 17. We can determine informally in the manner shown in Fig. 17 that $U(a, p)$ consists of at most two open intervals for any real number a and any positive real number p. Therefore, by Proposition 3, f is continuous.

Example 23. Let T consist of all real numbers except 0. Then

$$f(x) = 1/x$$

defines a function f from T into R. The graph of f is given in Fig. 18. We also see from the figure that for any real number a and any positive real number p, $U(a, p)$ is the intersection of T and an open interval; hence f is continuous by Proposition 3.

Example 24. If we examine the function f of Example 19, we find that $U(0, 1)$ is the half-open interval $[0, 1)$. Therefore f is not continuous. We could, however, verify that f fails to be continuous only at 0.

The term "smooth" was used to describe the graphs in Example 16. We may informally define "smoothness" to be continuity together with an absence of "sharp edges" in the graph. Continuity alone does not always give "smoothness" as we can see from the following example.

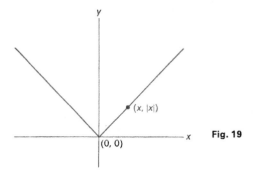

Fig. 19

Example 25. The graph of the function f defined by $f(x) = |x|$ is given in Fig. 19. It is left as an exercise to verify that this absolute value function is continuous; nevertheless, the graph of f has a sharp corner at $(0, 0)$. The graph, therefore, is not smooth at $(0, 0)$.

We shall make the notion of "smoothness" more precise in the next chapter.

EXERCISES

1. Prove that each of the following functions are continuous at the points indicated.
   a) $f(x) = x$ at 0 b) $f(x) = 4x$ at 5 c) $f(x) = -6x + 1$ at -2
   d) $f(x) = x^2$ at 0 e) $f(x) = |x|$ at 0

2. Prove that each of the following functions are continuous, that is, that they are continuous at each real number a for which the definition of the function makes sense.
   a) $f(x) = |x|$ b) $f(x) = -2x + 5$ c) $f(x) = x^2$
   d) $f(x) = x^{1/2}$ e) $f(x) = x^3$

3. Prove that each of the following functions are discontinuous at at least one point.
   a) $f(x) = \begin{cases} 1/x, & \text{if } x \neq 0, \\ 0 & \text{if } x = 0 \end{cases}$ b) $g(x) = \begin{cases} x & \text{if } x \neq 2, \\ -4 & \text{if } x = 2 \end{cases}$
   c) $f(x) = \begin{cases} 2x & \text{if } x < 0, \\ 1 & \text{if } 0 \leq x \leq 1, \\ x - 1 & \text{if } x > 1 \end{cases}$

4. A subset U of R is said to be open if, given any point u of U, there is an open interval (a, b) which contains u and is a subset of U. Prove that a function f from R into R is continuous if and only if, given any open subset U of R, $\{x \mid f(x) \text{ is in } U\}$ is an open subset of R.

5. Decide which of the following statements are true and which are false. Justify your answer in each case.

   a) If f and g are continuous functions from R into R, then $f + g$ is also continuous (see Exercise 3, Section 3.3).

   b) If f is continuous on $[0, \infty)$ and is also continuous on $(-\infty, 0]$, then f is continuous at 0.

3.5 SUMMATION NOTATION

The reader is of course familiar with the fact that mathematicians often use concise symbols to express notions that—if they were to be written out in full—would require a fair amount of space and effort. But the use of good notation does more than conserve space; for if notation is really good, it will be a positive aid to those who use it. Good notation not only represents, but also suggests. It is much more efficient than "long-hand" notation, because it helps one arrive at conclusions and see relationships faster. Alas, not all notation is good; and what is "good" notation is often largely determined by the particular temperament of the individual user. The author believes that the "summation notation" we now introduce is good notation. Regardless, however, of whether or not it is good, it is almost universally accepted among mathematicians (a fairly good sign that it is "good"), and it will save a great deal of space. We therefore feel more than justified in presenting it.

The Greek capital letter *sigma*,

$$\sum,$$

will indicate that a sum is to be taken. Written after, above, and below, the sigma will be certain instructions as to what sum is to be taken. We shall try to explain this notation through the use of examples.

Example 26. Consider

(1) $$\sum_{i=1}^{4} i.$$

The $\sum$ indicates that a sum is to be taken. The $i = 1$ below and the 4 above the $\sum$ indicate that i is to be allowed to take on all *integer* values between 1 and 4, inclusive, that is, 1, 2, 3, and 4. We take the sum of the terms obtained by substituting each allowable value of i for i in the term following the $\sum$. Since this term is merely i, the terms we sum are 1, 2, 3, and 4. Consequently, (1) is a shorthand way of writing $1 + 2 + 3 + 4$.

The letter i is used in (1) as an *index of summation*. It is the index of summation, the allowable values of the index of summation, and how this index is used in the expression following the $\sum$ that determine what terms we sum.

Example 27. Consider

(2) $\quad \sum_{i=-1}^{6} k_i.$

The $i = -1$ and the 6 together tell us that i is to be allowed to take on all integer values between -1 and 6, inclusive. It is the i that is the index of summation. Substituting -1 for i in the expression following the $\sum$, we obtain k_{-1}. There is no way of evaluating k_{-1} numerically from the information given; hence we leave it as k_{-1}. Substituting the allowable values for i in k_i, we find that we are to sum $k_{-1}, k_0, k_1, \ldots, k_6$. Thus, (2) is a short way of expressing

$$k_{-1} + k_0 + k_1 + k_2 + k_3 + k_4 + k_5 + k_6.$$

The symbol ∞ is used to denote *infinity*. This symbol has already been encountered in conjunction with open and closed half-lines. In the next example we show how the symbols ∞ and $-\infty$ are used with summation notation.

Example 28.

(3) $\quad \sum_{j=1}^{\infty} (\tfrac{1}{2})^j$

is another way of writing the geometric series

$$\tfrac{1}{2} + (\tfrac{1}{2})^2 + (\tfrac{1}{2})^3 + \cdots.$$

Since this geometric series has first term $\tfrac{1}{2}$ and ratio $\tfrac{1}{2}$, it sums to $(\tfrac{1}{2})/(1 - \tfrac{1}{2}) = 1.$* The $j = 1$, in conjunction with ∞ (and the $\sum$, of course), indicates that the sum is to be taken of all terms of the form $(\tfrac{1}{2})^j$ for any positive integer j, that is, for all integers greater than or equal to 1.

(4) $\quad \sum_{k=-\infty}^{-1} x^k$

represents the sum $x^{-1} + x^{-2} + x^{-3} + \cdots$ (This is a geometric series

---

* The general formula for the sum S of a geometric series with first term a and ratio r, that is, $a + ar + ar^2 + ar^3 + \cdots$, is given by $S = a/(1 - r)$. For this formula to be valid, we must have $|r| < 1$.

with first term $1/x = x^{-1}$ and ratio $1/x$. The index of summation is k. We use $k = -\infty$ and -1 in conjunction with x^k to indicate that we are summing all terms of the form x^k, where k has any integer value up to and including -1.

Notation is primarily for communication. Just as having learned a basic vocabulary for reading a language, we find we are able to deduce the meaning of many other words because they are related to the basic vocabulary, or the context in which they are used gives their meaning, so we find that having learned a basic mathematical notational device, we can often understand variations of that notation. At times certain special instructions are used in conjunction with the summation sign to indicate how a sum should be taken. The meaning of these variations should be clear from the context and the basic use of the summation notation. We illustrate two possible variants of the summation notation in the next example.

Example 29.

(5) $$\sum_{\substack{j=1 \\ j \neq 3}}^{5} f(x_j)$$

indicates that we are to sum all terms of the form $f(x_j)$, where j takes on all integral values from 1 to 5, except for 3. Thus the sum represented by (5) is

$$f(x_1) + f(x_2) + f(x_4) + f(x_5).$$

Let $A = \{1, 2, 3\}$ and consider

(6) $$\sum_{j \text{ in } A} (k, j).$$

From "j in A" we learn that j is the index of summation and j is to be allowed to take on all values in A. Consequently, (6) represents the sum

$$(k, 1) + (k, 2) + (k, 3).$$

In some instances it may be desirable to change the index of summation. A change of this sort must be made in such a way that the terms summed remain the same.* We illustrate a change of index of summation in each of the next two examples.

---

* In the case of sums of infinitely many terms, we usually must also keep the order in which the terms are summed the same. We shall discuss the sums of infinitely many terms (*series*) at greater length later in this book.

Example 30. Consider

(7) $$\sum_{n=0}^{8} x^n.$$

In (7) the index of summation n is to take on all integral values between 0 and 8, inclusive. We will express (7), using an index of summation k which takes on all integral values from 1 to 9. Using the new index of summation k, (7) will have the form

(8) $$\sum_{k=1}^{9} x^{f(k)},$$

where $f(k)$ is an expression involving k which will enable us to have the same terms to sum in (8) that we have in (7). Since n starts at 0 and ends at 8, while k starts at 1 and ends at 9, k is always 1 more than n. That is, $k = n + 1$, or $n = k - 1$. We replace $f(k)$ with $k - 1$ in (8) to obtain

(9) $$\sum_{k=1}^{9} x^{k-1}.$$

As k goes from 1 to 9, $k - 1$ goes from 0 to 8; hence we are summing the same terms in (9) as we are in (7), even though the indices of summation are different.

Example 31. Consider

(10) $$\sum_{k=1}^{10} x^k + \sum_{j=3}^{12} y^j.$$

The second summand of (10) can also be expressed using an index of summation k which takes on values from 1 to 10; specifically,

(11) $$\sum_{j=3}^{12} y^j = \sum_{k=1}^{10} y^{k+2}.$$

We can therefore rewrite (10) as

(12) $$\sum_{k=1}^{10} x^k + \sum_{k=1}^{10} y^{k+2} = \sum_{k=1}^{10} (x^k + y^{k+2}).$$

EXERCISES

1. Write out each of the following sums. If a sum involves an infinite number of terms, write out only the first six.

   a) $\sum_{n=1}^{4} n^2$

   b) $\sum_{j=0}^{7} f(x_j)$

c) $\displaystyle\sum_{\substack{k=-3\\k\neq 0}}^{5} 1/k$ d) $\displaystyle\sum_{x \text{ in } B} f(x^k)$, where $B = \{0, 7, 8, 15, 25\}$

e) $\displaystyle\sum_{n=0}^{\infty} (\tfrac{1}{3})^n$ f) $\displaystyle\sum_{\substack{k \text{ any positive}\\ \text{even integer}}} (\tfrac{1}{2})^{k-1}$ g) $\displaystyle\sum_{\substack{n=1\\n \text{ odd}}}^{15} 1/n^2$

2. Write each of the following sums using the summation notation developed in this section. There is more than one correct way of expressing each sum.

a) $1 + 2 + 3 + 4 + 5 + 6$
b) $x^4 + x^5 + x^6 + x^7 + x^9$
c) $\tfrac{1}{6} + (\tfrac{1}{6})^2 + (\tfrac{1}{6})^3 + (\tfrac{1}{6})^4 + \cdots$
d) $f(x) + f(x^{-1}) + f(x^{-2}) + f(x^{-3}) + \cdots$

3. Rewrite each of the following sums as directed.

a) $\displaystyle\sum_{n=1}^{4} (n+1)$; rewrite with an index of summation which takes values from 0 to 3.

b) $\displaystyle\sum_{k=-5}^{6} f(x_k)$; rewrite, using an index of summation which takes only positive values.

c) $\displaystyle\sum_{j=-\infty}^{-1} x^{-j}$; rewrite with an index of summation which takes only positive values.

4. Prove each of the following:

a) $\displaystyle\sum_{k=1}^{m} g(k) + \sum_{k=1}^{m} f(k) = \sum_{k=1}^{m} (g(k) + f(k))$, where m is any positive integer.

b) $q\left(\displaystyle\sum_{j=1}^{m} g(j)\right) = \sum_{j=1}^{m} qg(j)$, where q is any real number and m is any positive integer.

3.6 PROBABILITY DISTRIBUTIONS

The admissible range of a random variable X is only part of the picture of X. If meaningful results are to come from a study of X, then we must also be able to find the probabilities associated with X. In particular, suppose that X is a function from the sample space S into the set of real numbers. Recall that X is a function, while x is used to indicate a particular admissible value of X. If B is a subset of the admissible range of X, then we want some way of computing the probability that x will be in B. If $E_B = \{s \text{ in } S \mid X(s) \text{ is in } B\}$, then E_B is an event in S; moreover, $P(E_B)$ is the probability that x will be in B.

Definition 8. *If X is a random variable, then the admissible range of X, together with some means of finding the probabilities associated with X, is called the **probability distribution**, or merely the **distribution**, of X.*

Before a random variable can be of any practical value, its distribution must be determined. We now consider the problem of specifying the distribution of a random variable.

You may feel that the most natural means of specifying the distribution of a variate X (on a sample space S) is to specify the admissible range of X explicitly, and also provide a method—a table, rule, function, or some other device—for determining the probability that X will take on each of its admissible values. This procedure is illustrated in the following example.

Example 32. An unbiased die is rolled once. The sample space associated with this experiment is $\{1, 2, 3, 4, 5, 6\}$. We define

$$X(r) = r, \quad r = 1, 2, 3, 4, 5, 6.$$

Thus X is a random variable with admissible range $\{1, 2, 3, 4, 5, 6\}$. We can specify the probability of each admissible value of X by means of the function

$$f(x) = \tfrac{1}{6}, \quad x = 1, 2, 3, 4, 5, 6.$$

In this case, the function f and the admissible range are sufficient to determine the distribution of X.

In the case of a continuous variate, however, we have to contend with the fact that the probability that the variate will give any one admissible value is 0. We can illustrate this fact as follows: Suppose there are n apples in a barrel, and one apple is drawn from the barrel at random. Then the probability of drawing one particular apple is $1/n$. As n gets large, $1/n$ gets small. If n is infinite, then the probability of drawing one particular apple is 0. This does not mean that it is impossible for someone to draw that apple, but only that the chance is infinitesimally small. The point is also illustrated in the following example.

Example 33. A string is 3 feet long. A piece s is cut from the string at random. Let $X(s)$ be the length of this piece. Then $X(s)$ can take on any value from 0 to 3; that is, X is a random variable with admissible range $(0, 3)$. There are infinitely many admissible values of X, all of which can be assumed to have the same probability. The probability $P(x)$ for each admissible value x of X is therefore 0.

We can, nevertheless, specify the probability distribution of X by means of an appropriate function—not a function which gives the probability of each admissible value of X, however. Instead, we use a function which

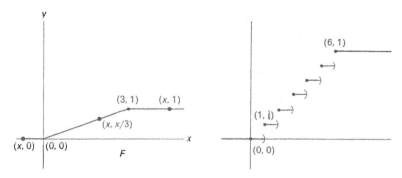

Fig. 20   Fig. 21

gives the probability that $x \leq y$ for any real number y, that is, the probability that X will give a value less than or equal to y. Specifically, if y is any real number, then we set $F(y) = P(x \leq y)$. If $y \leq 0$, then $F(y) = 0$; and if $y \geq 3$, then $F(y) = 1$. We may suppose from the way the variate X is defined that $F(y)$ for y in $(0, 3)$ will be directly proportional to y; that is, the greater y, the greater the chance of cutting a piece of string of length less than y, and $F(y) = ky$, where k is a suitable constant. Since $F(3) = k \cdot 3 = 1$, it follows that $k = \frac{1}{3}$. Hence, for y in $(0, 3)$, $F(y) = y/3$. The complete definition of $F(y)$ is

$$F(y) = \begin{cases} 0, & y \leq 0, \\ y/3, & y \text{ in } (0, 3), \\ 1, & y \geq 3. \end{cases}$$

The graph of F is given in Fig. 20.

The distribution of the variate X of Example 32 can also be specified by means of a function similar to that given in Example 33.

Example 34. If X is the variate of Example 32, then setting $F(y) = P(x \leq y)$, we find

$$F(y) = \begin{cases} 0 & \text{if } y < 1, \\ \frac{1}{6} & \text{if } 1 \leq w < 2, \\ \frac{2}{6} & \text{if } 2 \leq y < 3, \\ \frac{3}{6} & \text{if } 3 \leq y < 4, \\ \frac{4}{6} & \text{if } 4 \leq x < 5, \\ \frac{5}{6} & \text{if } 5 \leq x < 6, \\ 1 & \text{if } 6 \leq y. \end{cases}$$

The graph of F is given in Fig. 21.

Definition 9. *Let X be a random variable with admissible range A. The function F defined for each real number y such that $F(y) = P(x \leq y)$ is called the **distribution function of** X.*

*If x is discrete, then the function defined for each x in A such that $f(x)$ is the probability of x is called the **density function of** X. (The density function of a continuous variate will be defined later.)*

Let F be a distribution function, and f be a density function if the variate is discrete. It is clear that

1) f and F can never take on negative values since probabilities are always nonnegative;

2) f and F never take on values greater than 1 since a probability can never exceed 1; and

3) F is an increasing function. The proof of (3) is left as an exercise.

The next propositions give more fundamental properties of density and distribution functions.

Proposition 4. *If F is the distribution function of a continuous variate X and F is continuous, then*

$$P(x \leq y) = P(x < y).$$

Proof. By definition, $F(y) = P(x \leq y)$. But

$$P(x \leq y) = P(x < y) + P(x = y).$$

Therefore $F(y) = P(x < y) + P(x = y)$, or

(13) $P(x = y) = F(y) - P(x < y).$

Now if $P(x \leq y) \neq P(x < y)$, then

$$P(x = y) = P(x \leq y) - P(x < y) = p,$$

where p is some positive real number. Therefore given any positive number q, there will be at least one point w such that $|w - y| < q$ but

$$F(y) - F(w) \geq p$$

(see Fig. 22). (Looking at it another way, we find that if F is to be continuous at y, then the right-hand side of (13) must go to 0 as x approaches y since $P(x < y)$ will approach $F(y)$, but this will not happen if $P(x = y) \neq 0$.) Hence F would not be continuous. Since F is assumed to be continuous, we must have

$$P(x \leq y) = P(x < y).$$

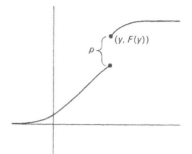

Fig. 22

Corollary: If F is the distribution function of a continuous variate X and F is continuous, then
$$P(y' \leq x \leq y) = P(y' \leq x < y) = P(y' < x \leq y)$$
$$= P(y' < x < y) = F(y) - F(y').$$

Proof. We prove that $P(y' \leq x \leq y) = F(y') - F(y)$ and leave the proof of the other equalities as an exercise. Now
$$P(x \leq y) = P(x < y') + P(y' \leq x \leq y)$$
$$= P(x \leq y') + P(y' \leq x \leq y)$$
by Proposition 4. Since $P(x \leq y) = F(y)$ and $P(x \leq y') = F(y')$, we have
$$F(y) - F(y') = P(y' \leq x \leq y).$$

The next proposition gives a fundamental relationship between the density and distribution functions of a discrete variate.

Proposition 5. *Let X be a discrete variate with admissible range A. Then if for any real number y, the number of admissible values of x less than or equal to y is finite, we have*

(14) $$F(y) = \sum_{\substack{x \leq y \\ x \text{ in } A}} f(x),$$

where F and f are the distribution and density functions of X, respectively. (The restriction that the number of values less than y is finite can be removed after we have studied infinite series.)

Proof. By definition, $F(y) = P(x \leq y)$. Denote the admissible values of x which are less than or equal to y by $x_1, \ldots, x_n$. Then $x \leq y$ is equivalent to $x_1 \cup x_2 \cup \cdots \cup x_n$. Since $x_1, \ldots, x_n$ are events which are mutually exclusive in pairs (since no two admissible values of X can be assumed

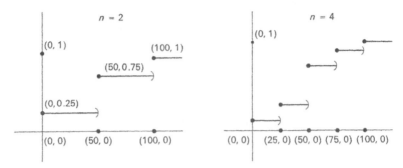

Fig. 23   Fig. 24

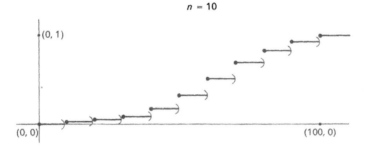

Fig. 25

simultaneously), we can apply Proposition 2 of Chapter 1 to obtain

(15) $\quad F(y) = P(x \leq y) = P(x_1) + P(x_2) + P(x_3) + \cdots + P(x_n).$

But $P(x_1) = f(x_1)$, $P(x_2) = f(x_2)$, etc.; hence (14) follows from (15).

Although continuous and discrete variates appear to be quite different in their properties, at times a discrete probability distribution can be "approximated" by a continuous distribution. This point is illustrated in the following example.

Example 35. A sample s of n people is chosen at random from a general population of which half are male. Let $X(s)$ be the percentage of males among the n people selected. As n increases, the number of admissible values of X also increases. For example, if $n = 4$, then X has only 0, 25, 50, 75, and 100 as admissible values; but if $n = 100$, then X has $\{0, 1, \ldots, 100\}$ as admissible range. The distribution function for X is given in Figs. 23, 24, and 25 for various values of n. As n increases, the

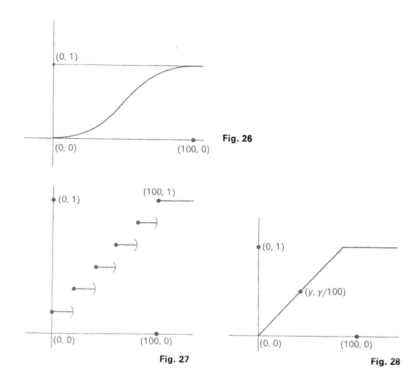

Fig. 26

Fig. 27

Fig. 28

graph of F, the distribution function for F, approaches the continuous function G whose graph is given in Fig. 26. Since working with continuous functions is often easier than working with functions which have discontinuities or summing a great number of terms (for example, as may be necessary in view of Proposition 5), such an approximation might be very useful from a computational viewpoint.

We close this section with one more example of a continuous "approximation" to a discrete distribution.

Example 36. A sample s of n people is chosen at random from a general population. Let $X(s)$ be the percentage of people who come from families whose yearly income exceeds $5000. We shall assume that all admissible values of X are equiprobable. As n increases, so does the number of admissible values of X, hence the closer X comes to being a continuous variate. The graph of the distribution function for X for a typical value of is shown in Fig. 27. As n increases, the distribution for x approaches that of the distribution of the continuous variate Y whose admissible range is $[0, 100]$ and whose density function is given by $F(y) = y/100$ (Fig. 28).

3.6   Probability Distributions   67

EXERCISES

1. Prove the equalities in the corollary to Proposition 4 that were left as exercises.

2. Find the density and distribution functions for each of the discrete random variables described below. The variates are described by indicating the values of the variates and giving the conditions under which the values are assumed.
   a) A fair coin is flipped fairly. Let x be 0 or 1 depending on whether the coin turns up heads or tails.
   b) Five slips numbered 1 through 5 are in an urn. One slip is drawn at random from the urn. Let x be the number of the slip drawn.
   c) Two fair coins are flipped together fairly. Let x be the number of heads that turn up.
   d) Five red and four green balls are in an urn. Three balls are drawn at random from the urn without replacement. Let x be the number of green balls drawn.
   e) A certain die has twice the probability of rolling a 3 as any other face. The die is rolled; let x be the value of the roll.

3. Verify in each part of Exercise 2 that
   $$\sum_{x \text{ in } A} f(x) = 1,$$
   where A is the admissible range of x and f is the density function of x. Why does this have to be true?

4. Let X be a continuous random variable with admissible range [0, 1]. Which of the following functions defined on [0, 1] are possible distribution functions for X?* Which of the possible distribution functions for X are continuous? Graph each of these distribution functions.
   a) $F(x) = x$
   b) $F(x) = x^3$
   c) $F(x) = x^2 + x$
   d) $F(x) = -x + 1$
   e) $F(x) = \begin{cases} x, & 0 \leq x < \frac{1}{2}, \\ x^2, & \frac{1}{2} \leq x < 1 \end{cases}$
   f) $F(x) = \begin{cases} 3x, & 0 \leq x \leq \frac{1}{3}, \\ 1, & \frac{1}{3} < x \leq 1 \end{cases}$
   g) $F(x) = \begin{cases} x/2, & 0 \leq x < \frac{1}{2}, \\ x - \frac{1}{4}, & \frac{1}{2} \leq x < \frac{3}{4}, \\ 2x + 3, & \frac{3}{4} \leq x < 1 \end{cases}$

---

* Since $F(y)$ will be 0 for any y less than all admissible values of X (that is, $y < 0$), and $F(y)$ will be 1 for any y greater than all admissible values of X (that is, $y > 1$), we need only define $F(x)$ for any admissible value of X (that is, for y in [0, 1]).

5. Let X be a continuous random variable with admissible range $[0, 3]$, and density function F defined by
$$F(x) = \begin{cases} x/2, & 0 \le x < 1, \\ (x+1)/4, & 1 \le x \le 3. \end{cases}$$
Find each of the following probabilities.
a) $P(x = 2)$ b) $P(2 \le x \le 3)$ c) $P(0 \le x < 2)$
d) $P(x < -1)$ e) $P(x > 2 \text{ or } x \le 1)$ f) $P(x > 1)$
g) $P(1 < x \le \frac{3}{2} \text{ or } x \ge 2)$ h) $P(x \not< 2)$

4 The Derivative

4.1 THE LIMIT OF A FUNCTION

Many words are used as technical terms in mathematics with meanings quite similar to their meanings in everyday speech; one such word is *limit*. According to common usage, *limit* is a boundary of something. For example, the "city limits" bound the city, the "speed limit" bounds the legal speed, and the "limit of our endurance" is the boundary of the strain that our minds or bodies can take.

Often a limit is dependent on, that is, "a function of," one or more factors. For example, the speed limit varies according to location and road conditions.

Consider the function f from R into R defined by $f(x) = 2x$. We feel intuitively that $f(x)$ should approach 4 as a limit as x approaches 2. The "closer" x gets to 2, the "closer" $f(x)$ gets to 4. From Section 3.4 we find that our intuition that $f(x)$ approaches 4 as a limit as x approaches 2 really amounts to saying that f is continuous at 2.

On the other hand, if we consider the function f of Example 19 of Chapter 3, then we may intuit that $f(x)$ approaches 2 as a limit if x approaches 0 from the negative side, but that $f(x)$ approaches 0 as a limit if x approaches 0 from the positive side. The fact that the limits are not the same as x approaches 0 from either side accounts for the discontinuity of f at 0.

What does it mean to say that "x approaches a"? It simply means that x assumes values closer and closer to a. If "L is the limit of $f(x)$ as x approaches a," then we expect $f(x)$ to assume values closer and closer to L as x assumes values closer and closer to a. This, stated quite informally, is what happens when a function approaches a limit. We must, however, formalize this notion considerably if it is to serve us mathematically. Once again, the absolute value will be used to measure distances. The informal notion of limit presented above is formalized in the following definition.

Definition 1. *Suppose f is a function from T, a subset of R, into R. Then L is said to be the **limit of f as x approaches** a if, given any positive number p, there is a positive number q such that if x is any point of T other than a such that $|x - a| < q$, then $|f(x) - L| < p$ (Fig. 1).*

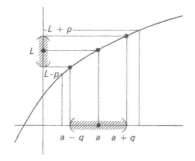

Fig. 1

We sometimes denote the phrase "L is the limit of $f(x)$ as x approaches a" by $\lim_{x \to a} f(x) = L$.

If L were replaced by $f(a)$ in Definition 1, we would have the equivalent of Definition 7 of Chapter 3. Consequently, we can say that if

$$\lim_{x \to a} f(x) = f(a),$$

then f is continuous at a. On the other hand, comparing Definition 7 of Chapter 3 and Definition 1 above, we see that if f is continuous at a, then $\lim_{x \to a} f(x) = f(a)$. We have therefore proved the following proposition.

Proposition 1. *If f is a function from T into R, then f is continuous at a if and only if $\lim_{x \to a} f(x) = f(a)$.*

In view of the close connection between limits and continuity, one may ask: How might there be a limit of f as x approaches a and yet f not be continuous at a?

In the first place, $f(a)$ might not be defined. For if f is a function from T into R and a is not a member of T, then $f(a)$ does not make sense.*

Second, $f(a)$ may be defined, but may not equal $\lim_{x \to a} f(x)$. This possibility is illustrated in the following example.

Example 1. Let f be the function from R into R defined as

$$f(x) = \begin{cases} 4, & x \neq 0, \\ 0, & x = 0. \end{cases}$$

Then $\lim_{x \to 0} f(x) = 4$ (see the graph of f in Fig. 2), but according to the definition of f, $f(0) = 0$. Therefore $\lim_{x \to 0} f(x) \neq f(0)$. Of course, f is not continuous at 0.

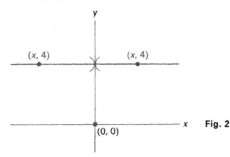

Fig. 2

Example 2. Let f be defined by $f(x) = 2x + 1$. (Recall that when the domain of f is not specified, it is to be taken to be as large a subset of R as is consistent with the definition of f. In this case, the domain of f will be R.) The function f is continuous; hence, by Proposition 1,

(1) $\lim_{x \to 0} f(x) = f(0) = 1.$

We now prove (1) directly from Definition 1.

Let p be any positive number and set $q = p/2$. Then if

$$|x - 0| = |x| < q,$$

---

* It may be that $f(a)$ makes sense formally, but not logically. For example, suppose f is the function from $(0, 1)$ into R defined by $f(x) = x$. By definition, we have restricted the values of x to $(0, 1)$. It follows from Definition 1 that

$$\lim_{x \to 0} f(x) = 0,$$

but $f(0)$ is not defined, even though we can formally substitute 0 for x in the rule for finding $f(x)$ and not arrive at any formal contradiction.

we have $|f(x) - 1| = |2x + 1 - 1| = |2x| = 2|x| < 2q = 2(p/2) = p$. Therefore (1) is proved.

Example 3. Let f be defined by $f(x) = 1/x$. Then f is defined for every real number x except 0. Even though $f(0)$ is not defined, we may ask whether $f(x)$ has a limit as x approaches 0. If there were a limit L of $f(x)$ as x approaches 0, then we could define a function g by

$$g(x) = \begin{cases} f(x), & x \neq 0, \\ L, & x = 0, \end{cases}$$

and g would be continuous at each real number x. But $f(x) = 1/x$ becomes arbitrarily large as x approaches 0 from the positive side, and $f(x)$ becomes arbitrarily small as x approaches 0 from the negative side (Fig. 3); hence $f(x)$ cannot have a limit as x approaches 0.

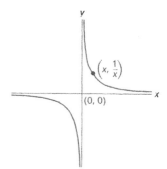

Fig. 3

At times a limit which at first glance seems to be undefined may be evaluated after an appropriate algebraic adjustment is made. This point is illustrated in the following examples.

Example 4. Let the function f be defined by

$$f(x) = \begin{cases} x^2/x, & x \neq 0, \\ 0, & x = 0. \end{cases}$$

We will evaluate $\lim_{x \to 0} f(x)$. If we merely substitute 0 for x, we obtain the undefined quantity $0/0$. However, if we first perform the division indicated, we obtain

$$\lim_{x \to 0} f(x) = \lim_{x \to 0} (x^2/x) = \lim_{x \to 0} x = 0.$$

Example 5. Let f be defined by $f(x) = x^2$. We now evaluate

(2) $$\lim_{h \to 0} \frac{f(3 + h) - f(3)}{h}.$$

Note first that setting $h = 0$ in the expression of which we are trying to take the limit merely gives $0/0$. Now if $f(x) = x^2$, then

$$f(3 + h) = (3 + h)^2 = 9 + 6h + h^2 \quad \text{and} \quad f(3) = 9.$$

Therefore $f(3 + h) - f(3) = 6h + h^2$. Consequently, $[f(3 + h) - f(3)]/h$ reduces to $6 + h$. Therefore (2) reduces to $\lim_{h \to 0} (6 + h) = 6$.

Limits such as (2) are of central importance in differential calculus. Because of this importance we present another example of evaluating a limit of this type.

Example 6. Let f be defined by $f(x) = x^{1/2}$, where x is any nonnegative number. We shall evaluate

(3) $\quad \lim_{h \to 0} \dfrac{f(1 + h) - f(1)}{h} = \lim_{h \to 0} \dfrac{(1 + h)^{1/2} - 1}{h}.$

Multiplying numerator and denominator of $[(1 + h)^{1/2} - 1]/h$ by $(1 + h)^{1/2} + 1$ and simplifying the resulting expression, we obtain

(4) $\quad \dfrac{1}{(1 + h)^{1/2} + 1}.$

Since $\lim_{h \to 0} (1 + h) = 1$, we have that the limit of (4) as h approaches 0, and hence (3), equals $1/(1 + 1) = \frac{1}{2}$.

In Exercise 3 of Section 3.3, we defined the sum of two functions. If the sum of two functions can be defined, then why not the product and quotient of two functions? In the following definition, we define the manner in which two functions should be added, multiplied, and divided. We then give a proposition which shows how this definition is pertinent to the study of limits.

Definition 2. *Suppose f and g are both functions from T, a subset of R, into R.*

*The **sum** $f + g$ of f and g is the function (from T into R) defined by*

$(f + g)(x) = f(x) + g(x)$ *for each x in T.*

*The **product** fg of f and g is the function defined by*

$(fg)(x) = f(x)g(x)$ *for each x in T.*

*The **quotient** f/g of f and g is the function defined by*

$(f/g)(x) = f(x)/g(x)$ *for each x in T for which $g(x) \neq 0$.*

If r is any real number, then the function rf is defined by

$(rf)(x) = rf(x)$ *for each x in T.*

[$rf(x)$ is the ordinary product of r and $f(x)$.]

Example 7. If f is defined by $f(x) = x^2$, and g is defined by $g(x) = 3x - 1$, then

$(f + g)(x) = x^2 + 3x - 1, \qquad (fg)(x) = x^2(3x - 1),$

$(f/g)(x) = x^2/(3x - 1),$

where $x \neq \frac{1}{3}$, and $(-2f)(x) = -2x^2$.

Example 8. The function h defined by $h(x) = x^{1/2} + 3x^2$, where x is any nonnegative number, is the sum of the functions f and g defined by $f(x) = x^{1/2}$ and $g(x) = 3x^2$.

Proposition 2. *Suppose f and g are functions from T into R and*

$\lim_{x \to a} f(x) = L, \qquad \text{while} \qquad \lim_{x \to a} g(x) = L'.$

Then:

a) $\lim_{x \to a} (f + g)(x) = L + L'.$

b) $\lim_{x \to a} (fg)(x) = LL'.$

c) $\lim_{x \to a} (f/g)(x) = L/L'$, provided that $L' \neq 0$.

d) *If r is any real number, then* $\lim_{x \to a} (rf)(x) = rL.$

If f and g are both continuous at a, then L and L' will be $f(a)$ and $g(a)$, respectively. Thus, for example, from (a), we find that if f and g are continuous, then $\lim_{x \to a} (f + g)(x) = f(a) + g(a) = (f + g)(a)$; hence by Proposition 1, $f + g$ is continuous at a. Using Proposition 1, we can establish the following important corollary to Proposition 2.

Corollary: *Suppose f and g are functions from T into R and f and g are continuous at a. Then $f + g, fg, rf$ for any real number r, and f/g if $g(a) \neq 0$, are continuous at a.*

Example 9. A *polynomial function* is a function f whose definition takes the form $f(x) = a_n x^n + a_{n-1} x^{n-1} + \cdots + a_1 x + a_0$, where n is some positive integer, and the a_i are real numbers. A typical polynomial function would be defined by $f(x) = 4x^3 + 3x^2 + (\frac{5}{4})x + 7$. A polynomial function can be "built" from the function g defined by $g(x) = x$ using simple algebraic operations. For $a_i x^i = a_i(g)^i(x)$, $i = 1, 2, \ldots, n$. Therefore, the general polynomial function f can be expressed as

(5) $\qquad f = a_n g^n + a_{n-1} g^{n-1} + \cdots + a_1 g + a_0.$

Since g is continuous, each term of the form $a_i g^i$ is continuous. Since any function which is constant is continuous, the function h defined by $h(x) = a_0$ is continuous. Therefore f is the sum of continuous functions, and hence is continuous. That is, any polynomial function is continuous.

If b is any real number and f is a polynomial function as in (5), then we have

(6) $\quad \lim\limits_{x \to b} f(x) = a_n b^n + a_{n-1} b^{n-1} + \cdots + a_1 b + a_0 = f(b).$

Example 10. Consider the function f defined by

(7) $\quad f(x) = \dfrac{x^3 + 7x + 4}{x^5 + 4x + 7}.$

Then f is the quotient of two polynomial functions. Therefore f is continuous wherever $x^5 + 4x + 7$ is not equal to 0. Since

$$\lim_{x \to 1} (x^3 + 7x + 4) = 12 \quad \text{and} \quad \lim_{x \to 1} (x^5 + 4x + 7) = 12,$$

it follows that $\lim_{x \to 1} f(x) = 12/12 = 1$.

Sometimes a limit may be simplified or put in a more desirable form by a change of variable. This procedure is illustrated in the following example.

Example 11. Consider

(8) $\quad \lim\limits_{h \to 0} \dfrac{f(a+h) - f(a)}{h},$

where a is a fixed real number. Set $h = x - a$. If we substitute $x - a$ for h in (8), we obtain

(9) $\quad \lim\limits_{h \to 0} \dfrac{f(x) - f(a)}{x - a}.$

If h is to approach 0, then x must approach a; in fact, $h \to 0$ if and only if $x \to a$. Consequently, (9), and therefore (8), can be rewritten as

(10) $\quad \lim\limits_{x \to a} \dfrac{f(x) - f(a)}{x - a}.$

EXERCISES

1. Suppose f is a constant function; that is, $f(x) = k$ for any real number x, where k is a constant. Prove that $\lim_{x \to a} f(x) = k$ for any real number a; hence f is continuous.

2. Prove (d) of Proposition 2.

3. Evaluate each of the following limits. You may use any method consistent with the results of this section.
   a) $\lim\limits_{x \to 2} (x + x^2)$
   b) $\lim\limits_{x \to 0} (3x^2 + 7x + 1)$
   c) $\lim\limits_{x \to 1} (x + 3)/(x^2 + 1)$
   d) $\lim\limits_{x \to -1} 7/(x^3 - 3x + 1)$

e) $\lim_{x \to 4} (x^{1/2} + x - (x-3)^{23})$ f) $\lim_{x \to -1} (x^2 + 2x + 1)/(x+1)$

g) $\lim_{h \to 0} \dfrac{(x+h) - x}{h}$

h) the limit of (8) with $f(x) = x^{1/2}$
i) the limit of (8) with $f(x) = x^3$
j) the limit of (8) with $f(x) = 1/x$ and $a \neq 0$.

4. In each of the following a limit is given together with a change of variable. Express the limit in terms of the new variable. Do not evaluate the limit.

   a) $\lim_{x \to 4} (x+2)/(x-3);\ y = x - 3$
   b) $\lim_{x \to -1} (x^2 + 1)/(x^2 + 2x + 1);\ y = x + 1$
   c) $\lim_{h \to 0} \dfrac{(x+h)^2 - h}{h};\ y = h - 5$

5. Let $f(x) = |x|$. Prove that

$$\lim_{h \to 0} \frac{f(0+h) - f(0)}{h}$$

does not exist even though f is continuous at 0. [*Hint.* Prove that the limit as h approaches 0 from the negative side is not the same as the limit as h approaches 0 from the positive side.]

6. Suppose f is a function from T into R. We shall say that

$$\lim_{x \to a^-} f(x) = L$$

if, given any positive number p, there is a positive number q such that if x is any point of T such that $x < a$ and $|x - a| < q$, then $|f(x) - L| < p$. Informally, $\lim_{x \to a^-} f(x)$ is what $f(x)$ approaches as x approaches a from the left.

   a) How would we define $\lim_{x \to a^+} f(x)$?
   b) Find $\lim_{x \to 0^+} f(x)$ and $\lim_{x \to 0^-} f(x)$ for the function f of Example 19 of Chapter 3. Do likewise for the function f of Example 1 of this chapter.

4.2 THE DERIVATIVE OF A FUNCTION

Recall that the equation of any line in the coordinate plane, except a line parallel to the y-axis, can be expressed in the form $y = mx + b$. The number m is called the *slope* of the line whose equation is $y = mx + b$. Recall too that the slope is a measure of how fast the line is "rising," that is, how fast y is increasing or decreasing relative to increases in x. For any real number x, y has the value $mx + b$. If x is increased by 1 to $x + 1$ (Fig. 4), then $y = m(x + 1) + b = (mx + b) + m$. Thus, as x increases 1 unit, y has a change of m units; this is true regardless of the value of x.

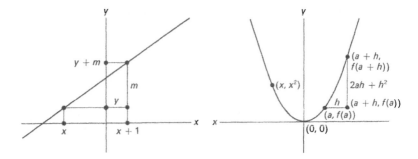

Fig. 4   Fig. 5

More generally, if a line has the equation $y = mx + b$ and x is increased by h units, then y increases by mh units.* The ratio

(11) $$\frac{\text{change in } y}{\text{change in } x} = \frac{mh}{h} = m,$$

is the slope of the line. Thus the slope of the line answers the question: How fast is y increasing relative to x?

We now wish to generalize the notion of *slope* to functions whose graphs are not necessarily straight lines. If we want to define the slope of an arbitrary function, however, we are immediately confronted by the following problem: When we are dealing with straight lines, the change in y relative to the change in x is uniform, that is, this change does not depend on x. However, this is not the case for arbitrary functions. For example, the function f defined by $f(x) = x^2$ is decreasing on $(-\infty, 0]$ and increasing on $[0, \infty)$; moreover, the larger the value of x, the more rapidly $f(x)$ seems to be increasing. Thus, we cannot expect a uniform rate of change of $f(x)$ relative to x.

Nevertheless, we may still be able to define the slope of a function as a function of x. Specifically, we try to answer the question: If the graph of the function f were to become a straight line at the point $(x, f(x))$, what would be the slope of that line? In other words, if we tried to approximate the graph of f at $(x, f(x))$ as closely as possible by a straight line, what would be the slope of that line? The slope of the approximating line will clearly depend on x.

Example 12. Let us examine the function f defined by $f(x) = x^2$ when $x = a$. An increase of a by h will mean a change of $2ah + h^2$ in $f(x)$, for

---

* We allow h to take any real number as a value; hence h may be negative. If h is negative, then an increase by h would actually be a decrease by $|h|$. We shall, however, use "increase" in the broadest possible sense; it will be synonymous with "change."

$f(a+h) - f(a) = (a+h)^2 - a^2 = 2ah + h^2$. See Fig. 5. Therefore the change in $f(x)$ relative to the change in x at a [compare with (11)] is

$$\frac{2ah + h^2}{h} = 2a + h.$$

The line passing through the point (a, a^2) with slope $2a + h$ will approximate the graph of f in the vicinity of (a, a^2). Moreover, the smaller h is, the better the approximation. The "limit slope" as h approaches 0 is $2a$; hence the line with slope $2a$ which passes through (a, a^2) will be the best approximation to the graph of f at (a, a^2). It would therefore be reasonable to assign a "slope" of $2a$ to f when $x = a$.

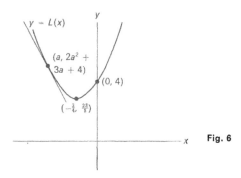

Fig. 6

Example 13. Let f be defined by $f(x) = 2x^2 + 3x + 4$. We will try to approximate f by a "linear function," that is, a function whose value at x is $mx + b$ for real numbers m and b when $x = a$. An increase in a of h (to $a + h$) will mean an increase of $4ah + 3h + 2h^2$ [that is, $f(a+h) - f(a)$] in $f(x)$. Therefore the ratio of increase in $f(x)$ to increase in x is given by

(12) $\qquad \dfrac{f(a+h) - f(a)}{h} = \dfrac{4ah + 3h + 2h^2}{h} = 4a + 3 + 2h.$

The limit of (12) as h approaches 0 is $4a + 3$; hence we may think of $4a + 3$ as the slope of f when $x = a$. If we wanted a straight line to approximate the graph of f at the point $(a, 2a^2 + 3a + 4)$, we would use the line with slope $4a + 3$ which passes through that point (Fig. 6). This approximating line is called the *tangent line* to the graph of f at $(a, f(a))$. Using the point-slope method for expressing the equation of a line, we find that the equation of the tangent line to the graph of f at $(a, f(a))$ is given by

(13) $\qquad \dfrac{y - f(a)}{x - a} = 4a + 3.$

The linear function which approximates f when $x = a$ is defined by $L(x) = (4a + 3)x - (a(4a + 3) + f(a))$. [We obtained this result by solving for y in (13) and relabeling y as $L(x)$. The graph of L approximates the graph of f "close to" $x = a$; the graphs are quite dissimilar, however, when x is not close to a.]

If the graph of a function f can be approximated near $(a, f(a))$ by a line with a positive slope, then it is reasonable to suppose that at least near $x = a$, f is an increasing function. Similarly, if the graph of f can be approximated near $(a, f(a))$ by a line with negative slope, then f is decreasing close to $x = a$. Now $f(x) = 2x^2 + 3x + 4$ in this example, and the slope of the tangent line to the graph of f at $(x, f(x))$ is given by the expression $4x + 3$. We now graph $y = 4x + 3$ and f together (Fig. 7). Note that for $x < -\frac{3}{4}$, $4x + 3$ is negative and hence f has negative slope, while for $x > -\frac{3}{4}$, the slope of f is positive. This means that f is decreasing on $(-\infty, -\frac{3}{4})$, increasing on $(-\frac{3}{4}, \infty)$, and at $-\frac{3}{4}$, f is neither increasing nor decreasing. At $-\frac{3}{4}$, the tangent line to the graph of f is parallel to the x-axis. It is clear from the graph of f that $f(-\frac{3}{4})$ is the minimum value of $f(x)$.

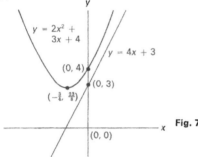

Fig. 7

Not all functions, not even all continuous functions, have a tangent line at every point. In the following example, we see a continuous function which fails to have a "slope" at $x = 0$.

Example 14. Let f be defined by $f(x) = |x|$. We shall show that the graph of f has no tangent line at $(0, 0)$. If we increase 0 by h, then the corresponding increase in $f(0)$ is $f(0 + h) - f(0) = |h| - |0| = |h|$. Therefore the change in $f(x)$ relative to the change in x is

(14) $|h|/h$.

If h is positive, then the value of (14) is 1, but if h is negative, then the value of (14) is -1. Then as h approaches 0, (14) does not approach a limit since the value of (14) depends on whether h approaches 0 from the positive or negative side.

We can illustrate this point more clearly by considering the graph of f shown in Fig. 8. We note that as x approaches 0 from the positive direction, $f(x)$ approaches 0 along the line whose equation is $y = x$; however, if x approaches 0 from the negative direction, then $f(x)$ approaches 0 along the line whose equation is $y = -x$. If the graph of f were to have a tangent line at $(0, 0)$, then we would assume it would have to be either $y = x$ or $y = -x$, but it can be neither.

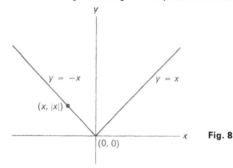

Fig. 8

The discussion of this section leads to the following definition.

Definition 3. *Let f be a function from T, a subset of R, into R. We define the **first derivative**, or **slope**, of f at a to be*

(15) $$\lim_{h \to 0} \frac{f(a + h) - f(a)}{h},$$

provided, of course, that this limit exists. We denote the first derivative of f at a by $f'(a)$.

*The line with slope $f'(a)$ which passes through the point $(a, f(a))$ is said to be the **tangent line** to the graph of f at $(a, f(a))$.*

Using the point-slope form for the equation of a line, we find that the tangent line to the graph of f at $(a, f(a))$ is given by

(16) $$\frac{y - f(a)}{x - a} = f'(a).$$

If f is not continuous at a, then f cannot have a first derivative at a. For f is continuous at a if and only if $f(a)$ is defined and $\lim_{x \to a} f(x) = f(a)$. Now $\lim_{x \to a} f(x)$ is the same as $\lim_{h \to 0} f(a + h)$. If this latter limit is not $f(a)$, then the numerator of (15) will not go to 0 as h goes to 0. The result then will be the division, by an expression which goes to 0, of an expression which does not go to 0; hence (15) will definitely not exist.

Even if a function is continuous, it need not have a derivative at every point (Example 14); even when the numerator and the denominator of the expression in (15) of which the limit is being taken both tend to 0 simultaneously, the limit may not exist.

Definition 4. *If a function f has a first derivative at a, then f is said to be* ***differentiable at*** *a. If f is differentiable at every point at which it is defined, we say that f is* ***differentiable***.

Thus, a function may be continuous but not differentiable. But no function can be differentiable and fail to be continuous. A function which is differentiable has a "smooth" graph.

Example 15. Let f be defined by $f(x) = 1/x$, $x \neq 0$. We compute $f'(3)$:

$$f(3+h) - f(3) = \frac{1}{3+h} - \frac{1}{3} = \frac{3+h-3}{3(3+h)} = \frac{h}{(3(3+h))}.$$

Consequently,

$$f'(3) = \lim_{h \to 0} \frac{f(3+h) - f(3)}{h} = \lim_{h \to 0} \frac{1}{3(3+h)} = \frac{1}{9}.$$

EXERCISES

1. Let f be defined by $f(x) = 3x^2 + 2x + 7$. Compute each of the following, using Definition 3 directly.

   a) $f'(1)$ b) $f'(0)$ c) $f'(10)$

   Do (a), (b), and (c) for $f(x) = x^3$.

2. Suppose $f(x) = k$, where k is a constant. Prove that $f'(x) = 0$ for any x.

3. Find the equation of the line tangent to the graph of f at $(1, f(1))$ for each of the following.

   a) $f(x) = 3x + 1$ b) $f(x) = mx + b$

   What can we say about the tangent line to the graph of a linear function at any point of the graph?

   c) $f(x) = -x^2 - 6$ d) $f(x) = x - x^{-1}$

4. For each of the following functions, find an expression for $f'(x)$ valid at every x for which the function has a first derivative. Indicate those points at which each function is differentiable.

   a) $f(x) = 1/x^2$ b) $f(x) = x^{1/2}$ c) $f(x) = x^3$
   d) $f(x) = 2x^2 - 3x + x^{-1}$ e) $f(x) = (x-1)^{1/2}$

5. Let F be the distribution function of a continuous random variate X with admissible range [0, 4], and suppose F is differentiable.

   a) Find $F'(0)$ and $F'(4)$.

   b) Provide an informal argument to show that $F'(x)$ is nonnegative for any x.

6. Let $f(x) = 3x^2 - 5$. Plot the graphs of f and f' together. Describe the behavior of f relative to the behavior of f'.

7. Do Exercise 6 with $f(x) = 1/x^2$.

4.3 BASIC RULES FOR FINDING A DERIVATIVE

In Definition 2 we learned how to form new functions from given functions by performing certain algebraic operations. In addition to the algebraic operations with functions defined in Definition 2, we have the operation of *raising a function to a power*. If r is any real number and f is any function from T into R, then

$(f^r)(x)$ is defined to be $(f(x))^r$

for each x in T for which $(f(x))^r$ makes sense.

Example 16. If $f(x) = x^2 + x$, then $f^{1/2}$ is defined by $f^{1/2}(x) = (x^2 + x)^{1/2}$. The function $f^{1/2}$ will only make sense where $x^2 + x$ is nonnegative.

In this section we answer the question: How is the derivative of a function formed from given functions using the algebraic operations we have defined, related to the derivatives of the given functions? Our basic tool in this section will be Proposition 2.

Proposition 3. *If f, g, and m are functions which are differentiable at a and $f = g + m$, then $f'(a) = g'(a) + m'(a)$. (That is, the derivative of a sum is the sum of the derivatives.)*

Proof: By definition of $f'(a)$ and the fact that $f = g + m$, we have

$$(17) \quad f'(a) = \lim_{h \to 0} \frac{f(a+h) - f(a)}{h}$$

$$= \lim_{h \to 0} \frac{(g(a+h) - g(a)) + (m(a+h) - m(a))}{h}$$

$$= \lim_{h \to 0} \left(\frac{g(a+h) - g(a)}{h} + \frac{m(a+h) - m(a)}{h} \right).$$

Applying Proposition 2 to (17), we find that (17) is equal to

$$\lim_{h \to 0} \frac{g(a+h) - g(a)}{h} + \lim_{h \to 0} \frac{m(a+h) - m(a)}{h} = g'(a) + m'(a),$$

which is what we wanted to show.

Example 17. If $g(x) = x^3$ and $m(x) = x^2$, then $g'(x) = 3x^2$ and $m'(x) = 2x$. If $f = g + m$, that is, if $f(x) = x^3 + x^2$, then

$$f'(x) = g'(x) + m'(x) = 3x^2 + 2x.$$

Proposition 4. *If k is any real number and $f'(a)$ exists, then $(kf)'(a) = kf'(a)$.*

Proof:

$$(kf)'(a) = \lim_{h \to 0} \frac{(kf)(a+h) - (kf)(a)}{h}$$

$$= \lim_{h \to 0} \frac{kf(a+h) - kf(a)}{h} \text{ (by definition of } kf\text{)}$$

$$= \lim_{h \to 0} \frac{k(f(a+h) - f(a))}{h}$$

$$= k \lim_{h \to 0} \frac{f(a+h) - f(a)}{h} \text{ (by Proposition 2)}$$

$$= kf'(a).$$

Example 18. If $f(x) = 3x^2$, then $f = 3g$, where $g(x) = x^2$. Since $g'(a) = 2a$, we have $f'(a) = 3(2a) = 6a$.

We accept the following proposition without proof.

Proposition 5. *Let $f(x) = x^r$, where r is a fixed real number and x is any real number for which x^r makes sense. Then if a^r and a^{r-1} are both defined, $f'(a) = ra^{r-1}$.*

Example 19. If $f(x) = x^5$, then $f'(a) = 5a^4$. If $g(x) = x^{-7}$, then $g'(a) = (-7)a^{(-7)-1} = -7a^{-8}$, $a \neq 0$.

If $m(x) = x^{6/7}$, then

$m'(a) = (\frac{6}{7})a^{-1/7}$.

Note that even though $m(0)$ is defined, $m'(0)$ is not defined.

If f is any function, f' is used to denote that function such that $f'(x)$ is the first derivative of f at x.

Example 20. If f is a polynomial function as in Example 9, then combining Propositions 3, 4, and 5, we find that

(18) $\quad f'(x) = na_n x^{n-1} + (n-1)a_{n-1} x^{n-1} + \cdots + a_1$.

As a particular case, suppose $f(x) = 4x^{45} + 7x^{23} + 2x + 76$. Then $f'(x) = 180x^{43} + 161x^{22} + 2$.

It may be possible to take the derivative of a derivative. Consider the following example.

Example 21. Suppose $f(x) = 3x^4 + 2x^3 + 7x + 1$. Then
$f'(x) = 12x^3 + 6x^2 + 7$,
$(f')'(x) = f''(x) = 36x^2 + 12x$,
$(f'')'(x) = f'''(x) = 72x + 12$, etc.

Note that after taking one more derivative all subsequent derivatives will be 0.

We now discuss the derivative of the product and quotient of two functions. Using Propositions 3, 4, and 5 as a guide, one might be tempted to say that the first derivative of the product of two functions is the product of the first derivatives; that is, if $f = gm$, then $f' = g'm'$. This, however, is not the case. The correct relationship is given in the next proposition.

Proposition 6. *Suppose $f = gm$ and g and m are differentiable at a. Then* $f'(a) = g(a)m'(a) + m(a)g'(a)$.

Proof: We shall use an algebraic trick to prove this proposition. Set $g(a+h) - g(a) = Dg(a)$ and $m(a+h) - m(a) = Dm(a)$. It can be verified directly that

$$g(a+h)m(a+h) = f(a+h) = (g(a) + Dg(a))(m(a) + Dm(a))$$
$$= g(a)m(a) + g(a)Dm(a) + m(a)Dg(a)$$
$$+ m(a)Dg(a) + Dg(a)Dm(a).$$

Consequently,

(19) $\quad f'(a) = \lim_{h \to 0} \dfrac{f(a+h) - f(a)}{h}$

$\qquad\qquad = \lim_{h \to 0} \dfrac{g(a+h)m(a+h) - g(a)m(a)}{h}$

$\qquad\qquad = \lim_{h \to 0} \left(g(a) \dfrac{Dm(a)}{h} + m(a) \dfrac{Dg(a)}{h} + Dg(a) \dfrac{Dm(a)}{h} \right).$

Applying Proposition 2, we find that (19) becomes

(20) $\quad g(a) \lim_{h \to 0} \dfrac{Dm(a)}{h} + m(a) \lim_{h \to 0} \dfrac{Dg(a)}{h} + \lim_{h \to 0} Dg(a) \lim_{h \to 0} \dfrac{Dm(a)}{h}.$

But

$\lim_{h \to 0} \dfrac{Dm(a)}{h} = m'(a), \qquad \lim_{h \to 0} \dfrac{Dg(a)}{h} = g'(a), \quad \text{and} \quad \lim_{h \to 0} Dg(a) = 0.$

Consequently, (20) is equal to $g(a)m'(a) + m(a)g'(a)$, which is what we wanted to prove.

Example 22. Suppose $f(x) = (x^2 + 3)(x^3 + 7)$. Setting $g(x) = x^2 + 3$, and $m(x) = x^3 + 7$, we have $f = gm$. Now $g'(x) = 2x$ and $m'(x) = 3x^2$. Hence

$f'(x) = g(x)m'(x) + m(x)g'(x) = (x^2 + 3)(3x^2) + (x^3 + 7)(2x)$
$\qquad = 5x^4 + 9x^2 + 14x.$

We confirm the accuracy of this conclusion by expanding the product which defines f and computing f' directly. Thus,

$$f(x) = (x^2 + 3)(x^3 + 7) = x^5 + 3x^3 + 7x^2 + 21.$$

Hence $f'(x) = 5x^4 + 9x^2 + 14x$, precisely what we obtained from the product formula.

Note that $(g'm')(x) = g'(x)m'(x) = 6x^3$, which is not equal to $f'(x)$.

Example 23. Suppose $f(x) = (g(x))^2$. Applying Proposition 6, we find $f'(x) = g(x)g'(x) + g(x)g'(x) = 2g(x)g'(x)$. Suppose $f(x) = (g(x))^n$ for some positive integer n. We could use finite induction to prove that $f'(x) = n(g(x))^{n-1}g'(x)$ (cf. Exercise 6 below). We shall actually prove a more general result after considering the *Chain Rule* in the next section.

The following proposition concerning the quotient of two functions is presented here without proof. We shall see in the next section that Proposition 7 follows from the Chain Rule and Proposition 6.

Proposition 7. *If $f(x) = g(x)/m(x)$, $m(a) \neq 0$, and $g'(a)$ and $m'(a)$ are both defined, then*

$$f'(a) = \frac{m(a)g'(a) - g(a)m'(a)}{(m(a))^2}.$$

Example 24. Suppose $f(x) = (x^2 - 3)/(x - 6)$. Then setting

$$g(x) = x^2 - 3 \quad \text{and} \quad m(x) = x - 6,$$

and applying Proposition 7, we obtain

$$f'(x) = \frac{(x-6)(2x) - (x^2 - 3)(1)}{(x-6)^2}.$$

Note that $f'(x)$, like $f(x)$ itself, is not defined for $x = 6$. We may also note that $g'(x) = 2x$ and $m'(x) = 1$; hence $g'(x)/m'(x) = 2x$, which bears little resemblance to what $f'(x)$ actually is.

We close this section with an example which combines several of the differentiation rules in one problem.

Example 25. We shall find $f'(x)$ when

$$f(x) = \frac{(x^2 + 7x)(x^3 + 4)^2}{(9x + x^{1/3})}.$$

By this time the reader may be able to find the derivative of a function which is the sum, product, or quotient of other functions without explicitly expressing the given function as a sum, product, or quotient. For the sake of completeness, however, we shall carry out explicitly most of the steps required to find $f'(x)$ using the rules developed in this section.

Let $g(x) = (x^2 + 7x)(x^3 + 4)^2$ and $m(x) = 9x + x^{1/3}$. Then $f(x) = g(x)/m(x)$.

Therefore

(21) $\quad f'(x) = \dfrac{m(x)g'(x) - g(x)m'(x)}{(m(x))^2}.$

Applying Propositions 3 and 5, we obtain $m'(x) = 9 + (\frac{1}{3})x^{-2/3}$. In order to evaluate $g'(x)$, we set $c(x) = x^2 + 7x$ and $d(x) = (x^3 + 4)^2$. Then $g(x) = c(x)d(x)$; hence $g'(x) = c(x)d'(x) + d(x)c'(x)$. Now $c'(x) = 2x + 7$, and applying Example 23, we find $d'(x) = 2(x^3 + 4)(3x^2)$. Therefore

$$g'(x) = (x^2 + 7x)(2(x^3 + 4)(3x^2)) + (x^3 + 4)^2(2x + 7).$$

All terms necessary to evaluate (21) have now been computed. The determined reader can complete the computation, if he wishes.

EXERCISES

1. Find the first derivatives of each of the following functions.
   a) $f(x) = x^2 + 1$ \qquad b) $f(x) = x^{4/5}$
   c) $g(x) = x(x - x^3)$ \quad (Do this two ways.)
   d) $m(x) = 3x^5 + 4x^2 + 10x + 156$
   e) $p(x) = (3x^2 + 7x + 3)(9x^4 + 4x + 1)$
   f) $q(x) = (x - 3)/(x - 7)$
   g) $r(x) = x^{1/2} + x^{1/3} + x^{-17} + x^{-0.67}$
   h) $t(x) = ax^{-2} + b(x^4 + 3x + c)$, where a and b are constants
   i) $f(x) = \dfrac{(x-3)(x-5)}{x^{1/2}}$

   (Where does $f'(x)$ make sense? Where does $f(x)$ make sense?)

   j) $g(x) = (x^2 + x^{-1/2})(3x^2 + 4x + 1)^{-1}$
   k) $h(x) = a^2 + b^2 + c^{-4} + x$, where a, b, and c are constants

2. Find the equation of the line tangent to the graph of $f(x) = (x + x^5)^2$ at (1, 4).

3. Find the equation of the line tangent to the graph of $f(x) = (x^2 + x^{1/2})/(x - 3)$ at (4, 18).

4. a) If f and g are both functions from T into R, how is $f - g$ defined? Suppose $f'(a)$ and $g'(a)$ both exist. Prove that $(f - g)'(a) = f'(a) - g'(a)$.
   b) You may assume that the derivative of a function is constantly 0, that is, $m'(x) = 0$ for all x for which $m(x)$ is defined, if and only if the function is constant, that is, $m(x) = k$, for some constant k. What can then be said about two functions which have the same derivative; for example, what could be said about f and g if $f'(x) = g'(x)$ for each x in T?

5. For each of the following, find a function which has the given function as its first derivative.
   a) $f(x) = 3$ b) $f(x) = 2x$ c) $f(x) = x^2$
   d) $f(x) = 2x + 7$ e) $g(x) = x^{-2}$

6. a) Suppose $f(x) = (g(x))^3$. Use Proposition 6 and Example 23 to find $f'(x)$.
   b) Use finite induction to show that if n is any positive integer and $f(x) = (g(x))^n$, then $f'(x) = n(g(x))^{n-1}g'(x)$.
   c) Suppose $f(x) = (g(x))^{-1} = 1/g(x)$. Use Proposition 7 to find $g'(x)$.
   d) Suppose n is any positive integer. Use finite induction, as well as any results from this section that may prove useful, to prove:

   If $f(x) = (g(x))^{-n}$, then $f'(x) = (-n)(g(x))^{-n-1}g'(x)$.

   e) Use the results obtained in (a) through (d), as well as any other results from this section that may prove useful, to compute the first derivatives of each of the following functions.
      i) $f(x) = (x^3 + 4)^{17}$
      ii) $g(x) = (x^2 + 3x + 1)^{-4}$
      iii) $h(x) = (x^2 + 3)^{18}(x^3 + 1)^{-15}$
      iv) $m(x) = (x^{1/2} + x^{1/3})^{23}(x + 1)^{1,678}$

4.4 THE CHAIN RULE. IMPLICIT DIFFERENTIATION

We have already learned to add, multiply, and divide functions, as well as raise a function to a power. In this section we introduce yet another operation between two functions, *composition*.

Suppose $f(x) = x^2$ and $g(x) = x + 6$. Then

$$f(g(x)) = f(x + 6) = (x + 6)^2 \quad \text{and} \quad g(f(x)) = g(x^2) = x^2 + 6.$$

The function defined by $f(g(x)) = (x + 6)^2$ is called the *composition of f with g*. The function defined by $g(f(x)) = x^2 + 6$ is the composition of g with f. In general, we make the following definition.

Definition 5. *If f and g are any two functions, then the **composition of f with g**, which we denote by $f \circ g$, is defined by $(f \circ g)(x) = f(g(x))$, for all x for which this definition makes sense.*

(For most functions we shall consider, the image of g is contained in the domain of f. This will mean $f(g(x))$ will be defined.)

Example 26. Suppose $f(x) = x^{25}$ and $g(x) = x^3 + x^{1/2}$. Then

$$(f \circ g)(x) = f(g(x)) = f(x^3 + x^{1/2}) = (x^3 + x^{1/2})^{25}.$$

Note that even though $f(x)$ makes sense for any real number x, $(f \circ g)(x)$ only makes sense for nonnegative values of x. The composition of g with f, that is, $g \circ f$, is defined by $(g \circ f)(x) = g(f(x)) = g(x^{25}) = x^{75} + x^{25/2}$.

We see from Example 26 that rather complicated functions can be formed by composing relatively simple functions. Since this is true, the following question is of interest: If f and g are differentiable, can $(f \circ g)'$ be expressed in terms of f, g, f', and g'? An affirmative answer to this question would allow us to compute the derivative of certain complicated functions more readily. The answer to the question is indeed affirmative; the relationship between the derivative of the composition of two functions and the derivatives of the component functions is given by the so-called *Chain Rule*. We make no attempt to prove the Chain Rule, but merely state it in the following proposition.

Proposition 8. *(Chain Rule).* *If $f = g \circ m$ and $g'(a)$ and $m'(a)$ are both defined, then*

$$f'(a) = g'(m(a))m'(a).$$

The application of Proposition 8 is best explained by examples.

Example 27. Suppose $f(x) = (x^3 + 2x + 4)^{1/2}$. Let $g(x) = x^{1/2}$ and $m(x) = x^3 + 2x + 4$. Then $f = g \circ m$; that is, $f(x) = g(m(x))$. Therefore $f'(x) = g'(m(x))m'(x)$. Now $g'(x) = (\frac{1}{2})x^{-1/2}$; consequently,

$$g'(m(x)) = (\tfrac{1}{2})(x^3 + 2x + 4)^{-1/2}.$$

Since $m'(x) = 3x^2 + 2$, we obtain $f'(x) = (\frac{1}{2})(x^3 + 2x + 4)^{-1/2}(3x^2 + 2)$.

Example 28. Consider $f(x) = (x^5 + 3)^{-1}$. Let $g(x) = x^{-1}$ and $m(x) = x^5 + 3$. Then $f(x) = g(m(x))$. Since $g'(x) = -x^{-2}$ and $m'(x) = 5x^4$, we have $f'(x) = g'(m(x))m'(x) = -(x^5 + 3)^{-2}(5x^4)$.

Example 29. Suppose $f(x) = (g(x))^r$, where r now is allowed to be any real number whatsoever. If we set $h(x) = x^r$, then $f(x) = h(g(x))$; hence $f'(x) = h'(g(x))g'(x)$. Since $h'(x) = rx^{r-1}$, we have

$$f'(x) = r(g(x))^{r-1}g'(x).$$

We have therefore generalized the results of Exercise 6 of the preceding section.

As an example of the use of the Chain Rule, we now prove Proposition 7.

Example 30. (Proof of Proposition 7). Suppose

$$f(x) = g(x)/m(x) = g(x)(m(x))^{-1},$$

$g'(a)$ and $m'(a)$ are defined, and $m(a) \neq 0$. Set $n(x) = (m(x))^{-1}$. Then $f(x) = g(x)n(x)$. Therefore $f'(a) = g(a)n'(a) + n(a)g'(a)$. Using the results of Example 29, we find $n'(a) = -(m(a))^{-2}m'(a)$. Hence

(22) $\qquad f'(a) = (m(a))^{-1}g'(a) - g(a)(m(a))^{-2}m'(a).$

Expressing the right-hand side of (22) as a fraction with denominator $(m(a))^2$, we obtain the form of $f'(a)$ given in Proposition 7.

Example 31. We shall study the *exponential function* in a later chapter. All we shall say at present about the exponential function E is that it is defined by $E(x) = e^x$, for each real number x, where e is a certain constant; the exponential function has the peculiar property that it is the same as its derivative, that is, $E'(x) = E(x) = e^x$.

Suppose $f(x) = e^{x^2+1}$. Let $g(x) = x^2 + 1$; then $f(x) = E(g(x))$. Using the Chain Rule, we find that $f'(x) = E'(g(x))g'(x) = e^{x^2+1}(2x)$.

The use of the Chain Rule may be repeated several times in the same problem as we see in the next example.

Example 32. Suppose $f(x) = e^{(2x+1)^3}$. Set $g(x) = (2x+1)^3$ and $E(x) = e^x$. Then $f(x) = E(g(x))$; hence $f'(x) = E'(g(x))g'(x) = e^{(2x+1)^3}g'(x)$. Applying the Chain Rule to g, we find $g'(x) = 3(2x+1)^2(2)$. This enables us to complete the evaluation of $f'(x)$.

A function f is said to be defined *implicitly*, or be an *implicit function* of x, if $f(x)$ is not given directly in terms of x but indirectly in the form of an equation to be solved. For example, the expression $f(x) = 1/x$ defines the function f explicitly, while the equation $xf(x) - 1 = 0$ is an implicit definition of the same function. A function which is defined implicitly can sometimes be differentiated implicitly in the manner to be described below. There are serious mathematical pitfalls involved in implicit functions and implicit differentiation. If what we are going to do strikes the reader as being too informal, he is right; it is. Nevertheless, implicit differentiation will usually work in the few situations in which the reader may have to use it.

The basic idea of implicit differentiation is this. We are given an equation of the form $F(x, f(x)) = 0$. We take the derivative of both sides of this equation and obtain an equation of the form $G(x, f(x), f'(x)) = 0$. We then solve $f'(x)$ in terms of x and $f(x)$. We illustrate this procedure with three examples.

Example 33. Suppose

(23) $\quad xf(x) - 1 = 0.$

We differentiate both sides of (23). Since (23) is an equality, the derivatives of both sides of (23) should also be equal. Of course, the derivative of 0 is 0. To differentiate $xf(x) - 1$, we note that $xf(x) - 1$ is the difference of two functions, one a constant function and the other the product of two functions. Using Proposition 6 to find the derivative of $xf(x)$, we find that the "derivative" of (23) is

(24) $\quad xf'(x) + f(x) = 0.$

Solving (24) for $f'(x)$, we find

(25) $\quad f'(x) = -f(x)/x$.

If we solve (23) for $f(x)$, we find $f(x) = 1/x$. Substituting for $f(x)$ in (25), we find $f'(x) = -1/x^2$, which is what we know $f'(x)$ to be. If it were not possible to find $f(x)$ explicitly, we would have to leave $f'(x)$ in the form given in (25).

Example 34. Suppose

(26) $\quad x^3 (f(x))^2 = 0$.

We consider $x^3(f(x))^2$ as the product of two functions and differentiate both sides of (26). In this way, we obtain

(27) $\quad 2x^3 f(x) f'(x) + 3x^2 (f(x))^2 = 0, \quad$ or
$\quad\quad\;\; 2xf'(x) + 3f(x) = 0$.

Therefore $f'(x) = -3f(x)/2x$.

This is one of the instances in which implicit differentiation gives a result which at first glance seems meaningful, but which is actually rather meaningless from a mathematical point of view. For if the product of two numbers is 0, at least one of the numbers must be 0. From the fact that $x^3(f(x))^2$ is always 0, we conclude that since x is 0 only for $x = 0$, $(f(x))^2$, and hence $f(x)$, must be 0 for every x except possibly 0. Thus we have gone to some trouble to find the derivative of a function which is essentially constant.

Example 35. The equation of the circle in the coordinate plane with center $(0, 0)$ and radius 1 is given by

(28) $\quad x^2 + y^2 = 1$.

We may consider y to be an implicit function in x. Taking the derivative of each side of (28), we obtain

(29) $\quad 2x + 2yy' = 0$.

Solving (29) for y', we obtain

(30) $\quad y' = -x/y$.

Suppose we wish to compute the slope of the tangent line to the circle at some point (a, b) on the circle. Then we would substitute a for x and b for y in (30). For example, $(1/\sqrt{2}, -1/\sqrt{2})$ is a point on the circle. We therefore conclude that the slope of the line tangent to the circle at this point is

$$\frac{-1/\sqrt{2}}{-1/\sqrt{2}} = 1.$$

If we consider the graph of the circle and the tangent line (Fig. 9), we see that 1 is a reasonable answer.

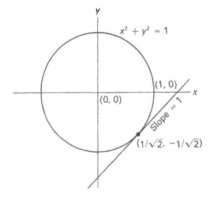

Fig. 9

EXERCISES

1. Find the first derivatives of each of the following functions.
   a) $f(x) = (x + 25)^{67}$ b) $g(x) = (x^3 + 5x + 78)^{96}$
   c) $h(x) = (x^3 + 2x^2 + 3x + 1)^{15}(x^4 + x^{1/2})^{10}$
   d) $k(x) = (x^{1/3} + 1)^{1/3}$ e) $f(x) = ((x^{1/2} + 1)^{1/2} + 1)^{1/2}$
   f) $f(x) = e^{x+4}$, where $E(x) = e^x$ is as introduced in Example 31
   g) $f(x) = e^{x^3 + (1-x^2)^2}$ h) $g(x) = e^{((x^{1/2}+1)^{1/2}+1)^{1/2}}$

2. Suppose $f(x) = g(m(n(x)))$. Prove that
   $f'(x) = g'(m(n(x)))m'(n(x))n'(x)$.

3. Suppose $f(x) = g(m(x))$ and g and m are differentiable. The Chain Rule tells us that $f'(x) = g'(m(x))m'(x)$. Find $f''(x) = (f')'(x)$. Find $f'''(x) = (f'')'(x)$. We call f'' and f''' the *second* and *third derivatives* of f, respectively. Use the results you have obtained to compute the second derivatives of each of the following functions.
   a) $f(x) = (x + 3)^{3.4}$ b) $f(x) = (x^3 + 4x + 5)^{6.7}$
   c) $g(x) = e^{(x^2 + 3v)}$ d) $h(x) = e^{e^x}$

4. Use implicit differentiation to $f'(x)$ for each of the following.
   a) $x^2 f(x) + 3 = 0$ b) $x + xf(x) + 2 = 0$
   c) $x(f(x))^2 + f(x) + 1 = 0$ d) $xf(x) + x^2(f(x))^2 + x^3 = 9$
   e) $e^x f(x) + e^{-x}(f(x))^{-1} = 1$

5. The equation of an ellipse in the coordinate plane with center (0, 0) has the form
   $$x^2/a^2 + y^2/b^2 = 1,$$
   where a and b are constants. Consider y to be a function of x and find y'.

6. The general form of the equation of a *conic section* is

$$Ax^2 + Bxy + Cy^2 + Dx + Ey + F = 0,$$

where A, B, C, D, E, and F are all constants. Consider y to be a function of x and find y'. Suppose $(3, 4)$ is a point on the conic section; find the equation of the tangent line to the conic at this point.

5 Applications of the Derivative

5.1 MAXIMA AND MINIMA

Thus far we have been concerned primarily with the computation of derivatives; we now put derivatives to use. First, however, some words of caution are in order.

First, it should be evident that we can gain information from the derivative of a function only where the function has a derivative. There are, however, perfectly decent functions (see, for instance, Example 14 of Chapter 4) which do not have derivatives at certain points. The absence of a derivative at a point may itself give us certain information about the function being investigated; but if a function does not have a derivative at a point, then the differential calculus will be a poor tool to investigate the behavior of the function at that point.

Second, we shall often be concerned with functions whose values we wish to consider only for a restricted portion of the real numbers. For example, if X is a random variable and F is the distribution function of X, then we might consider $F(x)$ only for admissible values of X. Quite often the formal definition of a function, for example, $f(x) = x^2$, will "make sense" even for values of x other than those to which we may restrict ourselves. When a value of x is found to have a formal property that we are interested in, we must be certain that that value of x lies in the subset of the real numbers that is of interest to us.

In this section we shall consider *maxima* and *minima*. In the context of this section, the admonitions above may be made concrete as follows:

1) When trying to find the maxima and minima of a function, we must take into account those points at which the function does not have a first derivative, as well as those points at which the function is differentiable; and

2) we will be interested only in those maxima and minima which occur for admissible values of the function variable.

Definition 1. *Let f be a function from T into R. The function f is said to have an **absolute minimum** at a if $f(a) \leq f(x)$ for each x in T.*

*We say that f has an **absolute maximum** at a if $f(x) \leq f(a)$ for each x in T.*

*If W is a subset of T, then f is said to have a **minimum** at a point a of W **relative to** W if $f(a) \leq f(x)$ for each x in W.*

*We say that f has a **maximum** at a in W **relative to** W if $f(x) \leq f(a)$ for each x in W.*

*If a is a point of T such that there is an open interval (u, v) containing a and such that f has a maximum at a relative to $\{x \mid x$ is both in T and $(u, v)\}$, then f is said to have a **relative maximum** at a.*

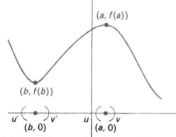

Fig. 1

Relative minimum at b | Relative maximum at a

*Similarly, if a is a point of T such that there is an open interval (u, v) which contains a and such that f has a minimum at a relative to $\{x \mid x$ is both in T and $(u, v)\}$, then f is said to have a **relative minimum** at a.*

Despite the imposing nature of Definition 1, the concepts involved should be fairly clear. A function f has an absolute maximum at a if $f(a)$ is the largest value that $f(x)$ assumes. The function f has a maximum

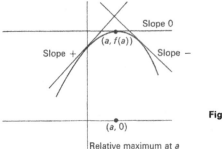

Fig. 2

Relative maximum at a

at a relative to W if $f(a)$ is the largest value that $f(x)$ assumes for any point x of W. And f has a relative maximum at a if $f(a)$ is the largest value of $f(x)$ for all the points x for which $f(x)$ is defined in some open interval which contains a.

Let us suppose now that f is a function which has a relative maximum at a and is differentiable at every point of some open interval (u, v) which contains a. We shall prove that $f'(a) = 0$.

Informally, if we interpret $f'(x)$ to be the slope of the tangent line to the graph of f at the point $(x, f(x))$, then showing that $f'(a) = 0$ if f has a relative maximum at a is geometrically equivalent to showing that the tangent line to the graph of f at $(a, f(a))$ is parallel to the x-axis. Since $f(a)$ is to be a maximum value, we intuitively see that the slope of the line tangent to the graph of f must be positive (or at least nonnegative) just before $(a, f(a))$ and nonpositve just after $(a, f(a))$ (Fig. 2).

More formally, consider

(1) $$\frac{f(a + h) - f(a)}{h}.$$

It is, of course, the limit of (1) as h approaches 0 that gives $f'(a)$. If h is negative and sufficiently close to 0,* then $f(a + h) - f(a)$ will be nonpositive because $f(a + h) \leq f(a)$ (Fig. 3). Consequently, if h is negative and close enough to 0, then (1) will be nonnegative. Therefore the limit of (1) as h approaches 0, and hence $f'(a)$, will be nonnegative. On the other hand, if h is positive and sufficiently close to 0, then $f(a + h) - f(a)$ is still nonpositive, but now (1) is nonpositive. Hence the limit of (1) as h approaches 0, that is, $f'(a)$, must be nonpositive. Since we have shown that $f'(a) \leq 0$ and $0 \leq f'(a)$, it must be that $f'(a) = 0$. We have therefore proved the following proposition.

---

* Specifically, if h is such that $a + h$ is in an open interval (u', v') which contains a and relative to which f has a maximum at a.

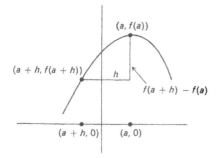

Fig. 3

Proposition 1. *If a function f has a relative maximum at a and f is differentiable at every point of some open interval which contains a, then $f'(a) = 0$.*

An argument similar to that used to prove Proposition 1 can be used to prove

Proposition 2. *If a function f has a relative minimum at a and f is differentiable at every point of some open interval which contains a, then $f'(a) = 0$.*

Definition 2. *A point a is said to be a **critical point** of the function f if $f'(a) = 0$.*

Thus a differentiable function will have a critical point at each point at which it has either a relative maximum or relative minimum. We shall see, however, that not all critical points represent relative maxima or minima.

We now consider some examples.

Example 1. Consider the function defined by

$$f(x) = x^3 - 6x^2 + 11x - 6 = (x-1)(x-2)(x-3).$$

The graph of this function is given in Fig. 4. Now $f'(x) = 3x^2 - 12x + 11$. By means of the quadratic formula, we determine that f has critical points at $x = 2 + (\sqrt{3}/3)$ and $x = 2 - (\sqrt{3}/3)$ (that is, we solve

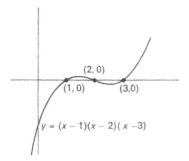

Fig. 4

$3x^2 - 12x + 11 = 0$). We see from Fig. 4 that f has a relative maximum at $2 - (\sqrt{3}/3)$ and a relative minimum at $2 + (\sqrt{3}/3)$. If we are considering $f(x)$ for all real numbers x, then f has no absolute maximum or absolute minimum since given any number M, there are always values of $f(x)$ both smaller and larger than M.

We may ask: If we did not have the graph of f before us, how could we tell whether f has a relative maximum, relative minimum, or neither relative maximum nor minimum at $2 + (\sqrt{3}/3)$? A function which has neither a relative maximum nor a relative minimum at one of its critical points is illustrated in the following example.

Example 2. Consider $f(x) = x^3$. Then $f'(x) = 3x^2$; hence f has a critical point at 0. We can see from the graph of f (Fig. 5), however, that f has neither a maximum nor a minimum at 0. We may also conclude that f does not have a relative maximum or minimum at 0 from the fact that $f'(x)$ is nonnegative for any x; therefore f is always an increasing function. If f is any differentiable function which has either a relative maximum or a relative minimum, then we would expect $f'(x)$ to take both positive and negative values.

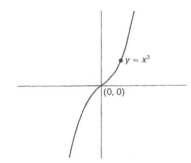

Fig. 5

A satisfactory answer to the question, "How do we tell what kind of point a critical point is?" must wait until later. For the moment, the only procedure we have is to test several points close to the critical point, taking care to choose some points from each "side" of the critical point. If the function values at the test points are consistently larger than, or at least as large as, the function value at the critical point, then we probably have a minimum. If the function values at the test points are no larger than the function value at the critical points, then we probably have a maximum. If some test values are larger and some are smaller than the value at the critical point, then we probably have neither a maximum nor a minimum. This method of testing, however, is not only tedious but often inaccurate. Much better methods will be discussed soon.

EXERCISES

1. Prove Proposition 2.

2. In each of the following, a function from R into R is defined and a value of x is given. Decide in each case which of the following the function has at the value of x: critical point, absolute maximum, absolute minimum, relative maximum, relative minimum. If a point is none of the latter, write NONE.
   a) $f(x) = x$; $x = 0$
   b) $f(x) = x^2$; $x = 0$
   c) $f(x) = x^2 + 4x + 4$; $x = -2$
   d) $f(x) = |x|$; $x = 0$
   e) $f(x) = x^3 - 4x^2 - 3x + 1$; $x = 3$
   f) $f(x) = x^3 - 4x^2 - 3x + 1$; $x = -\frac{1}{3}$
   g) $h(x) = 1/x^2$; $x = 0$

3. Find the critical points of each of the following functions.
   a) $f(x) = (x - 9)^{15}$
   b) $f(x) = (x - 3)^4(x - 7)^{65}$
   c) $g(x) = 4x^3 + 7x^2 + 5x + 6$
   d) $g(x) = x + x^{1/2}$
   e) $h(x) = (x - 3)/(3x + 5)$
   f) $f(x) = \bigl(h(x)\bigr)^3$, where $h(x)$ is as defined in (e)

4. In each of the following, f and g will be functions which are differentiable at every real number. If a statement below is true, write TRUE. If a statement is false, produce an example which illustrates the falsity of the statement.
   a) If $f'(x) \neq 0$ for any real number x, then f has no maxima and no minima.
   b) If a is a critical point of both f and g, then a is a critical point of $f + g$.
   c) If a is a critical point of f and g, then a is a critical point of f/g.
   d) If f and g each have no relative maxima, then $f + g$ has no relative maxima.
   e) If f and g each have a maximum at a, then fg also has a maximum at a.
   f) If f and g each have an absolute minimum at a, then $f(a) = g(a)$.
   g) If g has a critical point at a, then $f \circ g$ has a critical point at a.

5.2 MORE ABOUT MAXIMA AND MINIMA. INCREASING AND DECREASING FUNCTIONS

If we wish to find the (relative) maxima and minima of a function and there are certain points at which the function is not differentiable, then the behavior of the function must be investigated, if possible, at each of the exceptional points to determine whether the function has a maximum or minimum at any of them.

Example 3. We saw in Example 14 of Chapter 4 that the function defined by $f(x) = |x|$ is not differentiable at 0. Yet this function has an absolute minimum at 0.

In the event that we are only considering function values $f(x)$ for x in some closed interval, we must pay special attention to what happens at the endpoints of the interval. If, in fact, we are considering a continuous function defined on a closed interval, the function will always have an absolute maximum and an absolute minimum (relative to that interval for which it is defined); there may also be relative maxima and minima in addition to the absolute maximum and minimum. Often there will be either relative or absolute maxima and minima at the endpoints of the interval. We illustrate this in the next example.

Example 4. Consider

$$f(x) = 3x,$$

defined for x in $[0, \frac{1}{3}]$. The reader may recognize f as being a possible distribution function for a random variable with admissible range $[0, \frac{1}{3}]$. Since $f'(x) = 3$, there are no critical points for f in $[0, \frac{1}{3}]$; in fact, there would be no critical points for f even if we were considering $f(x)$ for all real numbers. Yet f has an absolute minimum at 0 and an absolute maximum at $\frac{1}{3}$. This endpoint maximum and this endpoint minimum could only be detected by direct testing or by noting that f is increasing on $[0, \frac{1}{3}]$; hence $f(0)$ is the least value of $f(x)$ for x in $[0, \frac{1}{3}]$ and $f(\frac{1}{3})$ is the greatest value.

The techniques of differential calculus can sometimes be used to find maxima and minima in practical situations. The following, a classical example of a maximum-minimum problem, is found in almost all calculus texts; its real practicality, at least to the average reader, is open to some question.

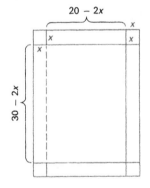

Fig. 6

Example 5. Find the volume of the largest box without a cover that can be made from a rectangular piece of cardboard 20 in. wide by 30 in. long.

We let x denote the height of the box to be made from the cardboard. Then the length of the box will be $30 - 2x$ and its width will be $20 - 2x$ (Fig. 6). Consequently, the volume $V(x)$ of the box is given by

(2) $\quad V(x) = x(30 - 2x)(20 - 2x) = 4x^3 - 100x^2 + 600x.$

We are trying to find an x for which $V(x)$ will be a maximum. Observe that x can only take values in the closed interval [0, 10], for if x were outside this interval, then at least one of the dimensions of the box would have to be negative. Examination shows, moreover, that for 0 and 10, the endpoints of [0, 10], the volume of the box would be zero; hence the endpoints of [0, 10] can be discarded as possible solutions to the problem.

Now $V(x)$ is differentiable at every point in the open interval (0, 10). Differentiating the expression for $V(x)$ in (2), we find

$$V'(x) = 12x^2 + 200x + 600.$$

Setting $V'(x) = 0$ and solving the resulting quadratic equation, we find that the critical points of $V(x)$ are

$$\frac{50 + 10\sqrt{7}}{6} \quad \text{and} \quad \frac{50 - 10\sqrt{7}}{6},$$

or approximately 12.6 and 3.9. The value 12.6 can be discarded immediately since it lies outside [0, 10]. It can be verified that $V(x)$ has a relative maximum at $(50 - 10\sqrt{7})/6$; hence the box will have greatest volume when x is approximately 3.9.

Thus far we have primarily been concerned with the behavior of a differentiable function at its critical points. Even when the derivative is not zero, however, we can still learn a fair amount about the function; moreover, we can use information about the derivative where it is not zero to give information about what is happening at critical points.

In Section 3.3 we defined the notions of an increasing and a decreasing function. We now apply the differential calculus to the further study of these notions.

Let f be any function which has a continuous first derivative in some open interval (u, v) which contains the point a and suppose that $f'(a)$ is positive. Consider

(3) $\quad \dfrac{f(a + h) - f(a)}{h}.$

The limit of (3), of course, as h approaches 0 is $f'(a)$. Since $f'(a)$ is positive and the limit of (3) as h approaches 0 is $f'(a)$, then if h is close enough to 0, (3) will have to be positive as well. [Or else the limit of (3) as h approaches 0 would have to be nonpositive, which would imply that $f'(a)$ was either not defined or was nonpositive.] More formally, there is some positive number p such that if h is in $(-p, p)$, (3) is positive. The limit of (3) as h approaches 0 is the same as

(4) $\quad \lim\limits_{x \to a} \dfrac{f(x) - f(a)}{x - a}$

(see Example 11, Chapter 4). And the condition that h be in $(-p, p)$ is equivalent to saying that $x - a$ be in $(-p, p)$ since $h = x - a$; hence the condition is equivalent to saying that x is in $(a - p, a + p)$. Therefore, if x is in $(a - p, a + p)$, then

(5) $$\frac{f(x) - f(a)}{x - a}$$

is positive.

Now if $x < a$, then $x - a$ is negative; hence in order for (5) to be positive, it must be that $f(x) - f(a)$ is negative, or $f(a) - f(x)$ is positive. This, in turn, implies that $f(x) < f(a)$. Similarly, if $a < x$, $x - a$ is positive; hence for (5) to be positive, we must have $f(x) - f(a)$ positive, or $f(a) < f(x)$.

Now $f'(x)$ is positive and continuous for every point x of

$$(a - p, a + p).$$

Hence we can show by an argument like that above that if x and y are points of $(a - p, a + p)$ with $x \leq y$, then $f(x) \leq f(y)$. That is, f is increasing on $(a - p, a + p)$. We summarize the results of this discussion in the following proposition.

Proposition 3. *Suppose f has a continuous derivative in some open interval which contains a, and $f'(a)$ is positive. Then f is increasing on some open interval which contains a.*

A discussion similar to that used to prove Proposition 3 could be used to prove the following proposition.

Proposition 4. *Suppose f has a continuous derivative in some open interval which contains a, and $f'(a)$ is negative. Then f is decreasing on some open interval which contains a.*

The following is a corollary to Propositions 3 and 4. Its proof is left as an exercise.

Corollary: *If f is differentiable at every point of some open interval (u, v) and $f'(x)$ is positive (negative) and continuous at each point of (u, v), then f is increasing (decreasing) on (u, v).*

We can use the corollary above to test a critical point. For suppose f is a function which has a continuous derivative in some open interval (u, v) which contains a and $f'(a)$ is 0. If it can be shown that $f'(x)$ is positive for $u < x < a$, and $f'(x)$ is negative for $a < x < v$, then this implies that f is increasing on (u, a) and decreasing on (a, v); therefore f has a relative maximum at a. Similarly, if $f'(x)$ is negative for $u < x < a$ and positive for $a < x < v$, then f has a minimum at a. We restate this result in the following proposition.

Proposition 5. *If f has a continuous derivative on an open interval (u, v) which contains a and $f'(a) = 0$, then if $f'(x)$ is positive (negative) for $u < x < a$ and negative (positive) for $a < x < v$, then f has a relative maximum (minimum) at a.*

We close this section by illustrating the use of the propositions proved above in several examples.

Example 6. The function $f(x) = x^2$ has a critical point at 0. Since $f'(x) = 2x$ is negative for $x < 0$ and positive for $0 < x$ (and is continuous), f has a relative minimum at 0. In fact, f has an absolute minimum at 0.

Example 7. Consider $f(x) = (x^2 + 3x + 1)^{27}$. Then
$$f'(x) = 27(x^2 + 3x + 1)^{26}(2x + 3).$$
Since 26 is an even number, $27(x^2 + 3x + 1)^{26}$ is nonnegative for any value of x. The sign of $f'(x)$ will therefore be determined by the factor $2x + 3$. We note that f has a critical point at $-\frac{3}{2}$, and that $2x + 3$ is negative for $x < -\frac{3}{2}$ and is positive for $-\frac{3}{2} < x$; therefore f must have a relative minimum at $-\frac{3}{2}$. The function f also has critical points at $-3 + \sqrt{5}$ and $-3 - \sqrt{5}$, the roots of $x^2 + 3x + 1$; f has neither maxima nor minima at these points. The graph of f is presented in Fig. 7.

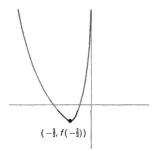

Fig. 7

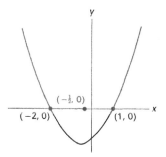

Fig. 8

Example 8. Consider $f(x) = 2x^3 + 3x^2 - 12x$. Then
$$f'(x) = 6x^2 + 6x - 12;$$
hence f has critical points at 1 and -2. The graph of f' is presented in Fig. 8. We note that $f'(x)$ (which is continuous) is negative for
$$-2 < x < 1$$
and positive for all other values of x except -2 and 1. Therefore f is decreasing on $(-2, 1)$ and increasing on $(-\infty, -2)$ and on $(1, \infty)$. The

function f has a maximum at -2 and a minimum at 1. The roots of f are

$$0, \quad \frac{-3+\sqrt{105}}{4}, \quad \text{and} \quad \frac{-3-\sqrt{105}}{4}.$$

We are thus able to draw a fairly accurate graph of f (Fig. 9) without having to compute a great many specific values of $f(x)$.

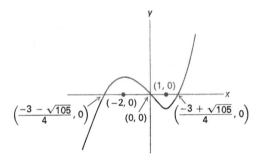

Fig. 9

EXERCISES

1. Prove Proposition 4; prove the corollary following Proposition 4.

2. Find the critical points of each of the following functions. Test each critical point to determine its nature. Determine where each function is increasing and where it is decreasing. Determine, at least approximately, where the graph of each function crosses the x-axis. Use all the information gained to draw a rough sketch of the graph of each function.
   a) $f(x) = x^3$
   b) $f(x) = -x^2 + 3$
   c) $f(x) = (x+3)^{43}$
   d) $f(x) = x^3 + 2x + 1$
   e) $f(x) = (3x+4)(x^2+1)$
   f) $f(x) = (3x+1)/(4x+7)$
   g) $f(x) = (3x+1)^{1/2}$
   h) $g(x) = (x^{1/2} + x)^2$
   i) $f(x) = x^2 + x^{-2}$

3. The general form of a cubic polynomial function is

   $$f(x) = Ax^3 + Bx^2 + Cx + D,$$

   where A, B, C, and D are constants. Find a necessary and sufficient condition involving A, B, C, and D in order for $f(x)$ not to have any relative maxima or minima.

4. The general form of a quadratic polynomial function is $f(x) = Ax^2 + Bx + C$, where A, B, and C are constants. Determine, in terms of A, B, and C, where f is increasing and where f is decreasing. Find the critical point of f and prove that its nature depends only on the sign of A.

5. Examine each of the following statements. If a statement is true, write TRUE. If a statement is false, supply an example which illustrates the false-

ness of the statement. Assume throughout that f and g are differentiable functions with continuous derivatives.

a) If f is a function which is increasing on the entire set of real numbers, then f' is also increasing on R.
b) If f and g are both increasing on (a, b), then $f + g$ is increasing on (a, b).
c) If f and g are both decreasing on (a, b), then fg is decreasing on (a, b).
d) If $f(x) = g(m(x))$, where m is also differentiable, g is increasing, and $m'(x)$ is negative for every x, then f is a decreasing function.
e) If $f(a) = f(b) = 0$ and $a \neq b$, then there is a point x' in (a, b) such that $f'(x') = 0$.

6. Prove that the composition of an increasing function with a decreasing function is a decreasing function. Prove that the composition of an increasing function with an increasing function is an increasing function.

5.3 SOME THEOREMS ABOUT CONTINUOUS FUNCTIONS

This text is not intended to be a rigorous treatment of functions of a real variable. Our primary purpose for developing the calculus is to be able to apply it to the study of continuous variates. This book omits a great deal of material that is found in other calculus books, simply because this material does not pertain to the use to which we want to put calculus. Likewise, we have left out a great deal of theoretical discussion, since this tends to confuse rather than aid comprehension. In sum, we are developing calculus as a tool and not as an end in itself. Nevertheless, it would be unwise to omit all "theory" which did not have an immediate computational application.

To omit too much theory would make this book like a book of recipes. Presented with some problem, one would find the proper recipe, follow the directions carefully, and cook up some sort of answer. The meaning of the intermediate steps leading from the statement of the problem to the solution would be lost, but an answer would be obtained. To be frank, this author is not opposed to the "recipe" method of solving problems—if it works. However, to be successful with this approach, one must have a recipe for each problem he may encounter and, when confronted with a problem, he must be able to pair the problem with the appropriate recipe for solving it. Since new problems will always arise and since our memory is generally rather limited and imperfect, the "recipe" method usually does not work very long. It is therefore important that one have some insight into the "theory" of the calculus in order to (1) recognize a problem to which the methods of the calculus should be applied, (2) avoid wasting time trying to use calculus to solve problems to which calculus is not applicable, and (3) be able to solve problems of types not previously encountered.

This section presents certain "theoretical" propositions which give frequently useful information about continuous and differentiable functions.

We shall also use some of the results of this section later in deriving certain fundamental results about integration. The first proposition we present is sometimes known as *Rolle's Theorem*.

Proposition 6. *Suppose f is continuous on $[a, b]$ and differentiable at every point of (a, b); assume too that $f(a) = f(b) = 0$. Then there is a point x' in (a, b) such that $f'(x') = 0$.*

Before we prove Proposition 6, we state a proposition which is needed to prove Proposition 6 and which is useful in other contexts as well. Its proof is beyond the scope of this text.

Proposition 7. *If f is a continuous function from $[a, b]$ into R, then f has both an absolute maximum and an absolute minimum at points of $[a, b]$.*

We now prove Proposition 6.

Proof of Proposition 6: If $f(x) = 0$ for every x in (a, b), then $f'(x') = 0$ for every x' in (a, b). If $f(x) \neq 0$ for some x in (a, b) (Fig. 10), then f has either an absolute maximum, absolute minimum, or both an absolute maximum and an absolute minimum at certain points of (a, b). Let x' be one of these points. Since an absolute maximum or minimum is also a relative maximum or minimum, by Propositions 1 and 2, we conclude that $f'(x') = 0$.

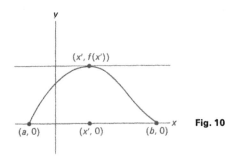

Fig. 10

Example 9. Suppose f is a polynomial function and a and b are roots of $f(x)$; that is, $f(a) = f(b) = 0$. If f is not constant, then f will have either a relative maximum or relative minimum at some point x' of (a, b); hence we will have $f'(x) = 0$. We therefore conclude that if f is a polynomial function, then f will have a critical point between any two of its roots.

Example 10. The function $f(x) = |x|$ is continuous on the closed interval $[-1, 1]$. We find that 1 is the largest value that $f(x)$ assumes for any x in

$[-1, 1]$, and 0 is the smallest value. None of these values can be detected using the critical points of f since f has no critical points.

We now apply Rolle's Theorem (Proposition 6) to prove the important *First Mean Value Theorem*.

Proposition 8. *Suppose f is continuous on $[a, b]$ and differentiable at every point of (a, b). Then there is a point x' of (a, b) such that*

(6) $\quad f'(x') = \dfrac{f(b) - f(a)}{b - a}.$

Before proving Proposition 8, we shall discuss its geometric significance. Consider the graph of f presented in Fig. 11. Then the right-hand side of (6) is nothing but the slope of the straight line determined by the endpoints of this graph. On the other hand, $f'(x')$ is the slope of the line tangent to the graph of f at $(x', f(x'))$. Therefore what Proposition 8 says, interpreted geometrically, is that there is a point x' in the open interval (a, b) such that the line tangent to the graph of f at $(x', f(x'))$ is parallel to (has the same slope as) the line determined by the endpoints of the graph over $[a, b]$.

Intuitively, as we "slide" a line along the graph of f keeping the line tangent to the graph, when we come to the point (or points) which are farthest from the line determined by the endpoints of the graph, the tangent line will be parallel to that line. Looking at the situation in another way, we see that Proposition 8 is just Rolle's Theorem with the line determined by $(a, f(a))$ and $(b, f(b))$ substituting for the x-axis (compare Figs. 10 and 11). We now give a formal proof of Proposition 8.

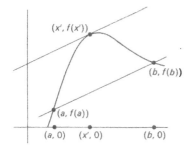

Fig. 11

Proof of Proposition 8: Set

$$F(x) = f(x) - \frac{f(b) - f(a)}{b - a}(x - a) - f(a).$$

Since f is continuous on $[a, b]$, so is F; and since f is differentiable on (a, b), so is F. Specifically,

$$F'(x) = f'(x) - \frac{f(b) - f(a)}{b - a}$$

for each x in (a, b). Now $F(a) = F(b) = 0$ and F satisfies the hypotheses of Proposition 6. Therefore there is a point x' in (a, b) such that

$$F'(x') = 0 = f'(x') - \frac{f(b) - f(a)}{b - a}.$$

Solving this last equation for $f'(x')$, we obtain (6).

The last proposition we present in this section is sometimes called the *Intermediate Value Theorem*; its proof is beyond the scope of this book.

Proposition 9. *Suppose f is a continuous function (f need not be differentiable) and $f(a) < y < f(b)$ for some numbers a and b. Then, if $f(x)$ is defined for every x between a and b, there is a number x' between a and b such that $f(x') = y$. In other words, if f is defined for the interval between a and b, then $f(x)$ will not only assume every value between $f(a)$ and $f(b)$, but will do so for the interval between a and b.*

As an example of the application of Proposition 9, we shall prove that every positive real number has an nth root for any positive integer n.

Example 11. Let $f(x) = x^n$, where n is any positive integer, and $0 \leq x$. The graph for a typical value of n is presented in Fig. 12. Since f is differentiable, it is certainly continuous. Moreover, given any positive number y, there is a positive number b such that $0 = f(0) < y < f(b) = b^n$. Therefore, by Proposition 9, there is a positive number x' such that $x'^n = y$. Therefore x' is an nth root of y.

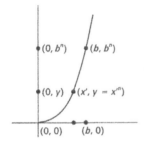

Fig. 12

Example 12. If f is a polynomial function such that $f(a) > 0$ and $f(b) < 0$, then there is a point x' between a and b such that $f(x') = 0$, that is, $f(x)$ has a root between a and b.

EXERCISES

1. The function defined by $f(x) = |x|$ is not differentiable at 0. Show that Proposition 8 is not true for f in the closed interval $[-1, 1]$.

2. Use Proposition 9 to prove that any positive real number has precisely two nth roots (which are real numbers) if n is a positive even integer, and exactly one nth root if n is a positive odd integer.

3. In each of the following f and g are differentiable functions. If a statement is true, write TRUE; supply a proof, if you can. If a statement is false, supply an example which demonstrates the falseness of the statement.
   a) If f has m relative maxima and n relative minima, then f has exactly $n + m$ critical points.
   b) The equation $x^{3/2} = a$ has a solution for any positive number a.
   c) If $(f + g)(a) = (f + g)(b) = 0$, then there is a number x' between a and b such that both $f'(x')$ and $g'(x')$ are 0.
   d) If $f(x)$ takes on every integer as a value, then $f(x)$ takes on every real number as a value.
   e) If g is a function whose derivative is always positive, then $g(x)$ takes on every real number as a value.
   f) If g is a function whose derivative is always positive and if $g(x) = g(x')$, then $x = x'$.

4. Each of the functions given in the following satisfy the hypotheses of Proposition 8 with regard to the interval given with each. Find a point x' in the given interval which satisfies (6).
   a) $f(x) = x^2$; $[-1, 2]$
   b) $f(x) = (x - 3)(4x - 1)(5 - x)$; $[3, 5]$
   c) $f(x) = 3x + 4$; $[a, b]$
   d) $f(x) = x^{1/2}$; $[1, 2]$

5. Find the largest and the smallest value that each function given in Exercise 4 attains on the given interval.

5.4 HIGHER DERIVATIVES AND THEIR APPLICATIONS

If a function f is differentiable, then, as we have seen, the first derivative, f', of f can give us useful information about f. It follows that the first derivative of f', which is usually denoted by f'', can give information about f', and hence information about f at least indirectly. We may even be able to take the first derivative of f'', and then the first derivative of that derivative, and so on. Usually, however, the more removed from f we get, the less information we obtain about f itself. We shall primarily be concerned in this section with f'', the first derivative of the first derivative of f; f'' is called the *second derivative* of f. More generally, the nth *derivative* of f is the first derivative of the $(n - 1)$-derivative of f.

We have seen that the first derivative of a function can tell us where the function is increasing and where it is decreasing, that is, informally, how the function values are changing. The second derivative of a function gives us the same information about the first derivative as the first derivative does about the function; that is, the second derivative tells us how the values of the first derivative are changing. Consider the following example.

Example 13. The graphs of $f(x) = x^2 - 5x + 6$, $f'(x) = 2x - 5$, and $f''(x) = 2$ are plotted in Fig. 13. Observe that f has a minimum at its only critical point $\frac{5}{2}$. Although $f'(\frac{5}{2}) = 0$, $f''(\frac{5}{2}) = 2$, which is positive. Since f'' is continuous, f' is an increasing function in some open interval (u, v) which contains $\frac{5}{2}$ (Proposition 3). Thus $f'(x)$ must be negative for x a little smaller than $\frac{5}{2}$ and positive for x a little larger than $\frac{5}{2}$. From Proposition 5, then, we see that f does have a minimum at $\frac{5}{2}$, which is what we knew to be the case.

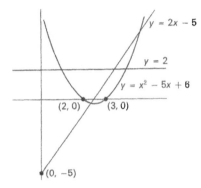

Fig. 13

The principle used to test the critical point of f in Example 13 is expressed in the following proposition.

Proposition 10. *If f and f' both have continuous derivatives on some open interval which contains a, $f'(a) = 0$, and $f''(a)$ is positive (negative), then f has a relative minimum (maximum) at a. If $f''(a) = 0$, then we cannot determine the nature of the critical point from $f''(a)$.*

Proof: If $f''(a)$ is negative, then by Proposition 4, f' is decreasing in some open interval which contains a. Therefore $f'(x)$ is positive for x a little smaller than a and negative for x a little larger than a. Hence by Proposition 5, f has a maximum at a.

The proof for $f''(a)$ positive is left as an exercise.

Example 14 presents a situation in which the second derivative is used successfully to test the nature of a critical point. Example 15 illustrates a situation in which the second derivative test fails.

Example 14. Consider $f(x) = x^3 - 3x + 6$. Then $f'(x) = 3x^2 - 3$; hence f has critical points at 1 and -1. Now $f''(x) = 6x$. Since $f''(1) = 6$, a positive number, and $f''(-1) = -6$, a negative number, f has a relative minimum at 1 and a relative maximum at -1.

Example 15. Consider $f(x) = (x + 3)^{25}$. Then $f'(x) = 25(x + 3)^{24}$ and $f''(x) = 600(x + 3)^{23}$. The sole critical point of f is -3. However, -3 is also a critical point of f'. We therefore cannot conclude the nature of -3 from $f''(-3)$. In this instance, f has neither a maximum nor a minimum at -3.

If $g(x) = (x + 3)^{24}$, then $g'(-3) = g''(-3) = 0$; here g has a maximum at -3. If $h(x) = -(x + 3)^{24}$, then again $h'(-3) = h''(-3) = 0$; and in this instance, h has a minimum at -3.

We might ask: If a function has neither a relative maximum nor a relative minimum at a critical point, then what happens geometrically at the critical point? Evidently, if a is a critical point of f at which f has neither a relative maximum nor a relative minimum, then if f is increasing just before a, it must also be increasing just after a; similarly, if f is decreasing just before a, it must also be decreasing just after a. We now examine a function which has a critical point at which the function has neither a maximum nor a minimum.

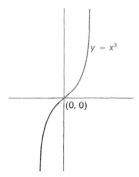

Fig. 14

Example 16. Consider $f(x) = x^3$; the graph of f is given in Fig. 14. Note that since $f'(x) = 3x^2$ is nonnegative for every x, f is always an increasing function. Note, too, however, that as x approaches 0 from the negative side, the graph of f rises less and less sharply; however, once x has passed 0, the graph begins to rise more and more sharply as x increases. This indicates that although the graph is always rising, it is rising least at $(0, 0)$. Since how sharply the graph is rising is measured by the first derivative, we would expect the first derivative to have a minimum at 0. Such is indeed the case. In this instance, the critical point 0 of f is also a critical point of f' and represents a point at which the rate of increase of f is at a minimum.

Definition 3. *A point a at which the first derivative f' of a differentiable function f has either a relative maximum or a relative minimum is said to be a* **point of inflection** *of f.*

Thus a critical point a of a differentiable function f may represent either a relative maximum, relative minimum, or point of inflection of f. Indeed, these possibilities for a critical point are exhaustive for those functions the reader is likely to encounter. A function may have a point of inflection, however, at some point which is not a critical point.

Example 17. Consider $f(x) = (x-3)^2(x-1)$. Then

$$f'(x) = (x-3)(3x-5) \quad \text{and} \quad f''(x) = 6x - 14.$$

Therefore f has critical points at 3 and $\frac{5}{3}$, while f' has a critical point at $\frac{7}{3}$ (Fig. 15). Since $f''(3)$ is positive and $f''(\frac{5}{3})$ is negative, f has a relative maximum at $\frac{5}{3}$ and a relative minimum at 3. The function f' has a relative minimum at $\frac{7}{3}$; hence f has a point of inflection at $\frac{7}{3}$. Although f is decreasing in $(\frac{5}{3}, 3)$, it is decreasing most rapidly at $\frac{7}{3}$, the point of inflection.

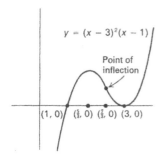

Fig. 15

EXERCISES

1. Find all critical points and points of inflection for each of the following functions. Clearly identify the nature of each point.

    a) $f(x) = x^3 + 1$ b) $f(x) = (5-x)^9$
    c) $f(x) = (x^2+1)(x-3)$ d) $f(x) = 1/(4+x^2)$
    e) $f(x) = x^2 + x^{1/2}$ f) $f(x) = 2x^3 - 30x^2 + 36x + 6$

2. Suppose $f(x)$ is a polynomial and a is a root of $f(x)$ such that

    $$f(x) = (x-a)^n g(x),$$

    where $g(x)$ is a polynomial function. Prove that if $2 \leq n$, then a is a critical point of f. If $n = 3$, does f necessarily have a point of inflection at a?

3. A differentiable function f is said to be *concave* at a point a if f' is increasing on some open interval which contains a; f is said to be *convex* at a if f' is decreasing on some open interval which contains a. The geometric difference between a convex and concave function is represented in Figs. 16 and 17. Determine where each of the following functions is convex and where each is concave. [*Hint:* Use f'' to test f'.]

a) $f(x) = x^3$
b) $f(x) = x^2 + 3x + 5$
c) $f(x) = 1/x$
d) $f(x) = 1/(x^2 + 1)$

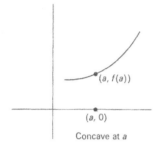

Fig. 16 — Concave at a
Fig. 17 — Convex at a

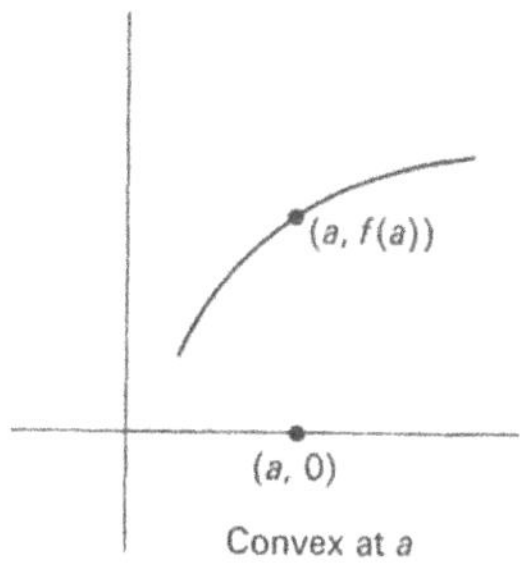

4. In each of the following statements, f and g represent differentiable functions with continuous derivatives. If a statement is true, write TRUE; supply a proof, if you can. If a statement is false, find an example which illustrates the falseness of the statement. The definitions of concave and convex functions are found in Exercise 3.

a) If $f(a) = g(a)$ and $f(b) = g(b)$ with $a < b$, and f is convex while g is concave for each point of (a, b), then $f(x) < g(x)$ for each x in (a, b).
b) If f and g are both convex at a, then $f + g$ is convex at a.
c) If f and g are concave at a, then fg is concave at a.
d) If a is a point of inflection of f, then f is convex on one side of a and concave on the other side.

5.5 THE DENSITY FUNCTION OF A CONTINUOUS DISTRIBUTION. THE MODE

Suppose X is a discrete random variable such that consecutive admissible values of X differ by 1. This would be the case, for example, if the admissible range of x were a set of consecutive integers. Let F be the distribution function of X and x_i and x_{i+1} be consecutive admissible values of X. Consider

(7) $\quad \dfrac{F(x_{i+1}) - F(x_i)}{x_{i+1} - x_i}.$

From Proposition 5 of Chapter 3 we find that

(8) $F(x_{i+1}) = \sum_{\substack{x \text{ in } A \\ x \leq x_{i+1}}} f(x),$

and

(9) $F(x_i) = \sum_{\substack{x \text{ in } A \\ x \leq x_i}} f(x),$

where A is the admissible range and f is the density function of X. Since x_i and x_{i+1} are consecutive, the only term which appears in (8) which does not appear in (9) is $f(x_{i+1})$. Using (8) and (9), we therefore find that

$$F(x_{i+1}) - F(x_i) = f(x_{i+1}).$$

Since

$$x_{i+1} - x_i = 1$$

by hypothesis, (7) reduces to

(10) $\dfrac{f(x_{i+1})}{1} = f(x_{i+1}).$

Suppose now that X is a continuous variate with admissible range A and distribution function F. Then the analog of (7) for this case if a is some admissible value of x is

(11) $\dfrac{F(y) - F(a)}{y - a},$

where y is some admissible value of X "close to" a. But no matter how close y is to a, if $y \neq a$, then there is an admissible value of X even closer to a. Instead of trying to set y arbitrarily, we might look at the limit of (11) as y approaches a. This limit, if it exists, is what we know as $F'(a)$.

The situation again is that described in the paragraph above. If h is a positive number, then

$P(x$ takes a value between a and $a + h) = F(a + h) - F(a)$

(Fig. 18). Thus, the ratio

(12) $\dfrac{F(a + h) - F(a)}{h}$

measures the "density" of the probability of x in the interval $[a, a + h]$. Informally speaking, the larger this ratio, the more the probability associated with x is concentrated in $[a, a + h]$. As we know, the limit of (12) as h approaches 0 is $F'(a)$.

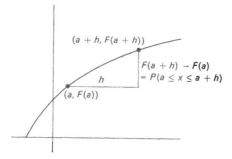

Fig. 18

We have already defined the density function of a discrete variate. We call upon the discussion above to motivate the following definition for the *density function* of a continuous variate.

Definition 4. *If F is the distribution function of a continuous random variable X and if F is differentiable at every point of every open interval contained in the admissible range of X, then the first derivative F' of F is said to be the **density function** of F.*

Example 18. Suppose $F(x) = 3x$ is the distribution function for a random variable X with admissible range $[0, \frac{1}{3}]$.* Then the density function of X is given by $F'(x) = 3$. The graphs of F and F' are compared in Figs. 19 and 20. We note first that $F'(x)$ is always nonnegative. This is to be expected since F is increasing on the admissible range of x (see the remarks following Definition 9 of Chapter 3).

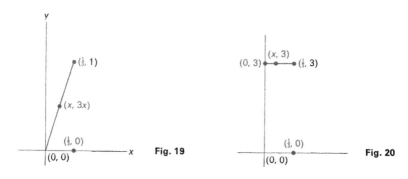

Fig. 19   **Fig. 20**

---

* Recall that we need only define the distribution function of a variate X for the admissible range of X since it is automatically assumed that $F(a) = 0$ for any a less than every admissible value of X, and $F(a) = 1$ for any a greater than every admissible value of X.

5.5   Density Function of a Continuous Distribution   115

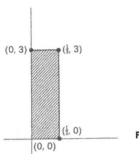

Fig. 21

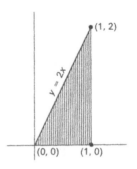

Fig. 22

In this instance we can also calculate the area between the graph of F' and the x-axis (the shaded area in Fig. 21). Since this area is that of a 3 by $\frac{1}{3}$ rectangle, it is 1. We can verify that the area between the graph of F' and the x-axis over the interval $[a, a + h]$ is

$$h(\tfrac{1}{3}) = F(a + h) - F(a) = P(a \leq x \leq a + h).$$

We conclude that, in this example at least, the area between the graph of F' and the x-axis has something to do with certain probabilities. We shall find that this is generally the case.

Example 19. Consider the random variable X with admissible range $[0, 1]$ and distribution function $F(x) = x^2$. The density function for X is $F'(x) = 2x$. The area between the graph of the density function and the x-axis, the shaded portion of Fig. 22, is $(\tfrac{1}{2})(2) = 1$. We could also show that the area between the graph of F' and the x-axis over the interval $[a, a + h]$ is $F(a + h) - F(a) = P(a \leq x \leq a + h)$.

Since the distribution of a continuous random variable is always an increasing function, we may say:

Proposition 11. *If X is a continuous random variable with a differentiable distribution function F, then the density function F' of X never takes on any negative values.*

The probability distribution of a discrete variate can be specified by giving its admissible range and density function; the same is true of a continuous variate. In fact, specifying the distribution of a continuous variate by means of its density function is often more useful than specifying the distribution by means of the distribution function. This point will be clearer after we have studied integration.

Suppose X is a random variable with admissible range $[a, b]$. What functions might be potential density functions for x? We have seen that if f is to be a density function for X, then $f(x)$ must be nonnegative for each admissible value x of X. Actually, not only must f take on only nonnegative values, but there must also be a differentiable function F, the distribution function of x, such that $F(a) = 0, F(b) = 1$, and $F'(x) = f(x)$ for each x in $[a, b]$. Again we find that a satisfactory discussion must await some knowledge of integration.

We now review briefly certain concepts associated with density functions. Some of these concepts will be discussed at greater length later in this book. Most of these concepts are applicable to density functions of discrete variates as well as of continuous variates.

Definition 5. *A density function f of a variate X with admissible range A is said to be* **symmetric with respect to** *a if for each real number y such that $a - y$ is an admissible value of X, $a + y$ is also an admissible value of x and $f(a - y) = f(a + y)$. One may also say that the distribution of x is* **symmetric with respect to** *a.*

Example 20. Assume that x is a continuous variate with admissible range $[-1, 1]$ and density function $f(x) = |x|$. Then f is symmetric with respect to 0 since $f(0 - y) = f(-y) = y = f(0 + y)$ for any admissible value y of x.

A function f will be symmetric with respect to a if the graph of f on one side of the line $x = a$ is the mirror image of the graph of f on the other side of that line (Fig. 23). This is because $a - y$ is the mirror image of $a + y$ with respect to a and $f(a - y)$ must be the same as $f(a + y)$ if f is to be symmetric with respect to a.

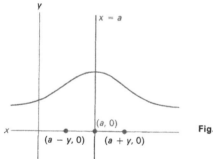

Fig. 23

The density function of a continuous variate may have one or more relative maxima; correspondingly, a discrete variate may have one or more admissible values for which its density function has a maximum, that is, whose probabilities are greatest.

Definition 6. *An admissible value of a continuous variate X at which the density function of X has a relative maximum is called a* **mode** *of X. An admissible value of a discrete variate X at which the density function of X has a maximum is also called a* **mode** *of X. It is quite possible for both discrete* and continuous variates to have more than one mode. A variate which has but one mode is said to be **unimodal**.

Example 21. The variate described in Example 20 has two modes, one at -1 and the other at 1. The variate of Example 19 has a single mode at 1. Every admissible value of the variate described in Example 18 is a mode.

The mode is one measure of the "normative" or "typical" value of a variate; we shall have more to say about this subject later.

The graph of the density function of a unimodal variate may be "lopsided" to the right or left. A distribution in which the longer "tail" goes to the right (Fig. 24) is said to be *skewed to the right*. If the longer tail goes to the left, then the distribution is said to be *skewed to the left* (Fig. 25). A more precise definition of *skewness* will be possible after the *mean* or *expected value* of a variate has been introduced.

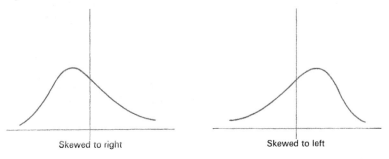

Skewed to right   Skewed to left

Fig. 24   **Fig. 25**

EXERCISES

1. Verify in both Example 18 and Example 19 that the area between the graph of F' and the x-axis over the interval $[a, a + h]$ is equal to $F(a + h) - F(a)$. It is implicit that a is an admissible value of X and h is a positive number such that $a + h$ is also an admissible value of X.

2. In each of the following, an admissible range and a distribution function are given for a variate X. Find the density function for X in each case. Furthermore in each case (i) find the modes of X; (ii) determine whether the distribution is symmetric or skewed; (iii) if the usual geometric formulas for areas are applicable, show that the area between the graph of the density function and the x-axis is 1; (iv) graph each density function.

   a) $F(x) = x$; $[0, 1]$ b) $F(x) = x^3$; $[0, 1]$
   c) $F(x) = x - 1$; $[1, 2]$ d) $F(x) = 1/x^2$; $(-\infty, -1]$
   e) $F(x) = 6x^3 + x^2$; $[0, \frac{1}{2}]$

3. Find an admissible range for the random variable X such that $f(x) = x^2$ becomes a density function for X which is symmetric with respect to 0.

[*Hint:* If f is the density function for X, what must the distribution function for X be if the admissible range of X is to have the form $[p, -p]$ for some positive number p?]

4. Find the distribution function of a continuous variate X with admissible range $[0, 1]$ and density function $f(x) = \frac{2}{3}(x + 1)$.

6 Sequences and Series

6.1 SEQUENCES AND SERIES

We introduce the notion of a sequence simply because it is a necessary prerequisite to the study of series; there are good reasons for studying series.

Definition 1. *A **sequence of real numbers** or, simply, a **sequence**, is a function from the set N of positive integers into R.*

Henceforth, N will denote the set of positive integers.

Example 1. The function s from N into R defined by $s(n) = 1 - n^2$ for each positive integer n is a sequence.

Example 2. The function s from N into R defined by $s(n) = 1/n$ for each positive integer n is a sequence.

If s is a sequence, it is common practice to write $s(n)$ as s_n, and the sequence itself as $\{s_n\}$, n in N. Although there is a "natural" ordering of the values that a sequence s assumes, that is, $s(1) = s_1$ is the *first term* of the sequence, s_2 is the *second term*, etc., a sequence should not be confused with an ordered set of real numbers. First, when writing a set, we do not, in general, repeat elements; for example, the set $\{1, 1, 1\}$ is the same as the set $\{1\}$. On the other hand, a sequence s can assume the same value for many different positive integers, or even the same value for every positive integer. The sequence s defined by $s(n) = 1$ for each n in N is not the same as the set $\{1\}$. Second, given a set T of numbers, the ordering of T is usually taken to be the usual "less than or equal to" ordering. The values of a sequence, however, are given their ordering from the positive integers; that is, s_n comes before s_m if $n < m$, even if numerically $s_m < s_n$. The ordering of the sequence of Example 2 inherited from the positive integers is, in fact, the opposite of the ordering of the values of s considered solely as real numbers.

Example 3. Let s be the sequence defined by $s(n) = (-1)^n$; this sequence may also be written $\{(-1)^n\}$, n in N. A third way of denoting this sequence is

$-1, 1, -1, 1, -1, 1, \ldots,$

where the three dots indicate that the pattern established by the terms shown continues indefinitely.

If we were to consider just the set of values of s, we would have $\{-1, 1\}$. Each odd term of this sequence (function value of an odd positive integer) is -1, and each even term is 1.

Intuitively, we see that the sequence of Example 2 seems to be approaching 0 as a "limit." More formally, suppose p is any positive number; then we can show that $|0 - s_n| = 1/n < p$ if n is large enough, that is, for all n greater than some positive integer M, $1/n < p$. For M take any positive integer greater than $1/p$. Then, since $M > 1/p$, we have $p < 1/M$; hence, if $n > M$, then $s_n = 1/n < 1/M < p$, which is what we wished to show.

The sequence in Example 1, on the other hand, does not seem to be approaching any limit since successive terms of the sequence get farther and farther apart as n gets larger. The first few terms of the sequence are

$0, -3, -8, -15, -24, -35, -48, -63, -80, \ldots$

The sequence in Example 3 does not seem to have a limit either. For if this sequence had a limit, we would expect that limit to be either 1 or -1. But this sequence never really stays close to either 1 or -1.

We formalize the notion of the limit of a sequence in the following definition.

Definition 2. A sequence $\{s_n\}$, n in N, is said to have a **limit** L if, given any positive number p, there is a positive integer M (which depends on p) such that if $M < n$, then $|s_n - L| < p$ (Fig. 1).

If the sequence $\{s_n\}$, n in N, has a limit L, then we say that the sequence **converges to** L, or merely that the sequence **converges**. We denote the fact that the sequence converges to L by writing $s_n \to L$.

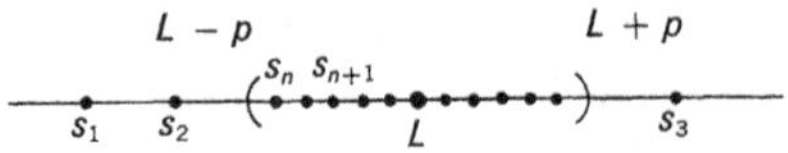

Fig. 1

Informally, a sequence $\{s_n\}$, n in N, has a limit L if the terms of the sequence ultimately get and stay arbitrarily close to L.

We already proved that the sequence in Example 1 converges to 0. We now prove that another sequence converges.

Example 4. Consider $\{1 + (-1/n)^n\}$, n in N. We shall prove that this sequence converges to 1. Now

$$|s_n - 1| = |1 + (-1/n)^n - 1| = |(-1/n)^n| = |-1/n|^n = (1/n)^n.$$

Since $(1/n)^n \leq 1/n$, we can say that if p is any positive number and M is any positive integer such that when $M < n$, $1/n < p$, we also have $|s_n - 1| < p$. Therefore $s_n \to 1$.

The next example formally proves that the sequence of Example 3 does not converge.

Example 5. Let L be any real number. Suppose $s_n \to L$, where s_n is as defined in Example 3. Suppose first that $L \neq 1$; set $p = |1 - L|$. Then for any positive integer M, there is a positive integer n' greater than M such that $s_{n'} = 1$, and hence $|s_{n'} - L| = |1 - L| = p \not< p$. Therefore the sequence cannot converge to L if $L \neq 1$. Assume now that $L = 1$, and set $p = 1$. Then for any positive integer M, there is an integer n' greater than M such that $s_{n'} = -1$; hence $|s_{n'} - 1| = |-2| = 2 \not< 1$. Therefore $s_n \not\to 1$; hence the sequence does not converge at all.

The following proposition tells us that if a sequence converges, then its limit is unique.

Proposition 1. Suppose $\{s_n\}$, n in N, is a sequence which converges to both L and L'. Then $L = L'$.

Proof: Suppose $L \neq L'$; then $|L - L'|$ is a positive number. Set
$$p = \frac{|L - L'|}{2}$$

(Fig. 2). Since $s_n \to L$ and $s_n \to L'$, there are positive integers M and M' such that if $M < n$, then $|s_n - L| < p$, and if $M' < n$, then $|s_n - L'| < p$.

Let M'' be the larger of M and M'. Then, if $M'' < n$, we have both $|s_n - L| < p$ and $|s_n - L'| < p$. Using the triangle inequality, we have

$$2p = |L - L'| \leq |L - s_n| + |s_n - L'| < p + p = 2p$$

if $M'' < n$. But $2p < 2p$ is a contradiction. This contradiction follows from the assumption that $L \neq L'$; hence it must be that $L = L'$, which is what we wanted to prove.

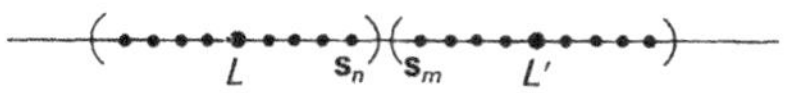

Fig. 2

Sequences can be added, multiplied, and divided in the following way.

Definition 3. *Let s and t be any sequences. Then:*

*The **sum** of s and t is defined by $s + t = \{s_n + t_n\}$, n in N.*

*The **product** of s and t is defined by $st = \{s_n t_n\}$, n in N.*

*If $t_n \neq 0$ for any n, then the **quotient** of s and t is defined by $s/t = \{s_n/t_n\}$, n in N.*

If r is any real number, then rs is defined to be $\{rs_n\}$, n in N.

Example 6. Let $s = \{1/n\}$, n in N, and $t = \{n^2\}$, n in N. Then

$$\begin{aligned} s + t &= \{1/n + n^2\}, & n \text{ in } N, \\ st &= \{n\}, & n \text{ in } N, \\ s/t &= \{1/n^3\}, & n \text{ in } N, \end{aligned}$$

and

$$4s = \{4/n\}, \qquad n \text{ in } N.$$

The following proposition is presented without proof.

Proposition 2. *If s and t are sequences such that $s_n \to L$ and $t_n \to L'$, then:*
a) *$s + t$ converges to $L + L'$;*
b) *st converges to LL';*
c) *if $L' \neq 0$ and s/t is defined, then s/t converges to L/L';*
d) *if r is any real nimber, then rs converges to rL.*

Now consider the sequence $s = \{(\frac{1}{2})^n\}$, n in N. What is the sum of all the terms of s; that is, what is

(1) $\quad \frac{1}{2} + (\frac{1}{2})^2 + (\frac{1}{2})^3 + (\frac{1}{2})^4 + (\frac{1}{2})^5 + \cdots ?$

The reader may recognize in (1) a geometric series with first term $\frac{1}{2}$ and ratio $\frac{1}{2}$ (see Example 28 of Chapter 3), and hence sum 1. However, although 1 is the "sum" asked for, this is not a "sum" in the usual sense.

The point is that although we can add two, or any finite number, of real numbers, the usual process of addition simply does not allow for adding infinitely many real numbers. In order to find the sum in (1), it is necessary to extend our concept of addition.

We can associate with (1) a sequence whose nth term is the sum of the first n summands of (1). If t_n is the nth term of this new sequence, then

$$t_n = \tfrac{1}{2} + (\tfrac{1}{2})^2 + (\tfrac{1}{2})^3 + \cdots + (\tfrac{1}{2})^n.$$

To find each term of $\{t_n\}$, n in N, we need add only finitely many real numbers; hence each t_n makes sense. If $t_n \to L$ for some number L, then it is reasonable to take (1) to be L, since this means that by adding enough terms of (1), we can bring (1) arbitrarily close to L and (1) will stay as close to L as we wish, even when more terms are added. Since t_n actually converges to 1, we take 1 to be the sum in (1).

In the case above, $\{t_n\}$, n in N, is said to be the *series formed from the sequence s*. More generally, we make the following definition:

Definition 4. *Given a sequence s, the sequence t defined by*

$$t_n = s_1 + s_2 + \cdots + s_n$$

*is said to be the **series formed from** s. The nth term of t, t_n, is called the nth **partial sum of** s. If $t_n \to L$, then L is said to be the **sum** of the series t, and the series is said to **converge**. A series which does not converge is said to **diverge**. The series t formed from s is often denoted by*

$$\sum_{n=1}^{\infty} s_n.$$

Example 7. The series t formed from $s = \{1/n\}$, n in N, does not converge; we prove this in the following argument. Now $t_2 = \frac{3}{2}$. Since

$t_4 = t_2 + (\frac{1}{3} + \frac{1}{4}) \quad$ and $\quad \frac{1}{4} < \frac{1}{3}$,

we have $t_4 < t_2 + (\frac{1}{4} + \frac{1}{4}) = t_2 + \frac{1}{2} = 2$. Now $t_{10} = t_4 + (\frac{1}{5} + \cdots + \frac{1}{10})$. Since $\frac{1}{5}, \frac{1}{6}, \ldots, \frac{1}{10}$ are all at least as large as $\frac{1}{10}$,

$t_{10} < t_4 + 5(\frac{1}{10}) = 2 + \frac{1}{2} = \frac{5}{2}$.

Given any positive real number K, we can, by continuing the pattern of the previous computations, find a positive integer n such that $K < t_n$. It follows then that $\{t_n\}$, n in N, diverges.

The proof of the following proposition is left as an exercise.

Proposition 3. *If t is the series formed from the sequence s and s does not converge to 0, then t diverges. Example 7 shows that t may diverge even though s converges to 0.*

In the next section we shall discuss the question: When does a series converge?

EXERCISES

1. Write out the first five terms in the sequences defined by each of the following:
   a) $s_n = n$
   b) $s_n = 5$
   c) $s_n = 1 - (1/n)^2$
   d) $s_n = (n+1)^{1/3}$
   e) $s_n = (3n+1)(2n-1)$
   f) $s_1 = 2$; $s_n = s_{n-1} + 1$
   g) $s_1 = 2$; $s_n = (s_{n-1})^2/2$
   h) $s_1 = 2$, $s_2 = 3$; $s_n = s_{n-1} + s_{n-2} + 1$

2. Discuss the convergence of the sequences defined by each of the following. If a sequence converges, find its limit and prove that it converges to that limit. If a sequence fails to converge, prove that it cannot converge to any limit L.
   a) $s_n = 3$
   b) $s_n = n$
   c) $s_n = (1/n)^{1/2}$
   d) $s_n = 1 - (1/n)^2$
   e) $s_n = 3 - n$
   f) $s_n = (-1)^n 3$

3. Prove parts (a) and (d) of Proposition 2.

4. Suppose s is a sequence such that $s_n = k$, where k is a constant, for each n in N. Prove $s_n \to k$.

5. Prove that a sequence in which every term is positive cannot converge to a negative limit.

6. Write out the first five terms of the series formed from the sequences defined by each of the following.
   a) $s_n = n$
   b) $s_n = 1$
   c) $s_n = (-1)^n$
   d) $s_n = (-1/n)^n$
   e) $s_n = ((n-1)/n)^2$

7. Prove Proposition 3; that is, prove that if s is a sequence which does not converge to 0, then the series t formed from s cannot converge.

8. A *geometric progression* with *first term* a and *ratio* r is an ordered set of numbers such that each successive term until the last (if there is a last term) is the product of the preceding term and r. Thus, a geometric progression has the

following form:

(2) $a, ar, ar^2, ar^3, \ldots, ar^{n-1}, ar^n$.

a) Find the sum of all the terms in (2). Proceed as follows: Let

$$S = a + \cdots + ar^n.$$

Find rS, and compute $S - rS$; then solve for S. If the computations were carried out properly, you should find that

(3) $S = \dfrac{a(1 - r^{n+1})}{1 - r}.$

b) Now suppose we have a sequence whose terms are in geometric progression, that is, each successive term of the sequence is found by multiplying the preceding term by a fixed ratio r. Consider the series formed from such a sequence; this series is called a *geometric series*. Let s be a geometric series formed from a sequence whose terms are in geometric progression and which has first term a and ratio r.

    i) Prove that s converges if $|r| < 1$.
    ii) Prove that if s converges, it will converge to

(4) $S' = \dfrac{a}{1 - r}.$

[*Hint:* Note that S in (3) is the nth term of s. What does S approach as n gets arbitrarily large?]

c) Find the sum of each of the following geometric series.

    i) $\displaystyle\sum_{n=1}^{\infty} \left(\tfrac{1}{3}\right)^n$ ii) $\displaystyle\sum_{n=1}^{\infty} \left(\tfrac{4}{5}\right)^n$ iii) $\displaystyle\sum_{n=1}^{\infty} \left(\tfrac{1}{2}\right)^n$

6.2 TESTS FOR CONVERGENCE OF SERIES

In this section we concern ourselves with the question: Given a series $\sum_{n=1}^{\infty} s_n$, can one tell from examining s_n whether or not the series converges? The answer to this question, as we shall see, is sometimes yes and sometimes no. Moreover, most tests which might tell us whether or not a series converges will not provide sufficient clues as to what the limit of the series is if it does converge. At best, a test may be able to tell us that the limit is not larger than some number. Nevertheless, it is often more important to know whether a series converges than what the series converges to. Once it is known that a series converges, a computer will readily compute the sum to any desired degree of accuracy.

Proposition 3 gave a necessary, but not sufficient, condition for a series to converge. Before applying any of the following tests for convergence of $\sum_{n=1}^{\infty} s_n$, we should first be certain that $s_n \to 0$; if s_n does not converge to 0, then we are certain that the series diverges.

Where a particular series is not specified, our statements will apply to the general series

(5) $$\sum_{n=1}^{\infty} s_n.$$

The first test we shall consider is the so-called *Comparison Test*.

Proposition 4. *Suppose all but finitely many s_n [in (5)] are positive. Then:*

a) *If each s_n, at least from some point on, is no larger than the corresponding term of a sequence of positive numbers whose series is known to converge, then (5) converges.*

b) *If each s_n, at least from some point on, is no smaller than the corresponding term of a sequence whose series is known to diverge, then (5) diverges.*

We shall neither presume the knowledge of the structure of the real numbers that would be prerequisite to proving Proposition 3 nor take the time to introduce such a discussion into this test; instead, we shall leave Proposition 3, as well as several other propositions in this section, unproved. We illustrate the use of Proposition 3 in the next example.

Example 8. Consider

(6) $$\sum_{n=1}^{\infty} (1/n)^n.$$

Then $s_n = (1/n)^n$ for the purposes of applying Proposition 3. For each positive integer n, s_n is positive, and for $n > 1$, $s_n \leq (\frac{1}{2})^n$. Since $\sum_{n=1}^{\infty} (\frac{1}{2})^n$ converges, Proposition 3 tells us that (6) also converges. Since $s_1 = 1$ and positive numbers are being added to s_1, the sum of (6) will be greater than 1. On the other hand, since $s_n \leq (\frac{1}{2})^n$ for $2 \leq n$, the sum of (6) will be no greater than

$$1 + \sum_{n=1}^{\infty} (\tfrac{1}{2})^n = 1 + 1 = 2.$$

The sum of (6) will therefore lie somewhere between 1 and 2.

Proposition 5. *Series (5) will converge if*

(7) $$\sum_{n=1}^{\infty} |s_n|$$

converges.

Example 9. Consider

$$\sum_{n=1}^{\infty} -(\tfrac{1}{3})^n.$$

Then $s_n = -(\frac{1}{3})^n$ and $|s_n| = (\frac{1}{3})^n$. Therefore

$$\sum_{n=1}^{\infty} |s_n| = \sum_{n=1}^{\infty} (\tfrac{1}{3})^n,$$

which is a convergent geometric series. The original series therefore converges.

We shall see that it is possible, however, for (7) to diverge even if (5) converges.

The following distinctions prove to be useful.

Definition 5. *The series (5) is said to be **alternating** if s_n has the opposite sign of s_{n+1} for all but finitely many n.*

*The series (5) is said to **converge absolutely** if (7) converges.*

*The series (5) is said to **converge conditionally** if it converges, but (7) diverges.*

The next proposition deals with the convergence of alternating series.

Proposition 6. *Assume that (5) is an alternating series. Then (5) converges if $s_n \to 0$ and if there is a positive integer M such that*

$$|s_M| > |s_{M+1}| > |s_{M+2}| > \cdots,$$

that is, from some point the absolute values of the s_n form a decreasing sequence.

Example 10.

$$(8) \quad \sum_{n=1}^{\infty} (-1)^n (1/n)$$

is an alternating series. Here $s_n = (-1)^n(1/n)$ and $|s_n| = 1/n$. The series (8) does not converge absolutely (Example 7), but according to Proposition 6, it does converge. Therefore (8) is an example of a series which converges conditionally.

Conditionally convergent series have certain almost paradoxical properties that we shall not discuss in this text. We shall just remark that by rearranging the s_n in a conditionally convergent series, we can change its limit into any number we want. If a series converges absolutely, then the s_n may be rearranged without fear that the sum of the series will be affected.

We next consider the *Ratio Test*.

Proposition 7. *Suppose all but finitely many s_n in (5) are positive and consider the sequence*

$$(9) \quad \{s_{n+1}/s_n\}, \; n \text{ in } N.$$

Suppose $s_{n+1}/s_n \to r$, where r will be taken to be ∞ if (9) does not converge. Then (5)

a) converges if $r < 1$;
b) diverges if $r > 1$ (which includes the case $r = \infty$);
c) the test gives no information about the convergence of (5) if $r = 1$.

The Ratio Test works because $s_{n+1}/s_n \to r$ implies that (5) "approximates" a geometric series with ratio r as n gets larger. If the ratio of the approximating geometric series is less than 1, then we can use Proposition 4 to prove the convergence of (5). Similarly, if $r > 1$, we can use Proposition 4(b) to show that (5) diverges. If $r = 1$, then we cannot draw any conclusions.

Example 11. Using the Comparison Test, we proved that the series (6) converges. We now prove that (6) converges, using the Ratio Test: For (6),

$$(10) \quad \frac{s_{n+1}}{s_n} = \frac{[1/(n+1)]^{n+1}}{(1/n)^n} = \left(\frac{n}{n+1}\right)^n \left(\frac{1}{n+1}\right).$$

Since $n/(n+1)$ is less than 1 for any positive integer 1, the limit of $[n/(n+1)]^n$ will be at most 1. On the other hand, $1/(n+1) \to 0$. Therefore $s_{n+1}/s_n \to 0$. Since $0 = r < 1$, we find that (6) converges.

Example 12. We have seen that if $s_n = 1/n$, then $\sum_{n=1}^{\infty} s_n$ does not converge. In this instance, $s_{n+1}/s_n = n/(n+1)$, which converges to 1. Although we know that the series in question does not converge, this conclusion must be arrived at by means other than the Ratio Test.

Example 13. We shall see that the series

$$(11) \quad \sum_{n=1}^{\infty} (1/n)^2$$

converges. In this case too though, $s_{n+1}/s_n \to 1$.

The following test works in many instances where the Ratio Test fails.

Proposition 8. *Suppose all but finitely many terms of (5) are positive and $s_{n+1}/s_n \to 1$. Suppose too that s_{n+1}/s_n is reducible in the form*

$$(12) \quad \frac{s_{n+1}}{s_n} = \frac{n^k + An^{k-1} + \cdots}{n^k + Bn^{k-1} + \cdots},$$

that is, s_{n+1}/s_n can be expressed as the quotient of two polynomials of degree k such that the coefficient of n^k in both polynomials is 1. Then the series converges if $B - A > 1$ and diverges if $B - A \leq 1$.

Example 14. Consider (11) once again. We cannot determine from the Ratio Test whether (11) converges. A straightforward computation shows, however, that

(13) $$\frac{s_{n+1}}{s_n} = \frac{n^2}{n^2 + 2n + 1},$$

the quotient of two polynomials in n of degree 2. Here $A = 0$ and $B = 2$; hence $B - A = 2$. According to Proposition 8, then, (11) converges.

Example 15. Consider

(14) $$\sum_{n=1}^{\infty} \frac{1}{2n + 3}.$$

Here $s_{n+1}/s_n = (2n + 5)/(2n + 3)$. Dividing numerator and denominator by n, we find

$$\frac{s_{n+1}}{s_n} = \frac{2 + 5/n}{2 + 3/n} \to 1$$

since $5/n \to 0$ and $3/n \to 0$. Therefore the Ratio Test does not tell us whether or not (14) converges. However, we also have

$$\frac{s_{n+1}}{s_n} = \frac{n + 5/2}{n + 3/2},$$

which is of the form needed to apply Proposition 8. Here $A = \frac{5}{2}$ and $B = \frac{3}{2}$. Since $B - A = \frac{2}{2} = 1$, we see that (14) diverges.

EXERCISES

1. Determine by any suitable test whether or not each of the following series converges. If a series converges, try to find two numbers m and M such that the sum of the series lies between m and M.

   a) $\sum_{n=1}^{\infty} n$

   b) $\sum_{n=1}^{\infty} \frac{n+1}{n^2}$

   c) $\sum_{n=1}^{\infty} (-1)^n (2n)$

   d) $\sum_{n=1}^{\infty} \frac{1}{n+1}$

   e) $\sum_{n=1}^{\infty} \frac{6}{(2n-1)^3}$

   f) $\sum_{n=1}^{\infty} \frac{n^2}{2^n}$

   g) $\sum_{n=1}^{\infty} \frac{n-1}{n+1}$

   h) $\sum_{n=1}^{\infty} (-1)^n (\tfrac{1}{2})^n$

   i) $\sum_{n=1}^{\infty} (1/n - 1/n^2)$

   j) $\sum_{n=1}^{\infty} [(1/n)^2 + (1/n)^3]$

   k) $\sum_{n=1}^{\infty} \frac{3n^2 + 4n + 5}{10n^3 + 2n^2}$

   l) $\sum_{n=1}^{\infty} \frac{-3n - 1}{3n + 1}$

2. Suppose

   $$S = \sum_{n=1}^{\infty} s_n \quad \text{and} \quad T = \sum_{n=1}^{\infty} t_n.$$

Define
$$S + T = \sum_{n=1}^{\infty} (s_n + t_n), \quad \text{and} \quad rS = \sum_{n=1}^{\infty} rs_n,$$
where r is any real number. Suppose $S \to L$ and $T \to L'$. Prove that $S + T \to L + L'$ and $rS \to rL$. [*Hint:* You will have to consider the series as sequences and apply Proposition 2.]

How would we define $S - T$? Is it true that if $S + T$ converges, then both S and T must also converge? Is it true that if $S + T$ converges and S converges, then T must also converge?

3. Suppose the series $\sum_{n=1}^{\infty} 1/kn$, where k is a nonzero constant, converges to a limit L. Prove that this implies that $\sum_{n=1}^{\infty} 1/n$ also converges.

4. Prove that the repeating decimal $0.637637637\ldots$ is expressible as the quotient of two integers. [*Hint:* This decimal is shorthand for a series
$$637(10^{-3} + 10^{-6} + 10^{-9} + \cdots).$$
Note that this is a geometric series.]

5. Express the repeating decimal $0.939393\ldots$ as the quotient of two integers.

6. Use the Comparison Test and the fact that $0 \leq |s_n| - s_n \leq 2|s_n|$ to prove Proposition 5.

6.3 POWER SERIES

Consider the series

(15) $$\sum_{n=1}^{\infty} x^{n-1} = \sum_{n=0}^{\infty} x^n = 1 + x + x^2 + x^3 + \cdots,$$

where x is a variable rather than a fixed real number. We recognize in (15) a geometric series with first term 1 and ratio x. If $|x| < 1$, then (15) will converge to $1/(1-x)$. We can also verify that if $|x| \geq 1$, then (15) will diverge. Even though (15) converges if and only if x is in the open interval $(-1, 1)$, when x is restricted to that interval, (15) represents the function

(16) $\quad f(x) = 1/(1-x);$

that is, (15) and (16) are equivalent for x in $(-1, 1)$.

The limit of (15) is the limit of a sequence of polynomials; specifically, (15) is by definition, the sequence

$$1, \ 1+x, \ 1+x+x^2, \ 1+x+x^2+x^3, \ 1+x+x^2+x^3+x^4, \ldots$$

Since the limit of (15) is the same as $f(x)$ for x in $(-1, 1)$, we see that for x in $(-1, 1)$, $f(x)$ can be approximated to an arbitrary degree of accuracy by a polynomial function.

In this instance, it would be both more cumbersome and more inaccurate to compute $f(x)$ from (15) rather than directly from the definition of $f(x)$. Nevertheless, there are times when a function can be represented in terms of a series, where the series is the most direct and computationally feasible means of arriving at specific function values.

In this section, we shall be concerned with the question: Given a function f, is it possible to represent f as a series of the form

(17) $\quad \sum_{n=0}^{\infty} a_n x^n = a_0 + a_1 x + a_2 x^2 + a_3 x^3 + \cdots$

such that if (17) converges for some value of x, it will converge to $f(x)$?

Definition 6. *A series of the form (17), where the a_i are constant real numbers, is called a **power series in** x.*

Given a function f, we want to find a power series in x which converges to $f(x)$ for those values of x for which it does converge.

The series (17) is the sequence of polynomials

$a_0, \; a_0 + a_1 x, \; a_0 + a_1 x + a_2 x^2, \; a_0 + a_1 x + a_2 x^2 + a_3 x^3, \ldots$

If this sequence is to converge to $f(x)$, then the farther out along the sequence we go, the better approximation to f the polynomial function we get should be. An "infinite" polynomial will give a perfect "representation" of f.

If a function is such that it can be approximated to any degree of accuracy by a polynomial function, it will have to be a very "nice" type of function. This is because polynomial functions are themselves very nice and cannot be made to behave very pathologically. Thus, if a function behaves pathologically, there will probably be no polynomial which can be used to approximate it very closely. We shall consider then only the best-behaved functions; specifically, we shall restrict our attention to those functions which have derivatives of all orders, that is, an nth derivative for any positive integer n.

Suppose now that f is a function with derivatives of all orders at least in some open interval which contains 0. If

(18) $\quad f(x) = a_0 + a_1 x + a_2 x^2 + a_3 x^3 + a_4 x^4 + \cdots,$

then $f(0) = a_0$, for all terms on the right-hand side of (18) except a_0 will vanish when $x = 0$.

If equality holds in (18), then equality should continue to hold if we take the first derivative of both sides. Therefore

(19) $\quad f'(x) = a_1 + 2a_2 x + 3a_3 x^2 + 4a_4 x^3 + \cdots$

Hence $f'(0) = a_1$. We now take the first derivative of both sides of (19) and obtain

(20) $\quad f''(x) = 2a_2 + 3 \cdot 2a_3 x + 4 \cdot 3a_4 x^2 + \cdots$

Thus, $2a_2 = f''(0)$; hence $a_2 = f''(0)/2$.

If we continued this process of taking successive derivatives and evaluating them at 0, we would arrive at the following expression for a_n: Letting $f^{(n)}$ denote the nth derivative of f, we have

(21) $\quad a_n = \dfrac{f^{(n)}(0)}{n \cdot (n-1) \cdot \ldots \cdot 2 \cdot 1}, \quad$ for $n \neq 0$, and $\quad a_0 = f(0)$.

The expressions in (21) are usually abbreviated by using a device called the *factorial* of a nonnegative integer.

Definition 7. *The **factorial** of 0, written 0!, is defined to be 1. If n is any positive integer, then n **factorial**, written $n!$, is defined to be the product of all positive integers less than or equal to n; that is*

$$n! = n \cdot (n-1) \cdot (n-2) \cdot \ldots \cdot 2 \cdot 1.$$

If we set $f^{(0)} = f$, then (21) reduces to

(22) $\quad a_n = \dfrac{f^{(n)}(0)}{n!}.$

Therefore (18) becomes

(23) $\quad f(x) = \sum\limits_{n=0}^{\infty} \dfrac{f^{(n)}(0)}{n!} x^n.$

Definition 8. *We call (23) the **Maclaurin series expansion** of f.*

Thus (15) is the Maclaurin series expansion of $f(x) = 1/(1-x)$.

Example 16. The exponential function $E(x) = e^x$ was introduced in Example 31 of Chapter 4. $E(x)$ has the property that $E'(x) = E(x)$ for each x. Since any nonzero number raised to the zeroth power is 1, we have

$$E(0) = E'(0) = E''(0) = \cdots = E^{(n)}(0) = \cdots = 1.$$

Consequently, the Maclaurin series expansion for $E(x)$ is

(24) $\quad E(x) = e^x = 1 + x + \dfrac{x^2}{2!} + \dfrac{x^3}{3!} + \dfrac{x^4}{4!} + \dfrac{x^5}{5!} + \cdots$

This series furnishes a method to compute $E(x)$ numerically, assuming that the series converges to e^x. We should aks though: For what values of x does (24) converge? Inasmuch as we can sum at most finitely many summands in (24), we might also ask: If we take only the nth partial sum, how accurate an approximation to $E(x)$ do we have? We first try to answer the first question.

Since a series converges if it converges absolutely, we shall determine where (24) converges absolutely; that is we shall find for which values of x

(25) $$\sum_{n=0}^{\infty} \left|\frac{x^n}{n!}\right| = \sum_{n=0}^{\infty} \frac{|x|^n}{n!}$$

converges. From (25) we obtain

(26) $$\frac{s_{n+1}}{s_n} = \frac{|x|^{n+1}/(n+1)!}{|x|^n/n!} = \frac{|x|}{n+1}.$$

For any value of x, the limit of s_{n+1}/s_n is 0. By the Ratio Test (25) converges for any value of x; hence the Maclaurin series for $E(x)$ converges for every value of x.

Suppose that instead of already having $E(x)$, we were trying to solve the problem: Find a function $f(x)$ such that $f(x) = f'(x)$ for any real number x. If f is a function with the desired property, then

$$f(0) = f'(0) = f''(0) = \cdots = f^{(n)}(0) = \cdots$$

Thus the Maclaurin series expansion for f becomes

(27) $$f(x) = f(0)\left(1 + x + \frac{x^2}{2!} + \frac{x^3}{3!} + \cdots\right) = f(0)E(x).$$

Thus a function f with the property that $f'(x) = f(x)$ is a constant times $E(x)$.

The question of how accurately the nth partial sum of the Maclaurin series expansion of a function approximates the function will be discussed in a later section. The remainder of this section will be devoted to the exponential and logarithmic functions.

We shall not explain how we knew that the function $E(x)$ having the property that $E(x) = E'(x)$ and $E(0) = 1$—this function is unique according to the preceding discussion—has the form e^x for some constant e. According to (24),

$$e^1 = e = 1 + 1 + \tfrac{1}{2}! + \tfrac{1}{3}! + \tfrac{1}{4}! + \cdots \doteq 2.71828.$$

The number e and the exponential function have extensive application in most branches of mathematics; we shall meet them again in some of the most important density functions.

Related to $E(x) = e^x$ is the function $\ln x$ which is defined as follows.

Definition 9. *If x is any real number, then $\ln x$ is that power to which e must be raised to obtain x, provided that such a power exists. That is, $y = \ln x$ if $e^y = x$; although we shall not prove it, $\ln x$ is unique for any number x for which it is defined. We call $\ln x$ the* **natural logarithm** *of x, or the*

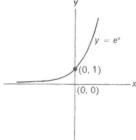

Fig. 3

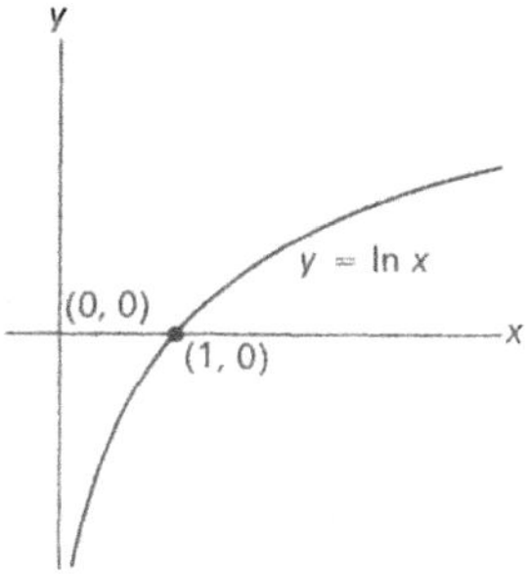

Fig. 4

logarithm to the base e *of* x. *The function defined by* $f(x) = \ln x$ *is called the logarithmic function.*

Since e^x is always positive, $\ln x$ makes sense only for positive values of x. The graphs of E and $\ln$ are presented in Figs. 3 and 4, respectively.

We now compute the first derivative of the logarithmic function.

Proposition 9. *Let* $L(x) = \ln x$. *Then* $L'(x) = 1/x$.

Proof: If $L(x) = \ln x$, then $x = e^{L(x)}$. Taking the derivative of both sides of the second equation, we find $1 = L'(x)e^{L(x)} = xL'(x)$. Thus, $L'(x) = 1/x$.

Observe that if we try to find a Maclaurin series expansion for $L(x) = \ln x$, we are stopped right away by the fact that $\ln 0$ is not defined. Is there nonetheless some method of expressing $\ln x$ as a series? This is a question that will be considered in the next section. We close this section by showing how what we know about e^x and $\ln x$ can be used to find the derivatives of other functions involving logarithms and powers.

Example 17. Consider $f(x) = 2^x$. Since $2 = e^{\ln 2}$, we have
$$f(x) = 2^x = (e^{\ln 2})^x = e^{x \ln 2}.$$
We can use the Chain Rule to find the derivative of $e^{x \ln 2}$; thus we obtain
$$f'(x) = \ln 2 e^{x \ln 2} = \ln 2 (e^{\ln 2})^x = 2^x \ln 2.$$
In general, if f is defined by $f(x) = a^x$, where a is any positive number, then $f'(x) = (\ln a)a^x$.

Definition 10. *If* a *is any positive number other than* 1, *and* x *is any positive number, then the number* y *such that* $a^y = x$ *is said to be the* **logarithm to the base** a *of* x; *the logarithm to the base* a *of* x *may also be written* $\log_a x$.

Example 18. Suppose $f(x) = \log_a x$. Since $a = e^{\ln a}$, we have
$$x = a^{\log_a x} = (e^{\ln a})^{\log_a x} = e^{\ln a \log_a x} = e^{\ln x}.$$
Consequently, $\ln x = \ln a \log_a x$, or
$$f(x) = \log_a x = (1/\ln a)\ln x.$$
Therefore $f'(x) = (1/\ln a)(1/x) = 1/(x \ln a)$. For example, if $f(x) = \log_3 x$, then $f'(x) = 1/x \ln 3$.

Example 19. Consider $f(x) = \ln(x^3 + 3x + 1)$. Then $f(x) = g(m(x))$, where $m(x) = x^3 + 3x + 1$ and $g(x) = \ln x$. Using the Chain Rule, we find
$$f'(x) = g'(m(x))m'(x) = \frac{1}{x^3 + 3x + 1}(3x^2 + 3) = \frac{3x^2 + 1}{x^3 + 3x + 1}.$$
More generally, if $f(x) = \ln m(x)$, then $f'(x) = m'(x)/m(x)$.

EXERCISES

1. Find the first derivatives of each of the following functions.
   a) $f(x) = \ln(2x)$
   b) $f(x) = e^{3x^2} + 1$
   c) $f(x) = \ln(x^3 + e^x)$
   d) $h(x) = 5^x$
   e) $f(x) = 5^{3x^4} + 5$
   f) $f(x) = \log_5(x + x^2)$
   g) $f(x) = 3^x + 4^x$
   h) $f(x) = e^{e^x}$
   i) $g(x) = \ln(\ln x)$
   j) $g(x) = x^x$
   k) $f(x) = e^{\log_3 x}$
   l) $h(x) = \log_3 e^{x+1}$

2. Let $P(x) = a_n x^n + a_{n-1} x^{n-1} + \cdots + a_1 x + a_0$. Find the Maclaurin series expansion for P. Why is the result obtained to be expected?

3. Prove that if $f(x) = \ln g(x)$, then $f'(x) = g'(x)/g(x)$.

4. Find the first four terms in the Maclaurin series expansion of each of the following functions
   a) $f(x) = e^{-x}$
   b) $f(x) = 1/(x - 2)$
   c) $f(x) = xe^x$
   d) $f(x) = 3x + e^x$
   e) $f(x) = 4^x$

5. Determine where each of the Maclaurin series found in Exercise 4 converge absolutely. Use the Ratio Test in the manner of Example 16. That is, suppose
$$f(x) = \sum_{n=0}^{\infty} s_n(x).$$
Find $|s_{n+1}|/|s_n|$ and determine for what values of x this ratio has $r < 1$ as a limit (as n gets arbitrarily large). If $r = 1$, then the convergence of the series must be tested by other means.

6. Using the laws of exponents, prove each of the following facts about logarithms.
   a) $\log_a(xy) = \log_a x + \log_a y$
   b) $\log_a(x^r) = r \log_a x$
   c) $\log_a x = \log_b x (1/\log_b a)$

6.4 TAYLOR'S SERIES. INTERVAL OF CONVERGENCE

We saw in the last section that $L(x) = \ln x$ does not have a Maclaurin series expansion. This is because ln 0 does not make sense. We know, however, that $L(2) = \ln 2$ is defined; moreover, $L(x)$ has derivatives of all orders at 2. Hence we might try to develop a series for $L(x)$ using 2 rather than 0. Of course, if we try to substitute 2 for x in (18), things do not fall out nicely like they do for $x = 0$. We would like to replace x in (18) with something which will be 0 at 2. The expression $x - 2$ works quite well. Therefore we shall try to represent $L(x)$ as a series of the form

$$(28) \quad L(x) = a_0 + a_1(x-2) + a_2(x-2)^2 + a_3(x-2)^3 + a_4(x-2)^4 + \cdots$$

Substituting $x = 2$ in (28), we find that $a_0 = L(2) = \ln 2$. Taking the first derivative of both sides of (28), we obtain

$$(29) \quad L'(x) = a_1 + 2a_2(x-2) + 3a_3(x-2)^2 + \cdots;$$

hence $a_1 = L'(2)$. Since $L'(x) = 1/x$, $L'(2) = \frac{1}{2}$. Consequently, $a_1 = \frac{1}{2}$. Now

$$(30) \quad L''(x) = 2a_2 + 3 \cdot 2a_3(x-2) + 4 \cdot 3a_4(x-2)^2 + \cdots;$$

hence $2a_2 = L''(2) = -(\frac{1}{2})^2$, or $a_2 = -(\frac{1}{2})^3$. We could continue in like fashion and evaluate all the a_i in (28). The series (29) is said to be the *Taylor series of $L(x)$ expanded about* 2. We could also have computed a Taylor series for $L(x)$ expanded about 1; that is, we could have used powers of $x - 1$ in (28) instead of powers of $x - 2$. Such an expansion would, in fact, be simpler than (28) since

$$L(1) = 0 \quad \text{and} \quad L^{(n)}(1) = (-1)^{n+1}(n-1)! \quad \text{for } n \geq 1.$$

The expansion in powers of $x - 1$ is given by

$$(31) \quad L(x) = \ln x = 0 + (x-1) - \frac{(x-1)^2}{2} + \frac{(x-1)^3}{3} - \frac{(x-1)^4}{4} + \cdots$$

Suppose now that f is a function which has derivatives of all orders in some open interval which contains a. If we wish to represent f as a series in powers of $x - a$, that is,

$$f(x) = a_0 + a_1(x-a) + a_2(x-a)^2 + a_3(x-a)^3 + \cdots,$$

then we find that

$$(32) \quad a_n = \frac{f^{(n)}(a)}{n!};$$

hence

(33) $$f(x) = \sum_{n=0}^{\infty} a_n(x-a)^n = \sum_{n=0}^{\infty} \frac{f^{(n)}(a)}{n!}(x-a)^n.$$

Definition 11. *Equation (33) is called the **Taylor series expansion of** $f(x)$ **about** a, or the **Taylor series of** f **expanded about** a. The set of values of x for which the series converges is called the **interval of convergence** of the series.*

Note that the Maclaurin series of f, if there is one, is the Taylor series expansion of $f(x)$ about 0.

Example 20. We have found the Maclaurin series for $f(x) = 1/(1-x)$. We now find the Taylor series expansion of this same function about 10. Straightforward computation shows

$f(10) = -\frac{1}{9},$
$f'(10) = (\frac{1}{9})^2,$
$f''(10) = -2(\frac{1}{9})^3,$
$\vdots$
$f^{(n)}(10) = n!(-1)^{n+1}(\frac{1}{9})^{n+1}$
$\vdots$

Therefore the Taylor series for f expanded about 10 is

(34) $$f(x) = -\frac{1}{9} + (\frac{1}{9})^2(x-10) - (\frac{1}{9})^3(x-10)^2 + (\frac{1}{9})^4(x-10)^3 - \cdots$$

The following proposition is of great importance in determining where the Taylor series expansion of a function converges and to what the series converges.

Proposition 10. *The Taylor series expansion of $f(x)$ about a given in (33) converges if $|x-a|$ is less than the limit of $|a_n/a_{n+1}|$. It will diverge if $|x-a|$ is greater than the limit of $|a_n/a_{n+1}|$. The series (33) may converge or diverge if $|x-a|$ is equal to this limit. If the series converges, it converges to $f(x)$.*

Proof: What we give now is not a formal proof, but rather a justification for the first part of Proposition 10. If (33) is to converge absolutely, then

$$\left| \frac{a_{n+1}(x-a)^{n+1}}{a_n(x-a)^n} \right| = |x-a| \left| \frac{a_{n+1}}{a_n} \right| \to r, \quad r \leq 1;$$

(33) will converge absolutely if $r < 1$, but the values of x where $r = 1$ must be tested separately. Now if

$$\lim_{n \to \infty} |x-a| \left| \frac{a_{n+1}}{a_n} \right| \leq 1,$$

then
$$\lim_{n\to\infty} \left|\frac{a_n}{a_{n+1}}\right| \leq |x - a|.$$

We now illustrate the use of Proposition 10 in several examples.

Example 21. Consider the series for $\ln x$ given in (31). Here
$$\left|\frac{a_n}{a_{n+1}}\right| = \left|\frac{n}{n+1}\right|$$
and the limit of this expression as $n \to \infty$ is 1. Therefore (31) will converge if $|x - 1| < 1$, and diverge if $|x - 1| > 1$. If $|x - 1| = 1$, then $x = 2$, or $x = 0$. If $x = 0$, then (31) diverges. If $x = 2$, then (31) converges by Proposition 6. Therefore (31) converges to $\ln x$ for all values of x in the interval $(0, 2]$.

Example 22. Consider the series in (34). The ratio $|a_n/a_{n+1}|$ in this case is
$$\left|\frac{(-1)^{n+1}(\frac{1}{9})^{n+1}}{(-1)^{n+2}(\frac{1}{9})^{n+2}}\right| = 9,$$
which has 9 as its limit as $n \to \infty$. Therefore (34) converges absolutely if $|x - 10| < 9$, that is, if x is in the interval $(1, 19)$. If $x = 19$, then the series diverges. If $x = 1$, then the series diverges since, in this case, (34) reduces to
$$-\tfrac{1}{9} - \tfrac{1}{9} - \tfrac{1}{9} - \tfrac{1}{9} - \cdots$$
Therefore (34) has interval of convergence $(1, 19)$. For x in this interval, however, the series converges to $1/(1 - x)$.

The last question we shall consider in this section is the following: If we take only the nth partial sum of the Taylor series expansion for $f(x)$ as in (33), how accurate will the approximation to $f(x)$ be? In other words, if $S_n(x)$ is the nth partial sum of (33), what is the largest that $|f(x) - S_n(x)|$ can be? A partial answer to this question is given in the following proposition.

Proposition 11.

a) *If the series (33) for f is alternating, then the sum of the series—provided, of course, that the series converges—will not differ in absolute value from the nth partial sum by more than*
$$|a_{n+1}(x - a)^{n+1}|.$$

b) *If the series (33) for f converges at y to f(y) and if x' is a point of the interval of convergence such that $x' \leq y$ and*

(35) $$\left| \frac{f^{(n+1)}(x')(y-a)^{n+1}}{(n+1)!} \right|$$

is a maximum, then the nth partial sum of (33) evaluated at y will not differ in absolute value from f(y) by more than (35).

We illustrate the use of Proposition 11 in the following examples.

Example 23. The series (31) for $L(x)$ can be used to compute $\ln(\frac{3}{2})$. Since (31) is an alternating series, its nth partial sum differs from $\ln(\frac{3}{2})$ by no more than

$$\left(\frac{1}{n+1}\right)\left(\frac{3}{2} - 1\right) = \frac{1}{2(n+1)}.$$

To be certain of having an error of no more than 0.01, we would have to take the 50th partial sum. We see from this that (31) is a series which converges rather "slowly."

Example 24. Consider the series expansion of $E(x) = e^x$ given in (24). If x is negative, then (24) is an alternating series; hence we can use (a) of Proposition 11 to find a bound for the error in computing e^x through the use of (24). If x is positive, then since e^x is an increasing function for all x, $E^{(n)}(x)$ has its maximum on the interval $(-\infty, x]$ at x itself. Therefore the error of stopping the summation at the nth partial sum is not greater than

$$\frac{e^x}{(n+1)!} x^{n+1}.$$

Consequently, the *relative error* (=error/true value) is

$$\frac{e^x}{e^x(n+1)!} x^{n+1} = \frac{x^{n+1}}{(n+1)!}.$$

When $x = 1$, we would only have to take the 5th partial sum of (24) to be sure of having a percentage of error of not more than 1%. We see then that series (24) converges more rapidly than (31) does.

EXERCISES

1. Find the Taylor series expansion of each of the following functions about the points indicated. Find the interval of convergence for each of the series found.
   a) $f(x) = 1/x$; 1
   b) $f(x) = e^{-x}$; 1
   c) $f(x) = \ln x + x^2$; 1
   d) $f(x) = xe^x$; 0
   e) $f(x) = x^3 + 3x + 7$; 0

2. Compute e to two decimal places. Compute e^{-2} to two decimal places. What partial sum of (24) would we have to take to find e^{10} with a percentage error of less than 1%?

3. Find a series which might be used to compute $\log_{10} 3$. What partial sum of this series would we have to take to ensure that we found $\log_{10} 3$ correct to five decimal places?

4. A certain function f has derivatives of all orders at all real numbers, never assumes a value greater than 1 or less than -1, and $f(0) = 0$, $f'(0) = 1$, $f''(0) = 0$, $f^{(3)}(0) = -1$, $f^{(4)}(0) = 0$, $f^{(5)}(x) = f(x)$. Find the Maclaurin series expansion for f. Where does this series converge? Find the maximum error in computing $f(1)$ using this series if only the 5th partial sum is taken; the 10th partial sum is taken; the 100th partial sum is taken.

5. Derive series (31).

6. The function defined by

$$f(x) = \begin{cases} e^{-\frac{1}{x^2}}, & x \neq 0, \\ 0, & x = 0 \end{cases}$$

has derivatives of all orders at 0. Try to find a Maclaurin series for f. This illustrates that not every function which has derivatives of all orders in an open interval containing 0 necessarily has a Maclaurin series expansion.

7 Integration

7.1 THE DEFINITE INTEGRAL. AREA UNDER A GRAPH

In Section 5.5 we had occasion to find the area between the graph of certain functions and the x-axis over particular intervals. In both instances (Examples 18 and 19 of Chapter 5), the areas concerned were standard geometric figures to which well-known area formulas could be applied. The area between the graph of $f(x) = x^2$ and the x-axis over the interval [0, 1], the shaded portion of Fig. 1, cannot be found, however, through the application of some standard area formula. If we are to find this area, then, we must do some trailblazing. That is, we must *define* what we mean by the area of a geometric configuration (such as that shown in Fig. 1) which does not fall into one of the "nice" categories for which we already have area formulas. This situation is analogous to our earlier problem of taking the sum of infinitely many numbers. Since we had only the sum of finitely many numbers defined to start with, we had to define what should be meant by the sum of infinitely many numbers. In that case, our definition of infinite sum had to be consistent with (that is,

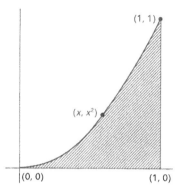 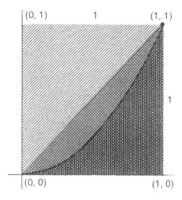

Fig. 1   Fig. 2

not contradict) what we knew about finite sums. Here, our extended definition of area must be such that it does not contradict the area formulas we already know.

Let the area presented in Fig. 1 be denoted by A. Then A lies inside a square with side 1 (Fig. 2). Since the area of this square is 1, then A ought to be less than 1 if it is to be consistent with the usual area formulas. Moreover, the shaded area also lies inside the triangle with vertices $(0, 0)$, $(1, 0)$, and $(1, 1)$ (Fig. 2). Since the area of this triangle is $\frac{1}{2}$, then A should be less than $\frac{1}{2}$.

We now partition $[0, 1]$ into four equal closed subintervals $[0, \frac{1}{4}]$, $[\frac{1}{4}, \frac{1}{2}]$, $[\frac{1}{2}, \frac{3}{4}]$, and $[\frac{3}{4}, 1]$. Over each of these subintervals we erect a rectangle of height $f(r) = r^2$, where r is the right-hand endpoint of the subinterval (Fig. 3). The sum of the areas of these rectangles, that is,

(1) $(\frac{1}{4})(\frac{1}{4})^2 + (\frac{1}{4})(\frac{1}{2})^2 + (\frac{1}{4})(\frac{3}{4})^2 + (\frac{1}{4})(1)^2$,

can be thought of as an approximation of A.

If we had used the left-hand endpoints of the subintervals to compute the height of the rectangles, then the rectangles obtained would be those presented in Fig. 4. The sum of the areas of these rectangles is

(2) $(\frac{1}{4})(0)^2 + (\frac{1}{4})(\frac{1}{4})^2 + (\frac{1}{4})(\frac{1}{2})^2 + (\frac{1}{4})(\frac{3}{4})^2$.

Since the area of Fig. 3 contains A while the area of Fig. 4 is contained in A, A should lie somewhere between (1) and (2).

We could get an even better approximation to A if we subdivided $[0, 1]$ into smaller subintervals. For example, if we subdivided $[0, 1]$ into 10 equal parts, then the subintervals would each have length $\frac{1}{10}$; in this instance, the subintervals would be

$[0, \frac{1}{10}], [\frac{1}{10}, \frac{2}{10}], \ldots, [\frac{8}{10}, \frac{9}{10}], [\frac{9}{10}, 1]$.

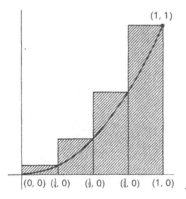

Fig. 3

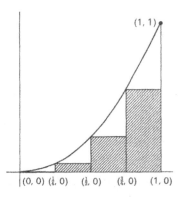

Fig. 4   **Fig. 5**

The approximations to A corresponding to (1) and (2), respectively, are given by

(3) $\quad (\tfrac{1}{10})(\tfrac{1}{10})^2 + (\tfrac{1}{10})(\tfrac{2}{10})^2 + \cdots + (\tfrac{1}{10})(\tfrac{9}{10})^2 + (\tfrac{1}{10})(1)^2$

and

(4) $\quad (\tfrac{1}{10})(0)^2 + (\tfrac{1}{10})(\tfrac{1}{10})^2 + \cdots + (\tfrac{1}{10})(\tfrac{8}{10})^2 + (\tfrac{1}{10})(\tfrac{9}{10})^2.$

The areas which (3) and (4) represent are shown in Fig. 5. Although (3) is larger than (4), the difference between (3) and (4) is only $(\tfrac{1}{10})(1)^2 = \tfrac{1}{10}$.

It seems then that the more subintervals into which we divide [0, 1], the better the approximation to A we shall obtain. Suppose [0, 1] is divided into n equal subintervals. These subintervals would be

$$\left[0, \frac{1}{n}\right], \left[\frac{1}{n}, \frac{2}{n}\right], \left[\frac{2}{n}, \frac{3}{n}\right], \ldots, \left[\frac{n-2}{n}, \frac{n-1}{n}\right], \left[\frac{n-1}{n}, 1\right].$$

The right endpoint of one interval is the left endpoint of the next interval. If the left endpoint of an interval is k/n, then its right endpoint will be $(k+1)/n$. The length of each of these subintervals is $1/n$. Using the right endpoints [as in (3) and (4)] to obtain the heights of the rectangles constructed to approximate A, we find the sum of the areas of the rectangles to be

(5) $$\left(\frac{1}{n}\right)\left(\frac{1}{n}\right)^2 + \left(\frac{1}{n}\right)\left(\frac{2}{n}\right)^2 + \left(\frac{1}{n}\right)\left(\frac{3}{n}\right)^2$$
$$+ \cdots + \left(\frac{1}{n}\right)\left(\frac{n-1}{n}\right)^2 + \left(\frac{1}{n}\right)\left(\frac{n}{n}\right)^2.$$

Using the left-hand endpoints, as in (2) and (4), the area approximating A is

(6) $$\left(\frac{1}{n}\right)(0)^2 + \left(\frac{1}{n}\right)\left(\frac{1}{n}\right)^2 + \cdots + \left(\frac{1}{n}\right)\left(\frac{n-2}{n}\right)^2 + \left(\frac{1}{n}\right)\left(\frac{n-1}{n}\right)^2.$$

Expressions (5) and (6) can be simplified by factoring $(1/n)^3$ out of each summand; thus (5) and (6) are equal to

(7) $$\left(\frac{1}{n}\right)^3 (1^2 + 2^2 + 3^2 + \cdots + n^2) = \left(\frac{1}{n}\right)^3 \sum_{k=1}^{n} k^2,$$

and

(8) $$\left(\frac{1}{n}\right)^3 (1^2 + 2^2 + 3^2 + \cdots + (1-n)^2) = \left(\frac{1}{n}\right)^3 \sum_{k=1}^{n-1} k^2,$$

respectively.

It can be shown that the sum of the squares of the first n positive integers is given by

(9) $\quad 1^2 + 2^2 + \cdots + n^2 = (\frac{1}{6})n(n+1)(2n+1)$.

Applying (9), we can reduce (7) and (8) to

(10) $$\left(\frac{1}{n}\right)^3 \left(\frac{1}{6}\right) n(n+1)(2n+1) = \left(\frac{1}{6}\right)\left(1 + \frac{1}{n}\right)\left(2 + \frac{1}{n}\right),$$

and

(11) $$\left(\frac{1}{n}\right)^3 \left(\frac{1}{6}\right) (n-1)(n)(2n-1) = \left(\frac{1}{6}\right)\left(1 - \frac{1}{n}\right)\left(2 - \frac{1}{n}\right),$$

respectively.

As n goes to infinity, that is, as $[0, 1]$ is divided into more and more equal subintervals, (10) and (11) each approach $(\frac{1}{6})(2) = \frac{1}{3}$ as a limit (since $1/n \to 0$). Thus we see that as we construct better and better approximations to A, we come arbitrarily close to $\frac{1}{3}$. It therefore makes sense to say that $A = \frac{1}{3}$.

Although we have been concerned in this rather lengthy example with finding a particular area, the technique used can be employed to find the area between the graph of any function f and the x-axis over the closed interval $[a, b]$, provided only that f is "nice" enough. We now summarize the steps that would be used in such a computation.

Problem. To find the area between the graph of the function f and the x-axis over the interval $[a, b]$. This area is often called the *area under the graph of f over $[a, b]$* (or, *from a to b*). This area is pictured in Fig. 6; we shall denote it by A. Moreover, we shall assume that $f(x)$ is nonnegative for each x in $[a, b]$.

We divide $[a, b]$ into n equal subintervals (which overlap only at endpoints). These intervals are

$$\left[a, a + \frac{b-a}{n}\right], \left[a + \frac{b-a}{n}, a + \frac{2(b-a)}{n}\right],$$
$$\left[a + \frac{2(b-a)}{n}, a + \frac{3(b-a)}{n}\right], \ldots,$$
$$\left[a + \frac{(n-2)(b-a)}{n}, a + \frac{(n-1)(b-a)}{n}\right],$$
$$\left[a + \frac{(n-1)(b-a)}{n}, a + \frac{n(b-a)}{n} = b\right].$$

We consider $a + (b - a)/n$ to be the first right-hand endpoint, and denote the kth right endpoint by r_k. We shall denote the kth left endpoint by s_k. There are n right-hand and n left-hand endpoints. Then the area A can be approximated by

$$(12) \quad \left(\frac{b-a}{n}\right) f(r_1) + \left(\frac{b-a}{n}\right) f(r_2) + \cdots + \left(\frac{b-a}{n}\right) f(r_n)$$
$$= \left(\frac{b-a}{n}\right) \sum_{k=1}^{n} f(r_k),$$

using righthand endpoints, or

$$(13) \quad \left(\frac{b-a}{n}\right) f(s_1) + \left(\frac{b-a}{n}\right) f(s_2) + \cdots + \left(\frac{b-a}{n}\right) f(s_n)$$
$$= \left(\frac{b-a}{n}\right) \sum_{k=1}^{n} f(s_k),$$

using lefthand endpoints.

If f is "reasonable," then the limits of both (12) and (13) as n goes to infinity should exist and be equal. It is this limit that we define to be A.

The sums (12) and (13) would still make sense even if $f(x)$ were negative for certain or all x in $[a, b]$. We can think of the area between the graph of

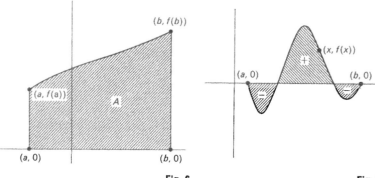

Fig. 6   Fig. 7

f and the x-axis as being positive if it lies above the x-axis, and as negative if it lies below the x-axis, since if $f(r_k)$ is negative, then

$$\left(\frac{b-a}{n}\right) f(r_k)$$

is negative.

The discussion above motivates us to make the following definition.

Definition 1. *Suppose f is any function from T into R and $[a, b]$ is a subset of T. If the limits of (12) and (13) above exist and are equal, call this limit A; then A is said to be the* **definite integral of f from a to b**. *The definite integral of f from a to b is generally denoted by*

$$\int_a^b f(x)\, dx.$$

If

$$\int_a^b f(x)\, dx$$

exists, we say that f is **integrable** *over $[a, b]$.*

Geometrically, $\int_a^b f(x)\, dx$ represents the area between the graph of f and the x-axis. This area is a "signed" area, that is, negative below the x-axis and positive above it. If there is exactly as much area between the graph of f and the x-axis over $[a, b]$ under the x-axis as there is above it (Fig. 7), then $\int_a^b f(x)\, dx$ will turn out to be 0.

The example which began this section showed that $\int_0^1 x^2\, dx = \frac{1}{3}$.

Definition 1 is actually a simplified version of the definition of the definite integral. It will work just as well as the more complicated definition in the case of "nice" functions. A nice function, for our purposes, will be one which has at most finitely many points at which it is discontinuous. Thus continuous functions and differentiable functions are "nice." If f is "nice" on $[a, b]$, then $\int_a^b f(x)\, dx$ will always exist. It is not likely that the reader will ever encounter many, if any, functions which are not "nice."

We should also note that Definition 1 is very awkward from a computational point of view. We certainly do not want to carry out the type of lengthy computation that was necessary to find $\int_0^1 x^2\, dx$ every time we have a definite integral to evaluate. Fortunately, in the next section we shall find a result which is very helpful in computing definite integrals. For the really perverse cases, there is the computer.

EXERCISES

1. Show that the difference between (7) and (8) has 0 as a limit (as n goes to infinity).

2. Use the formula

    (14) $\quad 1 + 2 + 3 + \cdots + n = (n/2)(n + 1)$

    to find the area between the graph of $f(x) = 2x$ and the x-axis over $[0, 3]$, using the method described in defining the definite integral. That is, evaluate

    $$\int_0^3 2x\, dx$$

    directly from Definition 1. Find the area in question, using the standard formula for the area of a triangle. The two results should be the same.

3. Find $\int_8^{10} 10x\, dx$, using Definition 1. Then find the area, using the formula for the area of a triangle. The two results should agree.

4. Let $f(x) = k$ for each real number x, where k is a constant. Show that

    $$\int_a^b f(x)\, dx = \int_a^b k\, dx = k(b - a).$$

    Why do we expect that this has to be true?

5. Using Definition 1 and what you know about limits, prove each of the following facts. We assume that f and g are both functions such that both

    $$\int_a^b f(x)\, dx \text{ and } \int_a^b g(x)\, dx \text{ exist.}$$

    a) $\displaystyle\int_a^b kf(x)\, dx = k \int_a^b f(x)\, dx$, where k is any real number.

    b) $\displaystyle\int_a^b (f + g)(x)\, dx = \int_a^b f(x)\, dx + \int_a^b g(x)\, dx$

    c) If M is the maximum value that $f(x)$ attains in $[a, b]$, then

    $$\int_a^b f(x)\, dx \leq (b - a)M.$$

6. Decide which of the following statements should be true on the basis of what we already know about the definite integral.

a) $\int_0^1 x^3 \, dx \geq \int_0^1 x^4 \, dx$ b) $\int_{-1}^1 x^2 \, dx = 2\int_0^1 x^2 \, dx$ c) $\int_0^1 (-x) \, dx < 0$

d) If $f(x) \geq g(x)$ for each x in $[a, b]$, then

$$\int_a^b f(x) \, dx \geq \int_a^b g(x) \, dx,$$

provided that both integrals exist.

e) If $a < c < b$, then

$$\int_a^b f(x) \, dx = \int_a^c f(x) \, dx + \int_c^b f(x) \, dx,$$

provided all the integrals in question exist.

f) $\int_a^b |f(x)| \, dx \geq \int_a^b f(x) \, dx$, if $\int_a^b |f(x)| \, dx$ exists.

g) $\int_0^1 (x - 1) \, dx = \int_{-1}^0 y \, dy$

7. Evaluate $\int_{-1}^1 |x| \, dx$. This integral exists even though $f(x) = |x|$ is not differentiable at every point of $[-1, 1]$.

7.2 THE FUNDAMENTAL THEOREM OF THE CALCULUS

If one reviews the somewhat lengthy computations which introduce the definite integral and realizes, moreover, that the example given was chosen because of its relative simplicity, then the hope must arise that there is a simpler way of evaluating the definite integral than taking the limit of a sum. Such is indeed the case for at least a very important class of functions, as we now see.

We wish to find

$$\int_a^b f(x) \, dx.$$

Assume that f is continuous on $[a, b]$ and that there is a function F such that $F'(x) = f(x)$ for each x in (a, b) and F is continuous on $[a, b]$. Consider a subinterval $[s_k, r_k]$ of $[a, b]$. Using the Mean Value Theorem (Proposition 8 of Chapter 5), we can find a point x_k in (s_k, r_k) such that

$$\frac{F(r_k) - F(s_k)}{r_k - s_k} = F'(x_k) = f(x_k);$$

hence

(15) $\qquad F(r_k) - F(s_k) = f(x_k)(r_k - s_k).$

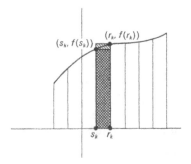

Fig. 8

If we have n intervals $[s_k, r_k]$, $k = 1, 2, \ldots, n$, which divide $[a, b]$ in accordance with the procedure for finding $\int_a^b f(x)\, dx$, and if f is continuous on $[a, b]$, then we expect that we can use $f(x_k)$ to replace $f(r_k)$ in (12), or $f(s_k)$ in (13), to obtain an approximation to the area sought (see Fig. 8); for as the intervals get smaller, that is, as n gets larger, the difference between $f(x_k)$ and $f(r_k)$ must get arbitrarily small as well, because of the continuity of f. Replacing $f(r_k)$ in (12) with $f(x_k)$, we obtain

(16) $\quad \left(\dfrac{b - a}{n}\right) \sum\limits_{k=1}^{n} f(x_k).$

The limit of (16) as n goes to infinity should be the same as the limit of (12) as n goes to infinity, namely, $\int_a^b f(x)\, dx$.

From (15), we also have

(17) $\quad F(r_k) - F(s_k) = f(x_k) \left(\dfrac{b - a}{n}\right).$

Consequently,

(18) $\quad \sum\limits_{k=1}^{n} \left(F(r_k) - F(s_k)\right) = \left(\dfrac{b - a}{n}\right) \sum\limits_{k=1}^{n} f(x_k).$

Let us examine $\sum_{k=1}^{n} \left(F(r_k) - F(s_k)\right)$ more closely. Written out, this sum is

(19) $\quad \left(F(r_1) - F(s_1)\right) + \left(F(r_2) - F(s_2)\right) + \cdots + \left(F(r_n) - F(s_n)\right).$

But $r_1 = s_2$, $r_2 = s_3$, etc.; that is, the right-hand endpoint of one interval is the same as the left-hand endpoint of the next interval. Hence we may rewrite (19) as

(20) $\quad \left(F(s_2) - F(s_1)\right) + \left(F(s_3) - F(s_2)\right) + \left(F(s_4) - F(s_3)\right) \\ + \cdots + \left(F(r_n) - F(s_n)\right).$

The terms of (20) may be regrouped and (20) rewritten as

(21) $\quad -F(s_1) + \left(F(s_2) - F(s_2)\right) + \left(F(s_3) - F(s_3)\right) \\ + \cdots + \left(F(s_n) - F(s_n)\right) + F(r_n).$

Since $s_1 = a$, and $r_n = b$, (21) reduces to

(22) $\quad F(b) - F(a)$.

Since (22) is the value of (18) for any value of n, (22) is the limit of (18) as n goes to infinity; this limit is $\int_a^b f(x)\,dx$.

We summarize the results of our discussion in the following proposition, a proposition which is so important that it is given the name the *Fundamental Theorem of Calculus*.

Proposition 1. *Suppose f is a function which is continuous on $[a, b]$ and F is a function such that F is continuous on $[a, b]$ and $F'(x) = f(x)$ for each x in (a, b). Then*

$$\int_a^b f(x)\,dx = F(b) - F(a).$$

We illustrate the use of this proposition in the following examples.

Example 1. Earlier we computed $\int_0^1 x^2\,dx$ the "long way." Set $F(x) = x^3/3$. Then $F'(x) = x^2$. The functions x^2 and F satisfy the hypotheses of Proposition 1 on the interval $[0, 1]$; therefore

$$\int_0^1 x^2\,dx = F(1) - F(0) = \tfrac{1}{3} - 0 = \tfrac{1}{3},$$

which is, of course, the result obtained before by lengthy computations.

Example 2. We shall compute $\int_0^5 x^5\,dx$. Observe that if we set $g(x) = x^6$, then $g'(x) = 6x^5 = 6f(x)$, where $f(x) = x^5$. Thus $f(x) = (\tfrac{1}{6})g'(x)$. If we set $F(x) = (\tfrac{1}{6})g(x) = x^6/6$, then we shall have $F'(x) = (\tfrac{1}{6})g'(x) = f(x)$. Therefore, by Proposition 1, $\int_0^5 x^5\,dx = F(5) - F(0) = 5^5/6$.

Example 3. We now compute $\int_1^2 (1/x)\,dx$. Because we remember it, can find it in a table of derivatives, or can derive it on our own, we observe that if we set $F(x) = \ln x$, then $F'(x) = 1/x$. Since $f(x) = 1/x$ and F both satisfy the hypotheses of Proposition 1 on $[1, 2]$, we have

$$\int_1^2 (1/x)\,dx = \ln 2 - \ln 1 = \ln 2.$$

If we had wanted to find $\int_0^1 (1/x)\,dx$, we could not do so—at least according to the definition of the definite integral as we now have it—since $1/x$ (as well as $\ln x$) is not defined for $x = 0$.

We shall learn more about the basic properties and evaluation of definite integrals in later sections. We now turn our attention to the density functions of continuous variates.

We shall assume that F is the distribution function of a continuous random variable X whose admissible range is $[a, b]$. We shall assume further that F is differentiable on $[a, b]$ (and hence is also continuous), and that F' is continuous on $[a, b]$. F' is the density function of F, to be denoted by f. By Proposition 1, then, $\int_a^b f(x)\, dx = F(b) - F(a)$. But $F(b) = 1$ and $F(a) = 0$; consequently, $\int_a^b f(x)\, dx = 1$. We therefore conclude that the area between the graph of f and the x-axis over $[a, b]$ is 1. This is a result which, in Section 5.5, we were able to verify only for special cases.

In general, suppose that we want to determine the probability that $y \leq x \leq y'$. This, as we proved in the corollary to Proposition 4 of Chapter 3, is $F(y') - F(y)$, which, in turn, is $\int_y^{y'} f(x)\, dx$, that is, the area under the graph of f from y to y'. This fact gives us greater justification for calling f the "density function" of X; for it is the area under the graph of f from y to y' which measures $P(y \leq x \leq y')$. Intervals of equal length will not necessarily have the same area over them; the greater the area, the "denser" the probability on the interval.

Suppose we are given a function f and are told that f is the density function of a continuous random variable X with admissible range $[a, b]$. How do we find the distribution function of X? If F is the distribution function and x is any point of $[a, b]$, then $F(x) - F(a) = F(x) - 0 = F(x)$. But since $F(x) - F(a) = \int_a^x f(x)\, dx$, we have

(23) $$F(x) = \int_a^x f(x)\, dx;$$

thus we have recovered the distribution function of x from f.

We summarize the results obtained about density functions in the following proposition.

Proposition 2. *Suppose f is the density function for a continuous random variable X with admissible range $[a, b]$. Suppose, too, that f is continuous on $[a, b]$. Then:*

a) $P(y \leq x \leq y') = \int_y^{y'} f(x)\, dx$, *where y and y' are in $[a, b]$ and $y < y'$. In particular, $\int_a^b f(x)\, dx = 1$.*

b) *If F is the distribution function for x, then $F(x) = \int_a^x f(x)\, dx$.*

Example 4. Evaluate the constant k such that $f(x) = kx^4$ can be the density function for a continuous random variate with admissible range $[-1, 1]$. We first observe that $f(x)$ will be nonnegative for each x in $[-1, 1]$ if and only if k is nonnegative. If f is to be the density function for a variate X with admissible range $[-1, 1]$, then $f(x)$ will have to be nonnegative for each x in $[-1, 1]$. If on calculating k, we find that k is negative, then we have to conclude that there cannot be a density function of the desired form for the admissible region given. By Proposition 2, we must have

(24) $$\int_{-1}^{1} kx^4\, dx = k \int_{-1}^{1} x^4\, dx = 1.$$

Now $F(x) = x^5/5$ is a function such that $F'(x) = f(x) = x^4$. Therefore $\int_{-1}^{1} x^4 \, dx = F(1) - F(-1) = \frac{1}{5} - (-\frac{1}{5}) = \frac{2}{5}$.

From (24), then, $k(\frac{2}{5}) = 1$; hence $k = \frac{5}{2}$.

We now find the distribution function G of X, using Proposition 2. By (b) of Proposition 2,

$$G(x) = \int_{-1}^{x} (\tfrac{5}{2}) x^4 \, dx = \tfrac{5}{2} \int_{-1}^{x} x^4 \, dx = (\tfrac{5}{2})(F(x) - F(-1))$$

(where F is defined as above)

$$= (\tfrac{5}{2})(x^5/5 - (-\tfrac{1}{5})) = x^5/2 + \tfrac{1}{2}.$$

EXERCISES

1. Evaluate each of the following definite integrals. The following facts—two of which were proved in Exercise 4 of the preceding section—may prove useful.

    i) $\int_{a}^{b} kf(x) \, dx = k \int_{a}^{b} f(x) \, dx$, where k is any constant.

    ii) $\int_{a}^{b} (f+g)(x) \, dx = \int_{a}^{b} f(x) \, dx + \int_{a}^{b} g(x) \, dx$.

    iii) $\int_{a}^{b} x^n \, dx = \dfrac{b^{n+1}}{n+1} - \dfrac{a^{n+1}}{n+1}$, $n \neq -1$.

    a) $\int_{1}^{2} x \, dx$ b) $\int_{-1}^{4} 5x \, dx$ c) $\int_{-3}^{-6} (-5x+1) \, dx$

    d) $\int_{-1}^{1} x^3 \, dx$. (Interpret geometrically why this integral is 0.)

    e) $\int_{0}^{1} e^x \, dx$ f) $\int_{7}^{16} (1/x) \, dx$ g) $\int_{2}^{3} (x^3 + 2x^2 + 4x + 1) \, dx$

    h) $\int_{-1}^{1} (x^4 + x^2 + 4) \, dx$

    (How do we know that this integral has the same as value as

    $2 \int_{0}^{1} (x^4 + x^2 + 4) \, dx$

    even before we calculate either of these integrals?)

    i) $\int_{5}^{6} x^{1/2} \, dx$ j) $\int_{1}^{2} (x^{1/3} + x^{-1/2}) \, dx$

    k) $\int_{4}^{5.1} (x^{0.9} + x^{-0.7} + 4x^{0.65}) \, dx$ l) $\int_{0}^{1} e^{x^2} 2x \, dx$

    [*Hint:* Note that $2x$ is the derivative of x^2 and use the Chain Rule.]

m) $\int_0^1 (x^3 + 4)^{17} 3x^2 \, dx$ n) $\int_0^1 (x^3 + 4)^{17} x^2 \, dx$

o) $\int_0^1 (x^2 + 3x + 1)(2x + 3) \, dx$

2. Prove formula (iii) of Exercise 1. What happens in the case that $n = -1$?

3. Try to evaluate k in each of the following so that the given function becomes a density function for a random variable with the given admissible range. It may be that no k can be found which satisfies this condition. In those instances where such a k can be found, find the distribution function of the variate.

a) $f(x) = kx$; $[0, 1]$
b) $f(x) = kx$; $[-1, 1]$
c) $f(x) = kx^3$; $[3, 4]$
d) $f(x) = kx^{1/2}$; $[1, 2]$
e) $f(x) = kx^2 + 2x + 1$; $[0, 1]$
f) $f(x) = ke^x$; $[0, 1]$
g) $f(x) = k/x$; $[0, 1]$

7.3 SOME BASIC INTEGRALS. THE INDEFINITE INTEGRAL

All but one of the facts about integrals presented in the following proposition have already been encountered in exercises. We present these facts again because of their importance and for the sake of ready reference.

Proposition 3. *Suppose that f and g are functions which are integrable over $[a, b]$. Then:*

a) $\int_a^b (f + g)(x) \, dx = \int_a^b f(x) \, dx + \int_a^b g(x) \, dx.$

b) $\int_a^b kf(x) \, dx = k \int_a^b f(x) \, dx$, *where k is any constant.*

c) *If $a < c < b$, then* $\int_a^b f(x) \, dx = \int_a^c f(x) \, dx + \int_c^b f(x) \, dx.$

d) $\int_a^b f(x) \, dx = -\int_b^a f(x) \, dx.$

Example 5. It was found in Exercise 1(d) of the preceding section that

(25) $\int_{-1}^1 x^3 \, dx = 0.$

This integral is 0 because there is just as much area under the graph of $f(x) = x^3$ below the x-axis (negative area) from -1 to 1 than there is area above the x-axis. This fact is illustrated in Fig. 9. Suppose we wish to find the "absolute" area of the shaded portion of Fig. 9. We can express (25) as

(26) $\int_{-1}^0 x^3 \, dx + \int_0^1 x^3 \, dx.$

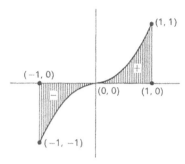

Fig. 9

The first summand will give us the negative area; the second summand the positive area. We can make both quantities positive either by taking the absolute value of each summand to obtain

(27) $\quad \left| \int_{-1}^{0} x^3 \, dx \right| + \left| \int_{0}^{1} x^3 \, dx \right|,$

or by reversing the order of the *limits of integration* in the first summand of (26) to change its sign, thus obtaining

$$\int_{0}^{-1} x^3 \, dx + \int_{0}^{1} x^3 \, dx.$$

Example 6. We will find

(28) $\quad \int_{-2}^{5} |x| \, dx.$

Since $f(x) = |x|$ is continuous on $[-2, 5]$, (28) exists. However, there is no obvious function F such that $F'(x) = f(x)$. We therefore express (28) as the sum of two integrals

(29) $\quad \int_{-2}^{0} |x| \, dx + \int_{0}^{5} |x| \, dx = \int_{-2}^{0} (-x) \, dx + \int_{0}^{5} x \, dx$

$\qquad\qquad\qquad\qquad\quad = -\int_{-2}^{0} x \, dx + \int_{0}^{5} x \, dx.$

If we set $F(x) = x^2/2$, then $F'(x) = x$; hence

$-\int_{-2}^{0} x \, dx = -(F(0) - F(-2)) = 2 \quad$ and $\quad \int_{0}^{5} x \, dx = \frac{25}{2}.$

Therefore (28) is $\frac{29}{2}$.

If we wish to evaluate a definite integral $\int_{a}^{b} f(x) \, dx$, where f is continuous on $[a, b]$, using the Fundamental Theorem of the Calculus, then we must be able to find a function F which is continuous on $[a, b]$ and $F'(x) = f(x)$ for each x in (a, b).

Definition 2. *A function F having the properties with regard to [a, b] and f referred to in the above paragraph is said to be the **indefinite integral of f**, or the **antiderivative of f**. Sometimes F is denoted by*

$$\int f(x)\, dx,$$

that is, the definite integral symbol without the limits of integration.

The antiderivative of f is not unique. For if F is one antiderivative and we define G by $G(x) = F(x) + k$, where k is any constant, then, since $G'(x) = F'(x) = f(x)$, G is also an antiderivative of f. Any two antiderivatives of f differ by at most a constant term. For if F and G are both antiderivatives of f, then $G'(x) - F'(x) = f(x) - f(x) = 0$ for at least all x in (a, b). Now a function has derivative 0 if and only if it is a constant function, that is, its value never changes; hence $G(x) - F(x)$ is a constant for each x in (a, b). From the viewpoint of evaluating $\int_a^b f(x)\, dx$, however, any antiderivative of f is as good as any other. When we present the general antiderivative of f, an undetermined constant is added on to indicate that no matter what value the constant has, we still have an antiderivative of f.

Example 7. Consider $f(x) = x^5 + 3$. Then one antiderivative of f is $F(x) = x^6/6 + 3x$. If k is any constant, then $G(x) = F(x) + k$ also gives an antiderivative of f; hence

$$\int f(x)\, dx = F(x) + k.$$

When we evaluate a definite integral, the k drops out; for

$$\begin{aligned}\int_a^b f(x)\, dx &= (F(b) + k) - (F(a) + k) \\ &= (F(b) - F(a)) + (k - k) \\ &= F(b) - F(a).\end{aligned}$$

If, however, certain conditions must be fulfilled by $\int f(x)\, dx$, it may be possible, or even necessary, to evaluate k. For example, if we demand that $F(0) = 0$, then k will have to be 0.

Example 8. Find k such that $f(x) = x + k$ can be the density function of a continuous variate with admissible range [6, 8]. We need

(30) $\quad \int_6^8 (x + k)\, dx = 1.$

We have

$$\int (x + k)\, dx = x^2/2 + kx + k',$$

where k' is a constant; hence $(32 + 8k) - (18 + 6k) = 14 + 2k = 1$. Therefore $k = -7$. Since with $k = -7$, $f(6) = -1$, $f(x)$ is not always nonnegative for x in $[6, 8]$; therefore there is no continuous variate with the given admissible range and density function f. If there were, and we wanted to find the distribution function F of the variate, then

$$F(x) = \int_6^x f(x)\, dx.$$

We could also find $F(x)$ by evaluating k' in

$$\int f(x)\, dx = x^2/2 - 7x + k',$$

in order for the value of this indefinite integral at 6 to be 0.

How do we find the indefinite integral of a function? There are certain tricks that can be employed, some of which will be introduced in later sections, but at least in the case of simple functions, we can find the indefinite integrals by reversing the process of differentiation and recalling the function which has the given function as a derivative. Some of the simplest indefinite integrals are presented in the following proposition.

Proposition 4.

a) $\int 0\, dx = k$, *where k is a constant.*

b) $\int a\, dx = ax + k$, *where a is a constant.*

c) $\int x^n\, dx = \dfrac{x^{n+1}}{n+1} + k$, *if $n \neq 1$.*

d) $\int (1/x)\, dx = \ln x + k$.

e) $\int e^x\, dx = e^x + k$.

The reader can check the accuracy of the claims of Proposition 4 by simply differentiating the right-hand side of each equality.

Example 9. We shall evaluate

(31) $\int (3x^5 + x^{-1} + e^x + 17)\, dx.$

Applying Proposition 3,* we may rewrite (31) as

(32) $3\int x^5\, dx + \int x^{-1}\, dx + \int e^x\, dx + \int 17\, dx.$

---

* Although Proposition 3 is stated for definite integrals, it works just as well in finding indefinite integrals.

Applying Proposition 4 to (32), we find that (32) is equal to
$3(x^6/6) + \ln x + e^x + 17x + k$, k some constant.

Example 10. Consider

(33) $\int e^{-x}\, dx.$

Set $f(x) = e^{-x}$. If we then set $F(x) = -f(x) = -e^{-x}$, then $F'(x) = -f'(x) = e^{-x}$.

Consequently, (33) is $F(x) = -e^{-x} + k$, where k is a constant. If we demand that $F(0) = 0$, then $0 = -e^{-0} + k = -1 + k$; hence k will have to equal $+1$. Specifying the value of $F(x)$ for some particular x will indirectly force k to assume a particular value.

EXERCISES

1. Find each of the following indefinite integrals. Be sure to include the constant. In each case, evaluate the constant if the value of the indefinite integral at 1 is to be 0.

   a) $\int x^4\, dx$
   b) $\int (x + x^{1/2})\, dx$
   c) $\int (x^{-1} + 3x^{-2} - x^{-3})\, dx$
   d) $\int (e^x + e^{-x})\, dx$
   e) $\int (a + bx + cx^2)\, dx$, a, b, and c constants
   f) $\int (ae^x - b/x + x^{78})\, dx$, a, b, and c constants
   g) $\int (e^{-3x} + e^{2x})\, dx$
   h) $\int (e^{-1} + e^8)\, dx$

2. Find the absolute area under the graphs of the following functions for the intervals given (see Example 5).

   a) $f(x) = x^3$; $[-5, 10]$
   b) $f(x) = |x|$; $[-10, 10]$
   c) $f(x) = (x - 2)(x - 3)$; $[-2, 3]$
   d) $f(x) = (x + 1)(x - 5)(2x + 1)$; $[-10, 10]$
   e) $f(x) = 1/x$; $[-3, -1]$

3. Evaluate each of the following definite integrals.

   a) $\int_0^2 |x - 1|\, dx$
   b) $\int_{-2}^2 f(x)\, dx$, where $f(x) = \begin{cases} \frac{1}{2}, & x \leq 0 \\ x + \frac{1}{2}, & 0 \leq x \end{cases}$
   c) $\int_0^{10} g(x)\, dx$, where $g(x) = \begin{cases} x, & x \leq 5 \\ x^2 - 20, & 5 \leq x \end{cases}$
   d) $\int_0^2 h(u)\, du$, where $h(u) = \begin{cases} 3, & u > 1 \\ 0, & 1 \geq u \end{cases}$

7.4 INTEGRATION BY PARTS AND CHANGE OF VARIABLE

Many of the computational manipulations introduced in this section should seem somewhat questionable to the reader. Putting these manipulations on a firm theoretical foundation, however, would involve us in a lengthy discussion that may be more confusing than enlightening. We do not wish to imply that the means justify the end, the end here being the right answer and the means being the questionable computations. In this instance, the means can be justified; we simply do not care to justify them in this text. With this warning, we proceed to study two of the most important integration tools, integration by parts and integration by change of variable. We begin with a definition.

Definition 3. *If $y = f(x)$, where f is differentiable, then dy defined by*

$$dy = f'(x)\,dx$$

*is said to be the **differential of** y **with respect to** x, or merely the **differential** of y.*

Example 11. If $y = x^3$, then $dy = 3x^2\,dx$. If $f(x) = 4x^2$, then $df(x) = 8x\,dx$. If $F(x)$ is such that $F'(x) = f(x)$, then $dF(x) = f(x)\,dx$.

From the last statement in Example 11, we see that if $F(x)$ is the antiderivative of $f(x)$, then

(34) $$\int_a^b dF(x) = \int_a^b f(x)\,dx = F(b) - F(a).$$

We have defined $dF(x)$ to make (34) true.

Example 12. Suppose $f(x) = g(m(x))$. Then (assuming everything is differentiable), we have $f'(x) = g'(m(x))m'(x)$. Set $y = m(x)$. Then $dy = m'(x)\,dx$. We also have $df(x) = f'(x)\,dx = g'(m(x))m'(x)\,dx$. Consequently, substituting y for $m(x)$, we obtain

$$df(x) = g'(y)m'(x)\,dx = g'(y)\,dy.$$

By a change of variable—that is, substituting y for $m(x)$—we have effected a simplification of $df(x)$.

Example 12 forms the basis for integration by transformation of variable. Using this technique, we try to change an integral of an unfamiliar form into a simple integral such as those given in Proposition 4. The method is best presented through the use of examples.

Example 13. Consider

(35) $$\int_2^3 (x-4)^3\,dx.$$

Set $f(x) = (x-4)^3$. Then $f(x) = g(m(x))$, where $g(x) = x^3$ and

$m(x) = x - 4$. Setting $y = m(x) = x - 4$, we obtain $f(x) = y^3$ and $dy = m'(x)\, dx = dx$. In terms of y, $(x - 4)^3 \, dx$ becomes $y^3 \, dy$. Now, if x varies from 2 to 3 and $y = x - 4$, then y varies from $2 - 4 = -2$ to $3 - 4 = -1$. In terms of y then, (35) can be written as

(36) $$\int_{-2}^{-1} y^3 \, dy.$$

Formula (36) is in a clearly recognizable form. Its evaluation yields

$$\frac{(-1)^4}{4} - \frac{(-2)^4}{4} = \frac{1}{4} - 4 = \frac{-15}{4}.$$

Example 14. Consider

(37) $$\int_0^2 2xe^{x^2} \, dx.$$

If we set $y = x^2$, then $dy = 2x\, dx$ and $e^{x^2} = e^y$. Consequently, $2xe^{x^2} \, dx = e^y \, dy$.

If x varies between 0 and 2, then $y = x^2$ varies between 0 and 4; hence, in terms of y, (37) becomes

(38) $$\int_0^4 e^y \, dy = e^4 - e^0 = e^4 - 1.$$

Example 15. Consider

(39) $$\int_0^1 (2x + 7)^{17} \, dx.$$

Set $y = 2x + 7$; then $dy = 2\, dx$ and $(2x + 7)^{17} \, dx = y^{17}(\frac{1}{2}) \, dy$. Therefore (39) becomes

(40) $$\int_{2(0)+7=7}^{2(1)+7=9} y^{17}(\tfrac{1}{2}) \, dy = (\tfrac{1}{2}) \int_7^9 y^{17} \, dy.$$

The definite integral (40) is in a simple form.

In using a change of variable, one should keep the following points in mind.

a) If an integral is of the form $\int_a^b g(m(x)) m'(x) \, dx$, then setting $y = m(x)$, the transformed integral will be $\int_{m(a)}^{m(b)} g(y) \, dy$.

b) Care must be taken to properly transform the limits of integration, as well as the function being integrated.

c) It is not always clear whether a transformation of variable will effect a simplification, or what transformation should be used. The examples above have been rather carefully chosen so that they will come out nicely. The real world is not nearly so cooperative. The reader should not feel disappointed or ignorant if the evaluation of an integral requires a fair amount of trial and error. Moreover he may expend a good deal of energy in trying to evaluate an integral by one means, only to find that the means simply will not work in the case being investigated.

We now proceed to *integration by parts*. If $f(x) = (gm)(x)$, then $f'(x) = g(x)m'(x) + m(x)g'(x)$;

therefore

(41) $\quad df(x) = f'(x)\, dx = \bigl(g(x)m'(x) + m(x)g'(x)\bigr)\, dx$
$\qquad\qquad = g(x)m'(x)\, dx + m(x)g'(x)\, dx.$

From (41), we obtain

(42) $\quad \int df(x) = f(x) = g(x)m(x) = \int g(x)m'(x)\, dx + \int m(x)g'(x)\, dx.$

If we have an integral of the form

$$\int g(x)m'(x)\, dx,$$

then from (42) we can obtain the equality

(43) $\quad \int g(x)m'(x)\, dx = g(x)m(x) - \int m(x)g'(x)\, dx.$

Equation (43) forms the basis for integration by parts. This method of integration might best be understood through the study of examples.

Example 16. Consider

(44) $\quad \int_1^2 x \ln x\, dx.$

If we set $m(x) = x^2/2$ and $g(x) = \ln x$, then $m'(x) = x$; hence (44) has the form $\int_1^2 g(x)m'(x)\, dx$. Therefore (44) is equal to

(45) $\quad (g(2)m(2) - g(1)m(1)) - \int_1^2 m(x)g'(x)\, dx.$

Since $g'(x) = 1/x$, the integral in (45) becomes

$$\int_1^2 (x^2/2)(1/x)\, dx = \int_1^2 (x/2)\, dx,$$

which is easily evaluated to be $\frac{3}{4}$. We therefore find that

Formula (44) = Formula (45) = $2\ln 2 - (\tfrac{1}{2})(0) - \tfrac{3}{4} = 2\ln 2 - \tfrac{3}{4}$.

Example 17. Consider

(46) $\quad \int_0^1 xe^x \, dx.$

If we set $m(x) = e^x$ and $g(x) = x$, then (46) has the form required to use (43). Therefore (46) equals

$$\bigl(m(1)g(1) - m(0)g(0)\bigr) - \int_0^1 m(x)g'(x)\, dx = e^1(1) - 0 - \int_0^1 e^x\, dx$$
$$= e^1 - e^1 + e^0 = 1.$$

At times an integral may seem to invite evaluation using integration by parts whereas it should really be evaluated using a change of variable. Unfortunately, an integral which looks like one for which a change of variable might be successful might really need an integration by parts. It is, alas, all too possible to encounter an integral for which neither a transformation of variable nor integration by parts will work. The following example illustrates this point.

Example 18. Consider

(47) $\quad \int_0^1 (x^2 + 1)^3 \, dx.$

If we set $y = x^2 + 1$, then $dy = 2x\, dx$. Consequently, $dx = dy/2x$. Thus (47) becomes

(48) $\quad \int_1^2 y^3(dy/2x).$

The x, being a variable rather than a constant, cannot be moved outside the integral sign. There is no way to evaluate (48) unless we can express everything entirely in terms of y. Since $y = x^2 + 1$, however,

$x = (y - 1)^{1/2}.$

Completely in terms of y, (48) is

(49) $\quad \int_1^2 y^3(y - 1)^{1/2} \, dy,$

which is substantially worse than (47). One way of successfully evaluating (47) is to expand $(x^2 + 1)^3$ and integrate the resulting polynomial.

EXERCISES

1. Express each of the following integrals in terms of the new variable given with each one. Be sure that the transformed integral is expressed entirely in terms of the new variable and that the appropriate limits of integration have been found. You need not evaluate the integrals.

a) $\int_0^1 3xe^{x^2}\,dx;\ y = x^2$

b) $\int_{-1}^{-2} (x^2 + 7x + 1)(2x + 7)\,dx;\ y = x^2 + 7x + 1$

c) $\int_4^{16} (x^3 + 7)^{1/2} x^2\,dx;\ y = x^3 + 7$

d) $\int_1^3 \left(\dfrac{2x+3}{x^2+3x}\right) dx;\ y = x^2 + 3x$

e) $\int_0^1 e^{e^x} e^x\,dx;\ y = e^x$

2. Evaluate the following integrals by any suitable means.

a) $\int_0^1 3xe^{x^2}\,dx$ b) $\int_{-1}^{-2} (x^2 + 7x + 1)(2x + 1)\,dx$ c) $\int_0^1 x^2 e^x\,dx$

[Hint: Sometimes integration by parts must be applied twice before a complete answer is arrived at.]

d) $\int_1^3 x^2 \ln x\,dx$

e) $\int_1^2 \ln x\,dx$

[Hint: Consider $\ln x$ as $1 \cdot \ln x$ and use integration by parts.]

f) $\int_0^1 x(x^2 + 3)^{-1}\,dx$ g) $\int_0^1 (x^2 + 2)^3\,dx$

h) $\int_{0.5}^1 e^{\ln x}\,dx$ i) $\int_0^1 xe^{-x}\,dx$

3. Let $C(x) = \cos x$ and $S(x) = \sin x$. (C and S are the *cosine* and *sine* functions, respectively; x is an angle measured in radians.) Then $C'(x) = -\sin x$ and $S'(x) = \cos x$. Find each of the following indefinite integrals.

a) $\int \cos x\,dx$ b) $\int 3\sin x\,dx$ c) $\int 4\sin x \cos x\,dx$

d) $\int \sin x/\cos x\,dx$ e) $\int \sin x(\cos x)^5\,dx$ f) $\int x \sin x\,dx$

g) $\int x \cos x\,dx$ h) $\int x^2 \sin x\,dx$ i) $\int e^x \sin x\,dx$

[Hint: Using integration by parts, obtain an equation that can be solved for $\int e^x \sin x\,dx$.]

j) $\int 5x \sin (x^2)\,dx$ k) $\int \sin^2 x\,dx$

7.5 IMPROPER INTEGRALS. TABLES OF INTEGRALS

If f is a function for which $\int_a^b f(x)\,dx$ exists, then $\int_a^b f(x)\,dx$, as we have seen, can be thought of geometrically as the area under the graph of f for the interval $[a, b]$. Because of the way in which $\int_a^b f(x)\,dx$ was defined, it was

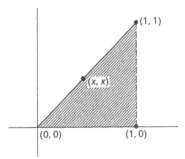

Fig. 10

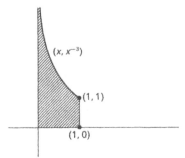

Fig. 11

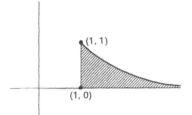

Fig. 12

important that $f(x)$ be defined for each x in $[a, b]$. There are situations in which a function f is not defined at one or both endpoints of $[a, b]$, yet it still seems reasonable to talk about the area under the graph of f for either (a, b), or $[a, b]$. We illustrate this point in several examples.

Example 19. Suppose $f(x) = x$ and we consider only values of x in $(0, 1)$. We are perfectly justified in talking about the area under the graph of f for $(0, 1)$ (Fig. 10), even though $(0, 1)$ is not a closed interval. We would expect this area to be the same as $\int_0^1 x \, dx$. Technically, however, $\int_0^1 f(x) \, dx$ cannot be found from Definition 1, since $f(0)$ and $f(1)$ are not defined.

Example 20. Consider $f(x) = 1/x^3$. Figure 11 depicts the area under the graph of this function for the interval $(0, 1)$. It is not clear whether this area is finite or infinite and $\int_0^1 (1/x^3) \, dx$ is not defined since $f(0)$ is not defined. If y is any point of $(0, 1)$, then

(50) $$\int_y^1 x^{-3} \, dx = -\tfrac{1}{2} - \frac{y^{-2}}{2}.$$

As y approaches 0, (50) gets arbitrarily large. Hence, in this instance, it appears as though the area under the graph in question for $(0, 1]$ were infinite.

Example 21. We could also consider the area under the graph of $f(x) = x^{-3}$ for $x \geq 1$ (Fig. 12). If y is any number greater than 1, then

(51) $$\int_1^y x^{-3}\,dx = -\frac{y^{-2}}{2} + \frac{1}{2}.$$

As y approaches infinity, (51) approaches $\frac{1}{2}$ (since $-y^{-2}/2 = -1/2y^2$ approaches 0 as y approaches ∞). Therefore the area represented in Fig. 12 appears to be finite.

The situations described in Examples 19, 20, and 21 are reminiscent of the situation which inspired the definition of the definite integral, and yet all the conditions necessary to apply Definition 1 were not present. In Examples 20 and 21, we reached conclusions by taking the limit of a definite integral. In Example 19, we could also talk of the limit of $\int_y^{y'} x\,dx$ as y approaches 0 from the right [that is, from inside $(0, 1)$] and y' approaches 1 from the left. This limit is the same as $\int_0^1 x\,dx$. But $\int_y^{y'} f(x)\,dx$ is defined (according to Definition 1) for each y and y' in $(0, 1)$, while $\int_0^1 f(x)\,dx$ is not defined. These considerations lead us to make the following definition.

Definition 4. *Suppose that f is a function defined for each point of (a, b), where a can be $-\infty$ and b can be ∞. Then*

$$\int_a^b f(x)\,dx$$

is defined to be

(52) $$\lim_{\substack{y \to a^+ \\ y' \to b^-}} \int_y^{y'} f(x)\,dx,$$

if this limit exists, where $y \to a^+$ and $y' \to b^-$ mean that y and y' approach a and b, respectively, from inside (a, b).

We call (52), if it exists, the **improper integral** of $f(x)$ from a to b. The improper integral is denoted in the same way as the definite integral, that is, $\int_a^b f(x)\,dx$.

Example 22. The area presented in Example 19 can be found by means of the improper integral

$$\int_0^1 x\,dx = \lim_{y \to 1^-} (y^2/2) - \lim_{y' \to 0^+} (y'^2/2) = \tfrac{1}{2}.$$

Example 23. The improper integral of Example 20 fails to exist, while the improper integral of Example 21 is $\tfrac{1}{2}$.

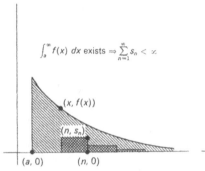
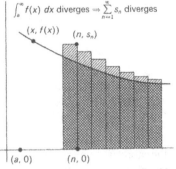

Fig. 13   Fig. 14

If an improper integral exists, we can also express this fact by saying that the integral *converges*. For an improper integral to fail to exist is the same as for the integral to *diverge*.

An improper integral can sometimes be used to obtain information about the convergence or divergence of a series; the following proposition tells how. The proof should be apparent from the figures given with the proposition.

Proposition 5.

a) *Suppose $\sum_{n=1}^{\infty} s_n$ is a series in which each s_n is positive and f is a function such that*

$$f(x) \geq s_n, \quad \text{for } x \text{ in } [n-1, n],$$

for each positive integer $n \geq a$ (Fig. 13). Then if $\int_a^{\infty} f(x)\,dx$ converges, $\sum_{n=1}^{\infty} s_n$ also converges.

b) *Suppose $\sum_{n=1}^{\infty} s_n$ is a series in which each s_n is positive and f is a function such that*

$$s_n \geq f(x) \geq 0, \quad \text{for } x \text{ in } [n-1, n],$$

for each positive integer $n \geq a$ (Fig. 14). Then if $\int_a^{\infty} f(x)\,dx$ diverges, $\sum_{n=1}^{\infty} s_n$ also diverges.

Example 24. Consider $\sum_{n=1}^{\infty} (1/n)^2$. Set $f(x) = (x-1)^{-2}$, $2 \leq x$. The graph of f in relation to the "graph" of the given series is depicted in Fig. 15. Since

$$\int_2^y (x-1)^{-2}\,dx = -(y-1)^{-1} - (-1) = 1 - (y-1)^{-1}$$

and the limit of this expression as $y \to \infty$ is 1, we see that $\int_2^{\infty} (x-1)^{-2}\,dx$ converges; hence $\sum_{n=1}^{\infty} (1/n)^2$ converges.

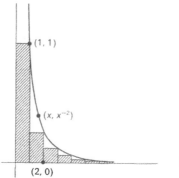

Fig. 15

The so-called *Integral Test*, Proposition 4, can be used to show that $\sum_{n=1}^{\infty} (1/n)^z$ will converge for any value of z greater than 1.

One important method of integration we have not yet considered is the use of tables of integrals. Compared to most calculus texts, this text presents very few methods of integration. We have not, for example, considered integration by partial fractions or the use of trigonometric transformations. There are good reasons for limiting our consideration of integration techniques. First, the integrals the reader will encounter in practical situations, particularly in statistics, are rarely cooperative enough to be readily evaluated using neat integration techniques. Second, the memory of human beings is limited; to use our memory to its best advantage, we should learn those things we will need most often and only know where to locate, if the need arises, those things we hardly use at all. To utilize our memory most efficiently, we shall omit most integration techniques in favor of material that we shall need more frequently. Third, "Tables of Integrals" have been compiled to aid integration. Why spend time deriving what can be looked up rather quickly?

A short table of integrals is given on the endpaper at the back of this text; more complete tables are available in various handbooks. In order to use a table of integrals, we must match the integral we wish to evaluate with the appropriate integral in the table. To obtain such a match a change of variable may be necessary. The following examples illustrate the use of a table of integrals.

Example 25. Consider

(53) $\quad \int_{8}^{9} (x^2 - 4)^{-1} \, dx.$

Looking through the table of integrals for something which resembles

(53), we encounter

(54) $$\int (u^2 - a^2)^{-1} \, du = \frac{1}{2a} \ln\left(\frac{u-a}{u+a}\right) + k.$$

Letting $x = u$ and $a = 2$, we see that (54) becomes the appropriate indefinite integral for evaluating (53). Thus, (53) is equal to $\frac{1}{4}(\ln \frac{7}{11} - \ln \frac{6}{10})$.

Example 26. Consider

(55) $$\int_0^\infty x^3 e^{-x} \, dx.$$

In the table of integrals, we find

(56) $$\int_0^\infty x^{n-1} e^{-x} \, dx = \Gamma(n),$$

where $\Gamma(x)$ is the so-called *Gamma function*. Thus, (55) is $\Gamma(4)$. Tables of values of $\Gamma(x)$ are found in most handbooks of probability and statistics.

Example 27. Consider

(57) $$\int_{-2}^{1} x \sin^2 (3x^2 - 1) \, dx.$$

There is no integral in the table of integrals which looks exactly like (57). However, if we let $u = 3x^2 - 1$, then $du = 6x \, dx$, and, in terms of u, (57) becomes

(58) $$\int_{11}^{2} (\tfrac{1}{6}) \sin^2 u \, du = -(\tfrac{1}{6}) \int_{2}^{11} \sin^2 u \, du.$$

We can now use the indefinite integral $\int \sin^2 u \, du$ found in the table to evaluate (58).

This text does not include much material on the trigonometric functions (sine, cosine, tangent, etc.) or inverse trigonometric functions ($\sin^{-1}$ (sometimes denoted by arcsin), $\cos^{-1}$, etc.), for three reasons: (1) it is assumed that the reader has already studied these functions in previous courses; (2) these functions are not encountered very often in work with probability and statistics; and (3) the inclusion of all topics that might somehow be of use under some circumstances would make for a prohibitively long book. One simply cannot expect to cover all that is usually covered in three semesters of calculus as well as learn a significant amount of statistics in a book designed for a two-semester course. While not apologizing for the omission, we recognize that the omission is there and suggest that the reader consult almost any standard text on trigonometry or calculus to learn more about the trigonometric functions if learning about them seems to be desirable.

EXERCISES

1. Evaluate each of the following improper integrals. If an integral fails to exist, simply write DIVERGES. Do *not* use tables to evaluate these integrals.

   a) $\int_1^\infty (1/x)\, dx$ b) $\int_0^1 x^{-1/2}\, dx$ c) $\int_0^1 \ln x\, dx$ d) $\int_3^\infty x^{-5}\, dx$

   e) $\int_0^1 (x-1)^{-2}\, dx$ f) $\int_{-\infty}^\infty e^{-x}\, dx$ g) $\int_0^\infty e^{-x}\, dx$

2. For what values of w does $\int_0^\infty x^{-w}\, dx$ converge?

3. For what values of w does the series $\sum_{n=1}^\infty (1/n)^w$ converge? For what values of w does this series diverge?

4. Use Proposition 4 to investigate the convergence of each of the following series. If a series converges, use what you know about the integral to find a number M such that the limit of the series is no larger than M.

   a) $\sum_{n=1}^\infty \dfrac{n}{n+1}$ b) $\sum_{n=1}^\infty \dfrac{3}{n^2+1}$ c) $\sum_{n=1}^\infty \dfrac{n}{e^n}$ d) $\sum_{n=1}^\infty n^2 e^{-n}$

5. Use the table of integrals found at the end of this text to evaluate each of the following integrals.

   a) $\int_0^1 2^x\, dx$ b) $\int_1^2 (x^2+9)^{-1}\, dx$

   c) $\int_0^1 (16-x^2)^{-1/2}\, dx$ d) $\int_{0.1}^1 (3x^2 - 0.01)^{-1/2}\, dx$

   e) $\int_{10}^{50} [(3x-1)^2 + 15]^{1/2}\, dx$ f) $\int_0^\infty e^{-x^2}\, dx$

   g) $\int_{-\infty}^\infty e^{-x^2}\, dx$ h) $\int_0^\pi \left(\dfrac{1}{3x\sqrt{x^2+5}}\right) dx$

7.6 NUMERICAL METHODS OF INTEGRATION

At times a definite integral cannot be evaluated by means of either a table of integrals or the usual integration techniques. In most such instances, however, we can make an approximate evaluation, using one of three techniques: *Simpson's Rule*, the *Trapezoidal Rule*, or *Integration by Series*.

The Trapezoidal Rule. Suppose we wish to evaluate

(59) $\int_a^b f(x)\, dx.$

We use the fact that geometrically (59) represents the signed area under the graph of f from a to b. Divide the interval into n equal subintervals each of length $(b-a)/n$. Let $[s_i, r_i]$ be the ith interval; then $s_1 = a$ and $r_n = b$. Over each such interval we build the trapezoid with sides

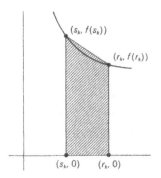

Fig. 16

$f(s_i)$ and $f(r_i)$ as shown in Fig. 16; the area of this trapezoid is

$$(\tfrac{1}{2})(f(s_i) + f(r_i))\left(\frac{b-a}{n}\right).$$

The sum of the area of all n trapezoids built over the n subintervals gives an approximation to (59); this sum reduces to

(60) $\quad \dfrac{1}{2}\left(\dfrac{b-a}{n}\right)(f(s_1) + 2f(r_1) + 2f(r_2) + \cdots + 2f(r_{n-1}) + f(r_n)),$

where we have used the fact that the right endpoint of one interval is the left endpoint of the next. We find that (59) is approximately equal to (60). Generally speaking, the larger the value of n, the better the approximation.

Example 28. We first consider an integral we can evaluate exactly in order to get some idea of how accurate an approximation (60) gives. Consider

(61) $\quad \displaystyle\int_0^1 x^2\, dx.$

We will use $n = 4$. Then

$$\frac{b-a}{n} = \frac{1-0}{4} = \frac{1}{4}.$$

The endpoints of the subintervals are $0, \tfrac{1}{4}, \tfrac{1}{2}, \tfrac{3}{4}$, and 1. Therefore (60) evaluates to

$\tfrac{1}{2} \cdot \tfrac{1}{4}[0^2 + 2(\tfrac{1}{4})^2 + 2(\tfrac{1}{2})^2 + 2(\tfrac{3}{4})^2 + 1^2] = \tfrac{11}{32}.$

The exact value of (61) is $\tfrac{1}{3}$. The error in the approximation is on the order of 3%.

Example 29. Consider

(62) $$\int_0^4 e^{x^2}\, dx.$$

If we let $n = 4$, then $(b - a)/n = \frac{4}{4} = 1$ and the endpoints of the subintervals are 0, 1, 2, 3, and 4. In this case, (60) evaluates to

$(\frac{1}{2})(e^0 + 2e^1 + 2e^2 + 2e^9 + e^{16})$.

Logarithms, or other useful computational devices, can then be used to evaluate this expression.

Simpson's Rule. Again, suppose we wish to evaluate (59), and $[a, b]$ has been divided into n equal subintervals as before, where n is even. Then (59) is approximately equal to

(63) $$\frac{1}{3}\left(\frac{b-a}{n}\right)[f(a) + 4\bigl(f(x_1) + f(x_3) + \cdots\bigr) \\ + 2\bigl(f(x_2) + f(x_4) + \cdots\bigr) + f(b)],$$

where x_1 is the first endpoint of a subinterval after a, x_2 is the second endpoint, x_3 is the third endpoint, etc. None of the x_i are either a or b; they are the endpoints between a and b. *Because of the way in which Simpson's Rule is derived, it is necessary that the number of subintervals be even.** If the same number of subintervals are used to approximate (59) by (60) and by (63), then Simpson's Rule, (63), will in general give the better approximation.

Example 30. We shall evaluate (61) approximately using Simpson's Rule with $n = 4$. In this case, (63) becomes

$\frac{1}{3} \cdot \frac{1}{4}\{0 + 4[(\frac{1}{4})^2 + (\frac{3}{4})^2] + 2(\frac{1}{2})^2 + 1^2\} = \frac{1}{3}$,

which is exactly the value of (61).

Example 31. Evaluating (62) using Simpson's Rule with $n = 4$, we obtain

$(\frac{1}{3})[e^0 + 4(e^1 + e^3) + 2e^2 + e^4]$.

Integration by Series. Suppose the function f of (59) has a power series representation

(64) $$f(x) = \sum_{n=0}^{\infty} a_n x^n,$$

---

* The Trapezoidal Rule was developed using straight-line approximations to the graph of f. Simpson's Rule uses parabolic arcs, each group of three consecutive endpoints determining one arc.

and a and b are both in an open interval in the interval of convergence of (64). Then

(65) $$\int_a^b f(x)\,dx = \sum_{n=0}^{\infty} \int_a^b a_n x^n\,dx,$$

that is, we can integrate the series (64) term by term to get a series which will converge to (59).

Example 32. Consider

(66) $$\int_0^1 e^x\,dx.$$

The Maclaurin series expansion for e^x is

(67) $\quad e^x = 1 + x + x^2/2! + x^3/3! + x^4/4! + \cdots$

Therefore

(68) $$\int_0^1 e^x\,dx = \int_0^1 1\,dx + \int_0^1 x\,dx + \int_0^1 x^2/2!\,dx + \cdots$$
$$= 1 + \tfrac{1}{2} + \tfrac{1}{3!} + \tfrac{1}{4!} + \cdots$$

The last expression in (68) can be recognized to be $e^1 - 1$, which is, in fact, the value of (66). We can sum (68) to as many terms as is necessary to obtain the degree of accuracy we desire.

Example 33. Consider

(69) $$\int_5^7 \frac{e^x}{x}\,dx.$$

Then $e^x/x = 1/x + (1 + x/2! + x^2/3! + \cdots)$ as a result of dividing (67) by x. Therefore

Formula (69) $= \int_5^7 \dfrac{1}{x}\,dx + \int_5^7 1\,dx + \int_5^7 \dfrac{x}{2!}\,dx + \cdots$
$\qquad = (\ln 7 - \ln 5) + 2 + \dfrac{1}{2}\cdot\dfrac{1}{2!}(7^2 - 5^2)$
$\qquad\quad + \dfrac{1}{3}\cdot\dfrac{1}{3!}(7^3 - 5^3) + \cdots$

The sum may be taken to as many terms as desired.

EXERCISES

1. Evaluate each of the following using
    i) the Trapezoidal Rule with $n = 4$;
    ii) Simpson's Rule with $n = 4$; and
    iii) integration by series, if possible, taking the fifth partial sum.

If possible, find the integral exactly and determine which of the methods
(i) through (iii) gives the best approximation.

a) $\int_{-1}^{1} x\, dx$ b) $\int_{-3}^{3} (4x + 1)\, dx$ c) $\int_{0}^{8} (3x^2 + 1)\, dx$

d) $\int_{-1}^{1} x^3\, dx$ e) $\int_{0}^{1} (2 - x)^{-1}\, dx$ f) $\int_{0}^{1} e^{-x^2}\, dx$

8 The Integral and Continuous Variates

8.1 MEASURES OF CENTRAL TENDENCY

Let X be any variate, either discrete or continuous. Then by the *norm* of X we mean the "expected," "typical," or "normative" value of X. Evidently this description of a norm is too imprecise to be of much use computationally. Moreover, we shall see that there is not just one way to measure a normative value of X, but several, and the results obtained by these different methods do not always agree. We shall investigate three measures of normative value: the *mean*, *mode* (or modes), and the *median*. The mean, mode, and median of X are also called *measures of central tendency*, since the distribution of X can be thought of as being "centered" about these values.

Given the distribution of a variate X, we may consider that value of X at which the density function of X has a maximum as being the normative, or typical, value.

Definition 1. *Any admissible value of a variate X at which the density function f of X has a maximum is called a **mode** of X.*

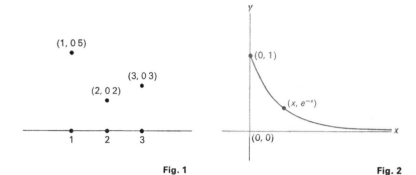

Fig. 1 Fig. 2

If X is a discrete variate, the density function measures the probability of each admissible value of X. A modal value m of X, then, in this instance, is an admissible value which has the greatest probability.

Evidently a random variable can have several modes. In fact, in Example 18 of Chapter 5 and Example 32 of Chapter 3 we find that every admissible value is a mode. The very fact that a random variable can have several modes makes the mode a rather unsatisfactory measure of normative value, or of central tendency. However, where there is only one mode and the value of the density function at that mode is large compared with its value at other admissible values of X, the mode may be a reasonably good measure of normative value.

Example 1. Assume a discrete variate X has admissible range $\{1, 2, 3\}$ and density function f defined by $f(1) = 0.5$, $f(2) = 0.2$, and $f(3) = 0.3$. The graph of f is presented in Fig. 1. The only mode of f is 1. In the long run, X can be expected to assume the modal value about half of the time.

Example 2. Suppose X is a continuous variate with admissible range $[0, \infty)$ and density function $f(x) = e^{-x}$. Then 0 is the only mode of X. Nevertheless, as one can see from the graph of f (Fig. 2), the modal value in this instance does not seem to be a very satisfactory measure of central tendency.

In instances in which there is more than one mode, the mode is rather useless as an indication of an "expected" value of the variate.

Another measure of normative value is the *mean*, or *average*. As motivation for the definition of the mean, we shall look at the way an average is usually obtained.

Example 3. On a certain test given to twenty-five students, five students achieved a score of 75, three students scored 80, ten students scored 85, and seven students scored 90. To compute the average test score, we divide the sum of the scores by the total number of tests. The average which in this case is $(5 \cdot 75 + 3 \cdot 80 + 10 \cdot 85 + 7 \cdot 90)/25$ can be rewritten as

(1) $\quad (\frac{5}{25})75 + (\frac{3}{25})80 + (\frac{10}{25})85 + (\frac{7}{25})90.$

The coefficient of each of the test scores in (1) is the relative frequency with which that particular score occurs.

Example 4. A discrete variate X has admissible values x_1, x_2, and x_3. A total of n trials are performed. In these n trials, x_1, x_2, and x_3 occur n_1, n_2, and n_3 times, respectively. The average value of X for the n trials is

(2) $\quad \dfrac{n_1 x_1 + n_2 x_2 + n_3 x_3}{n} = \left(\dfrac{n_1}{n}\right) x_1 + \left(\dfrac{n_2}{n}\right) x_2 + \left(\dfrac{n_3}{n}\right) x_3.$

Again, n_1/n, n_2/n, and n_3/n are the relative frequencies of x_1, x_2, and x_3, respectively.

If n is "sufficiently large," then the relative frequencies of x_1, x_2, and x_3 will be close to $f(x_1)$, $f(x_2)$, and $f(x_3)$, the respective probabilities of x_1, x_2, and x_3. Thus, as n gets "large," (2) should approach

(3) $\quad f(x_1)x_1 + f(x_2)x_2 + f(x_3)x_3.$

Examples 1 and 2 are intended to serve as motivation for the following definition.

Definition 2. *Let X be a variate with admissible range A and density function f. Then the **mean** (sometimes called the **mean value, expected value,** or **average**) of X is defined to be*

(4) $\quad \displaystyle\sum_{x \text{ in } A} f(x)x \quad$ *if X is discrete,*

and

(5) $\quad \displaystyle\int_A xf(x)\,dx \quad$ *if X is continuous.**

We shall denote the mean of X by m_x. [*The Greek letter μ (mu) is frequently used to denote the mean.*]

Note that (5) is the limiting situation of (4) as the discrete variate "becomes continuous." See Examples 35 and 36 of Chapter 3.

Example 5. The mean of X in Example 1 is

$$(0.5)1 + (0.2)2 + (0.3)3 = 1.8.$$

---

* The mean of a variate may not exist; in particular, m_x will not exist if (4) or (5) diverges. See Exercise 5.

The mean of X in Example 32 of Chapter 3 is

$$(\tfrac{1}{6})(1 + 2 + 3 + 4 + 5 + 6) = 3.5.$$

Example 6. The mean of X in Example 2 is $\int_0^\infty xe^{-x}\, dx = 1$. The mean of X in Example 18 of Chapter 5 is $\int_0^{1/3} x(3)\, dx = \tfrac{1}{6}$.

The mean of X need not be an admissible value of X; thus it is possible for a family to have an average of 2.5 children. The mean and the mode need not be the same, even when there is only one mode. In Example 1, the mean of X, 1.8, is closer to 2 than to the modal value 1.

Whereas a variate can have many modes, it can have only one mean.

Yet another measure of normative value is the *median*.

Definition 3.

a) *Suppose X is a discrete variate with distribution function F. If there is no admissible value for which $F(x) = \tfrac{1}{2}$, then the **median** of X is defined to be the least admissible value x_i of X such that $F(x_i) > \tfrac{1}{2}$. If there is an admissible value x_i of X such that $F(x_i) = \tfrac{1}{2}$, then the **median** of X is defined to be*

$$\frac{x_i + x_{i+1}}{2},$$

where x_{i+1} is the least admissible value of X greater than x_i.

b) *Suppose X is a continuous variate with distribution function F and density function f. Then the **median** of X is defined to be that admissible value M of X such that $F(M) = \tfrac{1}{2}$. Alternatively, M is the median of X if*

$$\int_a^M f(x)\, dx = \tfrac{1}{2},$$

where a is the left endpoint of the admissible range of X.

The median of a variate X is thus a number y such that X can be expected to assume values greater than or equal to y about half the time, and values less than or equal to y about half the time.

Example 7. The median of the variate in Example 1 is 1.5. The median of the random variable in Example 32 of Chapter 3 is $(3 + 4)/2 = 3.5$.

Example 8. Let X be a variate with the set of positive integers as its admissible range and density function f defined by $f(x) = (\tfrac{1}{2})^x$ for any admissible value of X. The distribution function F for X is given by

$$F(w) = \begin{cases} 0 & \text{if } w < 1, \\ \sum_{k=1}^{[w]} (\tfrac{1}{2})^k & \text{if } 1 \leq w, \end{cases}$$

where $[w]$ denotes the greatest integer less than or equal to w. The median of X is $(1 + 2)/2 = 1.5$. The mode of X is 1. The mean m_x of X is given by

(6) $$m_x = \sum_{k=1}^{\infty} k(\tfrac{1}{2})^k,$$

which is 2 (as we can see, using the moment-generating function introduced in Section 8.3).

Example 9. To find the median of X in Example 2, we must solve

(7) $$\int_0^y e^{-x}\, dx = 1 - e^{-y} = \tfrac{1}{2}$$

for y. From (7), we find that $e^{-y} = \tfrac{1}{2}$; multiplying both sides of this last equation by e^y, we find $e^y = 2$. Hence y, the median of x, is $\ln 2$, which is approximately 0.69315.

EXERCISES

1. In each of the following a random variable is given with its admissible range and density function. In each case:
    i) Find the mean, mode (or modes), and median of the variate.
    ii) Determine which measure of central tendency seems most appropriate. Clearly explain the reasons for your choice.
    iii) Draw a graph of the density function and indicate on the graph where the various measures of central tendency occur.
    a) X; admissible range $\{1, 2, 3\}$; $f(1) = 0.4$, $f(2) = 0.2$, $f(3) = 0.4$
    b) X; admissible range $\{1, 2, 3, 4, 5, 6, 7\}$; $f(x) = \tfrac{1}{7}$ for any admissible value of X
    c) X; admissible range $\{1, 2, 3, 4\}$; $f(x) = 0.x$, for each admissible value of X, for example, $f(1) = 0.1$
    d) X; admissible range $\{1, 10, 100\}$; $f(1) = 0.8$, $f(10) = f(100) = 0.1$
    e) Z; admissible range $\{0, 1, 2, 3, 4\}$; $f(z) = C(4, z)(\tfrac{1}{2})^4$ for any admissible value of Z
    f) X; admissible range $[0, 1]$; $f(x) = 1$
    g) X; admissible range $[0, 1]$; $f(x) = 3x^2$
    h) X; admissible range $(-\infty, -1]$; $f(x) = -2x^{-3}$
    i) X; admissible range $[1, e]$; $f(x) = 1/x$

2. The notions of symmetry and skewness were introduced in Section 5.5. We can define *skewness* more formally than we did in Section 5.5, as follows: If X is a variate with a single mode p_x, the X is *skewed to the right* if $p_x > m_x$; if $m_x < p_x$, then X (or the distribution of X) is *skewed to the left*.
    a) Suppose a variate X is symmetric with respect to K. Prove that the probability that X will be less than or equal to K is $\tfrac{1}{2}$.
    b) If a variate X is symmetric about 0, prove that $m_x = 0$.

3. Consider the variate X with admissible range $[-1, 1]$ and density function $f(x) = |x|$. Find the mean, modes, and median of X. Note that the value of the variate has little chance of falling "close" to the mean and median.

4. Suppose X is a continuous variate with admissible range $[a, b]$ and density function $f(x) = 1/(b - a)$. Prove that the mean and median of X coincide. What is the mean of the modes of X?

5. Prove that the mean is *not* defined for each of the following variates.
  a) X; admissible range $\{2^n \mid n = 1, 2, 3, \ldots\}$; density function defined by $f(2^n) = (\frac{1}{2})^n$
  b) X; admissible range $(0, \infty)$; density function defined by $f(x) = 1/(x+1)^2$

8.2 VARIATION FROM THE NORM

Once we select an expected, or normative, value of a random variable X, we are faced with the question: How "typical" is the norm? In other words, are the values of X likely to stay close to, or wander far from, the norm? From a mathematical point of view, we are interested in knowing whether the tendency of a variate to deviate from its norm can be studied quantitatively as well as qualitatively.

In the following examples we see one variate which seems to stay "close" to its mean, and one variate which seems quite "dispersed" about its mean.

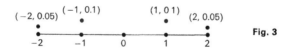

Fig. 3

Example 10. Let X be a variate with admissible range $\{-2, -1, 0, 1, 2\}$ and density function f defined by

$$f(-2) = f(2) = 0.05, \qquad f(-1) = f(1) = 0.1, \qquad f(0) = 0.7.$$

The graph of f is given in Fig. 3. In this example, the mean, mode, and median of X are all 0; hence there can be no debate over which of these norms is best. We see that X is "concentrated" rather heavily at 0; in fact the probability is only 0.1 that X will assume either of the extreme admissible values -2 and 2.

Example 11. Let X be a discrete variate with the set of integers from 1 to 100 as its admissible range and density function f defined by $f(x) = \frac{1}{100}$ for any admissible value of X. Then

$$m_x = (\tfrac{1}{100})(1 + 2 + \cdots + 99 + 100) = 50.5.$$

Since all admissible values of X have the same probability and there are 100 admissible values, we expect the values of X to vary rather widely from the mean.

Let X be a variate with admissible range A, density function f, and norm $\bar{x}$. One measure of the concentration of X about $\bar{x}$ can be obtained as follows: Define $Z = |\bar{x} - X|$. For each admissible value x_i of X, $z = |\bar{x} - x_i|$ is the distance of x_i from $\bar{x}$. Since X is a random variable, so is Z. We can therefore take the mean of Z, m_z (the mean distance of x from $\bar{x}$), as a measure of concentration.

Proposition 1. *If X is discrete, then*

(8) $\qquad m_z = \sum\limits_{x \text{ in } A} |\bar{x} - x| f(x).$

Proof: The admissible range of Z is $\{|\bar{x} - x| \mid x \text{ is in } A\}$. For each admissible value z_i of Z, there are at most two admissible values x_i and $x_{i'}$ of X such that $z_i = |\bar{x} - x_i| = |\bar{x} - x_{i'}|$. Let g be the density function of Z. If x_i is the only admissible value of X such that $z_i = |\bar{x} - x_i|$, then $g(z_i) = f(x_i)$. For, in this case, z_i will occur if and only if x_i occurs. If x_i and $x_{i'}$ are distinct values of x which give z_i, then

$$g(z_i) = f(x_i) + f(x_{i'}).$$

By definition of m_z, we have

(9) $\qquad m_z = \sum\limits_{\substack{z \text{ in} \\ \{|\bar{x}-x| \mid x \text{ in } A\}}} zg(z).$

It is left to the reader to show the equivalence of (8) and (9) from the observations that have been made.

Example 12. In Example 10 all measures of normative value are 0. In that example, the measure of concentration about 0 given by (8) is

$$|0 - (-2)|(0.05) + |0 - (-1)|(0.1) + |0 - 0|(0.7) + |0 - 1|(0.1)$$
$$+ |0 - 2|(0.05) = 0.3.$$

Taking the continuous case as the limit of discrete cases, we have that if X is a continuous variate with admissible range A and density function f, then

(10) $\qquad m_z = \int_A |\bar{x} - x| f(x)\, dx.$

Using (8) or (10) as a measure of concentration about $\bar{x}$ has the disadvantage that the absolute value is rather distasteful from a computational point of view [particularly in (10)]. To avoid using the absolute value, we might define $W = \bar{x} - X$, and use the mean of W to measure the concentration about $\bar{x}$. The mean of W, m_w, is given by

(11) $\quad m_w = \sum_{x \text{ in } A} (\bar{x} - x) f(x) \quad$ (discrete case),

or

(12) $\quad m_w = \int_A (\bar{x} - x) f(x)\, dx \quad$ (continuous case),

where A is the admissible range and f is the density function of X. Unfortunately, (11) and (12) have the most unpleasant property that in the case of the most important norm, the mean, they always give the same answer, 0. We prove this in the following proposition.

Proposition 2. If $\bar{x} = m_x$, then m_w as in (11) and (12) is 0.

Proof: If $\bar{x} = m_x$, then (11) becomes

(13) $\quad \sum_{x \text{ in } A} (m_x - x) f(x) = \sum_{x \text{ in } A} (m_x f(x) - x f(x))$

$\qquad = \sum_{x \text{ in } A} m_x f(x) - \sum_{x \text{ in } A} x f(x)$

$\qquad = m_x \sum_{x \text{ in } A} f(x) - \sum_{x \text{ in } A} x f(x).$

But

$\sum_{x \text{ in } A} f(x) = 1 \quad$ and $\quad \sum_{x \text{ in } A} x f(x) = m_x.$

Therefore (11) reduces to

$m_x(1) - m_x = 0.$

Likewise, with $\bar{x} = m_x$,

$\int_A (\bar{x} - x) f(x)\, dx = \int_A \bar{x} f(x)\, dx - \int_A x f(x)\, dx$

$\qquad = \bar{x} \int_A f(x)\, dx - m_x$

$\qquad = m_x(1) - m_x = 0.$

The most important measure of concentration about a norm $\bar{x}$ of a variate X is the mean, or expected, value of $(\bar{x} - X)^2$. Since $(\bar{x} - X)^2$ is nonnegative for each admissible value of X, its mean value will be nonnegative, and the mean value will be 0 if and only if X has only the norm as an admissible value. The less concentrated X is about $\bar{x}$, the larger the mean of $(\bar{x} - X)^2$ will be. If A is the admissible range and f is the density function of X, then the mean value of $(\bar{x} - x)^2$ is given by

(14) $\quad \sum_{x \text{ in } A} (\bar{x} - x)^2 f(x) \quad$ (discrete case)

and

(15) $\quad \int_A (\bar{x} - x)^2 f(x)\, dx \quad$ (continuous case).

The case where $\bar{x} = m_x$ is particularly important.

Definition 4. *If X is a variate with admissible range A and density function f, then*

(16) $\quad \sum_{x \text{ in } A} (m_x - x)^2 f(x) \quad$ (discrete case),

or

(17) $\quad \int_A (m_x - x)^2 f(x)\, dx \quad$ (continuous case),

*is called the **mean square deviation from the mean**, or the **variance** of X. We shall denote the variance of X by σ_x^2 (σ is the Greek lower case sigma).*

*The positive square root of the variance is called the **standard deviation**. We shall use σ_x to denote the standard deviation.*

As was the case with the mean, there are variates whose variance is not defined. Clearly, if the mean of a variate is not defined, neither will its variance be defined.

Before giving specific examples of the variance, we derive a formula which helps us to compute it.

Proposition 3. *Let X be a variate with admissible range A and density function f. Then*

(18) $\quad \sigma_x^2 = \sum_{x \text{ in } A} x^2 f(x) - m_x^2 \quad$ (discrete case),

and

(19) $\quad \sigma_x^2 = \int_A x^2 f(x)\, dx - m_x^2 \quad$ (continuous case).

(It should be noted that

$$\sum_{x \text{ in } A} x^2 f(x) \quad \text{and} \quad \int_A x^2 f(x)\, dx$$

are the mean of X^2; hence Proposition 3 states that $\sigma_x^2 = m_{x^2} - m_x^2$.)

Proof: We supply the proof for the discrete case. The proof for the continuous case is left as an exercise. Since

$$(m_x - x)^2 = m_x^2 - 2m_x x + x^2,$$

(18) becomes

$$\sum_{x \text{ in } A} (m_x^2 - 2m_x + x^2)f(x)$$

$$= \sum_{x \text{ in } A} m_x^2 f(x) - \sum_{x \text{ in } A} 2m_x x f(x) + \sum_{x \text{ in } A} x^2 f(x)$$

$$= m_x^2 \sum_{x \text{ in } A} f(x) - 2m_x \sum_{x \text{ in } A} x f(x) + \sum_{x \text{ in } A} x^2 f(x)$$

$$= m_x^2 (1) - 2m_x(m_x) + \sum_{x \text{ in } A} x^2 f(x)$$

$$= \sum_{x \text{ in } A} x^2 f(x) - m_x^2.$$

Example 13. Consider Example 32 of Chapter 3 again; we previously found $m_x = 3.5$. Now

$$\sum_{x \text{ in } A} x^2 f(x) = (\tfrac{1}{6})(1^2 + 2^2 + 3^2 + 4^2 + 5^2 + 6^2) = \tfrac{91}{6}$$

and $m_x^2 = \tfrac{49}{4}$. Therefore $\sigma_x^2 = \tfrac{91}{6} - \tfrac{49}{4} = \tfrac{35}{12}$ and σ_x is approximately 1.71. Since the mean value of $(m_x - X)^2 = (3.5 - X)^2$ is approximately 3, we see that X is not concentrated very closely about 3.5. Although σ_x^2 is the mean value of $(m_x - X)^2$, σ_x is not the mean value of $m_x - X$ (recall that we proved that the mean value of $m_x - X$ is always 0).

In the next example we see a variate with the same mean and admissible range as in Example 13, but with a smaller variance.

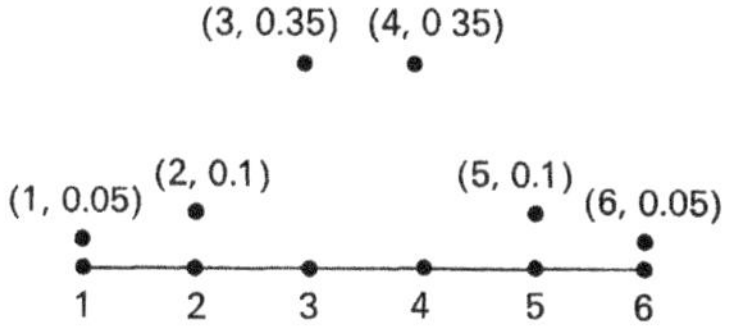

Fig. 4

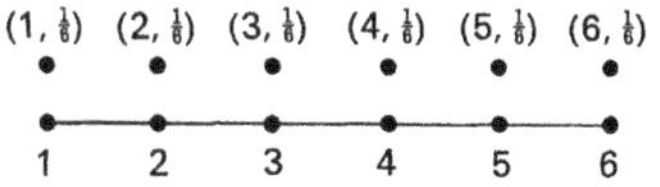

(1, 0.05), (2, 0.1), (3, 0.35), (4, 0.35), (5, 0.1), (6, 0.05)

Fig. 5

$(1, \tfrac{1}{6})$ $(2, \tfrac{1}{6})$ $(3, \tfrac{1}{6})$ $(4, \tfrac{1}{6})$ $(5, \tfrac{1}{6})$ $(6, \tfrac{1}{6})$

Example 14. Let X have admissible range $\{1, 2, 3, 4, 5, 6\}$ and density function defined by

$f(1) = f(6) = 0.05$, $f(2) = f(5) = 0.1$, and $f(3) = f(4) = 0.35$.

Straightforward calculation reveals that $m_x = 3.5$. The graph of f is given in Fig. 4, and the graph of the density function for the variate of Example 13 is given in Fig. 5. From these graphs alone we should expect the variate of this example to be concentrated more closely about 3.5. In this case,

$$\sigma_x^2 = \sum_{x=1}^{6} x^2 f(x) - (3.5)^2 = 0.88.$$

(The variance found in Example 13 was 2.9175.) Here the standard deviation is approximately 0.94 as compared with 1.71 in Example 13.

Example 15. Let X be a continuous variate with admissible range $[0, 1]$ and density function $f(x) = 2x$. Then $m_x = \int_0^1 2x^2 \, dx = \frac{2}{3}$. Therefore

$$\sigma_x^2 = \int_0^1 x^2(2x) \, dx - (\tfrac{2}{3})^2 = \tfrac{1}{2} - \tfrac{4}{9} = \tfrac{1}{18},$$

and $\sigma_x = (\tfrac{1}{18})^{1/2}$.

EXERCISES

1. In each of the following, an admissible range and density function are given for a random variable X. In each case, compute the mean, mode, and median of X. Denote these three values by m_x, p_x, and q_x, respectively. In each case, find the mean values of $m_x - X$, $p_x - X$, and $q_x - X$. For the discrete variates, find the mean value of $|m_x - X|$. In each case, also find the variance and standard deviation.
   a) Admissible range $\{1, 2, 3\}$; $f(1) = f(3) = 0.3$, $f(2) = 0.4$
   b) Admissible range $\{1, 2, 3\}$; $f(1) = f(3) = 0.01$, $f(2) = 0.98$
   c) Admissible range $\{1, 2, 3\}$; $f(1) = 0.2$, $f(2) = 0.3$, $f(3) = 0.5$
   d) Admissible range $\{-2, -1, 0, 1, 2\}$;
      $f(-2) = f(2) = 0.01$, $f(1) = f(-1) = 0.04$, $f(0) = 0.90$
   e) Admissible range $[0, 1]$; $f(x) = 4x^3$
   f) Admissible range $[0, 1]$; $f(x) = nx^{n-1}$, where n is any positive integer
   g) Admissible range $(0, \infty)$; $f(x) = e^{-x}$
   h) Admissible range $[1, e)$; $f(x) = 1/x$

2. Prove Proposition 3 for the continuous case.

3. Let X be a discrete variate with admissible values $1, 2, \ldots, n$, and density function f defined by $f(x) = 1/n$ for each admissible value of X. Find the mean, modes, variance, and standard deviation of X.

4. In (a)–(d) of Exercise 1, find the probability that X will have a value between $m_x - \sigma_x$ and $m_x + \sigma_x$.

5. Assume that X is a variate with mean m_x, and define t_x to be the least positive real number with the property that the probability that X will lie between $m_x - t_x$ and $m_x + t_x$ is greater than or equal to $\frac{1}{2}$. Find t_x for each of the variates in Exercise 1. Does t_x seem to measure the concentration of x about m_x? In particular, answer the following questions about t_x:
   a) Is $t_x = 0$ if and only if X always assumes the value m_x?
   b) For random variables which are concentrated rather closely about their mean, is t_x small (thus indicating that x does not vary much from its mean)?
   c) If X and X' have the same mean m_x, but $t_x \leq t_{x'}$, is X more closely concentrated about m_x than is X'?

8.3 PROBABILITY OF EXTREME VALUES. MOMENT-GENERATING FUNCTIONS

If the variance and mean of a variate can be either computed exactly or estimated, then their values can be used to estimate the probability that the variate will assume a value which deviates from the mean by more than some specified amount. One proposition which gives an estimate of this sort is *Tchebycheff's Lemma*.

Proposition 4. Let X be a variate (either discrete or continuous) with mean m_x and variance σ_x^2. Then if $b > 0$, we have

(20) $\qquad P(|x - m_x| \geq b\sigma_x) \leq 1/b^2$.

We omit the proof of Proposition 4.

Example 16. Suppose X is a variate with $m_x = 0$ and $\sigma_x = 1$. Then $P(|x - 0| \geq 4(1)) = P(|x| \geq 4) = P(x \leq -4 \text{ or } x \geq 4) \leq \frac{1}{16}$.

Example 17. In Example 15, we had $m_x = \frac{2}{3}$ and $\sigma_x = (\frac{1}{18})^{1/2} \doteq 0.24$. By Proposition 4,

$P(|x - \frac{2}{3}| \geq 2(0.24)) \leq \frac{1}{4}$.

In actual fact, $P(|x - \frac{2}{3}| \geq 2(0.24))$ is considerably smaller than $\frac{1}{4}$. Specifically, this probability is $\int_0^{2/3 - 2(0.24)} 2x \, dx \doteq \frac{1}{36}$. The area under the graph of f which represents this probability is pictured in Fig. 6.

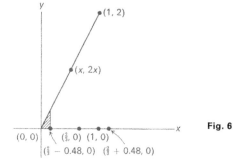

Fig. 6

As a rule, the probability mentioned in Proposition 4 is considerably smaller than $1/b^2$; $1/b^2$ is the very largest that it will be. In the case of certain distributions, we can find a better estimate of the probability than that given by (20) that X will give some "extreme" value. One such instance in which a better estimate is sometimes possible occurs when the variate has only one mode.

Suppose now that X is a variate with a single mode p_x. If $m_x \neq p_x$, then the distribution of X is *skewed* (that is, not symmetric; see Section 5.5). If $m_x > p_x$, then the distribution is *skewed to the right*; if $p_x > m_x$, then the distribution is *skewed to the left*. The degree of skewness can be measured by the quantity

(21) $\quad \dfrac{m_x - p_x}{\sigma_x} = S_k,$

where S_k is known as *Pearson's measure of skewness*. In the event that $m_x = p_x$, we have $S_k = 0$.

We now present another proposition related to the extreme values of a random variable.

Proposition 5. *Suppose X is a unimodal variate with mode p_x. Then, if $b > |S_k|$,*

(22) $\quad P(|x - m_x| \geq b\sigma_x) \leq \dfrac{4(1 + S_k^2)}{9(b - |S_k|)^2}.$

The proof of this proposition is beyond the scope of this text. In the event that $S_k = 0$, that is, if the mean and mode coincide, then we have

(23) $\quad P(|x - m_x| \geq b\sigma_x) \leq 1/(2.25 b^2).$

Expression (23) is known as the *Camp-Meidell inequality*.

Example 18. Consider once more the distribution presented in Example 2. The variate X has a mode at 0 and $m_x = 1$ (Example 6). The variance of x is $\int_0^\infty x^2 e^{-x}\, dx = 2$; hence $\sigma_x = 1$. Then $S_k = (1 - 0)/1 = 1$. Consequently, applying Proposition 4, we have

$P(|x - 1| \geq 5) \leq \frac{1}{25}.$

Applying Proposition 5, we obtain

$P(|x - 1| \geq 5) \leq \dfrac{4(1 + 1)}{9(5 - 1)^2} = \dfrac{8}{9(4^2)} = \dfrac{1}{18}.$

Whether or not (22) will give a better estimate than (20) depends on how large $|S_k|$ is—the smaller $|S_k|$, the better estimate (22) will give—and the size of b—if b is large in relation to $|S_k|$, then (22) will give the better estimate.

Although of some theoretical importance, Propositions 4 and 5 were introduced into this text because of their value in estimating certain probabilities. *Moment-generating functions* are also useful computational devices, and, for this reason, we introduce them now.

Definition 5. Let X be a variate with density function f and admissible range A. Define $M(t; x)$ by

$$M(t; x) = \sum_{x \text{ in } A} e^{tx} f(x) \quad \text{(discrete case)},$$

and

$$M(t; x) = \int_A e^{tx} f(x) \, dx \quad \text{(continuous case)}.$$

(Note that $M(t; x)$ is a function in the variable t.) $M(t; x)$, if it exists, is called the **moment-generating function** of x.

The moment-generating function of a variate and its admissible range completely determine the density function of the variate if the moment-generating function is defined for all t in some open interval which contains 0; that is, no two distinct density functions give the same moment-generating function. The following proposition gives one of the key properties of the moment-generating function.

Proposition 6. If a variate X has moment-generating function $M(t; x)$, then $M^{(n)}(0; x) = m_{x^n}$; that is, the nth derivative of $M(t; x)$ evaluated at 0 gives the mean value of X^n.

Example 19. Consider again the variate X with admissible range $[0, \infty)$ and density function $f(x) = e^{-x}$. Then

$$M(t; x) = \int_0^\infty e^{tx} e^{-x} \, dx = \int_0^\infty e^{(t-1)x} \, dx = 1/(1 - t).$$

We see that

$$M'(0; x) = -(1 - 0)^{-2}(-1) = 1 = m_x \quad \text{and}$$

$$M''(0, x) = m_{x^2} = 2.$$

Consequently, $\sigma_x^2 = 2 - 1 = 1$. This agrees with the results of Example 18.

Example 20. Let X be the discrete variate with the set of nonnegative integers as admissible range and density function $f(x) = (\frac{1}{2})^{x+1}$. Then

$$(24) \quad M(t; x) = \sum_{x=0}^\infty e^{tx}(\tfrac{1}{2})^{x+1} = \sum_{x=0}^\infty \left(\frac{1}{2}\right)\left(\frac{e^t}{2}\right)^x$$

$$= \left(\frac{1}{2}\right) \sum_{x=0}^\infty \left(\frac{e^t}{2}\right)^x$$

$$= \left(\frac{1}{2}\right) \frac{1}{1 - e^t/2} = \frac{1}{2 - e^t}.$$

The fourth equality in (24) comes from the fact that the summation is a geometric series with first term 1 and ratio $e^t/2$. Since
$$M'(0;x) = (2-e^0)^{-2}e^0 = 1,$$
it follows that $m_x = 1$. Since $M''(0;x) = 3 = m_{x^2}$, it follows that $\sigma_x^2 = m_{x^2} - (m_x)^2 = 2$.

The next proposition gives another important property of moment-generating functions.

Proposition 7. *Suppose X is a variate with admissible range A and density function f. Let W be a variate defined by $W = aX + b$, where a and b are constants and $a \neq 0$. Then*
$$M(t;w) = e^{bt}M(at;x).$$

Proof: We prove this proposition for the continuous case. The proof for the discrete case is left as an exercise. By definition,
$$M(t;w) = \int_A e^{t(ax+b)}g(ax+b)\,dx,$$
where g is the density function of W. But $g(ax+b) = f(x)$; hence this integral equals
$$\int_A (e^{tax})(e^{tb})f(x)\,dx = e^{tb}\int_A e^{(at)x}f(x)\,dx = e^{tb}M(at;x).$$

Example 21. Let X be a variate with admissible range A and density function f and set $W = (X - m_x)/\sigma_x = (1/\sigma_x)X + (-m_x/\sigma_x)$. Then
$$M(t;w) = e^{-m_x t/\sigma_x}M(t/\sigma_x;x).$$
Then
$$M'(0;w) = (1/\sigma_x)e^{-m_x 0/\sigma_x}M'(0;x) + M(0;x)(-m_x/\sigma_x)e^{-m_x 0/\sigma_x}$$
$$= (1/\sigma_x)(1)m_x + (1)(-m_x/\sigma_x)(1) = 0.$$
(The proof that $M(0;x) = 1$ is left as an exercise.) This means that, regardless of what the mean of X is, the mean of $(X - m_x)/\sigma_x = 0$. It can also be shown that the variance of $(X - m_x)/\sigma_x = 1$; the proof of this fact is left as an exercise. A variate with mean 0 and variance 1 is said to be a *standard score*; the substitution $W = (X - m_x)/\sigma_x$ is said to be a *reduction to standard score*.

EXERCISES

1. Find the moment-generating functions of each of the following variates with admissible ranges and density functions as given. If possible, use the moment-generating function to find the mean and variance of each variate. For each of the variates which are unimodal, find Pearson's measure of skewness (21).

Use either Proposition 4 or Proposition 5 to obtain an estimate of the probability that the variate and the mean do not differ by more than 5.
a) X; admissible range $(0, 1)$; $f(x) = 1$
b) X; admissible range $(-\infty, 0)$; $f(x) = e^x$
c) X; admissible range = the non-negative integers; $f(x) = (\frac{1}{3})^x$, $x > 0$; $f(0) = \frac{1}{2}$

2. Prove Proposition 6 for the discrete case.

3. Prove that $M(0; x) = 1$ for any variate X for which the moment-generating function exists. [*Hint:* Apply the definition directly with $t = 0$.]

4. Prove directly that if X is a variate with mean m_x and variance σ_x^2, then $W = (X - m_x)/\sigma_x$ is a variate with mean 0 and variance 1. For example,

$$m_w = \int_A ((x - m_x)/\sigma_x)f(x)\,dx,$$

where A is the admissible range and f is the density function of X, and X is assumed to be continuous. Show that the value of this integral is 0. We showed in Example 21 that $m_w = 0$, using moment-generating functions. Use moment-generating functions to show that $\sigma_x = 1$.

5. Find a quadratic equation to be used in finding how small S_k must be for a given b to have the estimate obtained by means of (22) at least as small as that obtained by means of (20).

6. If $S_k = 1$ in (22), will the right-hand side of (22) ever be as small as the right-hand side of (20). If so, how large must b be for this to happen?

9 Some Basic Discrete Distributions

9.1 THE RECTANGULAR AND HYPERGEOMETRIC DISTRIBUTIONS

In this chapter we shall consider certain probability distributions involving discrete variates. The particular distributions we shall consider are important because of the frequency of their occurrence in practical situations. Any variate discussed in this chapter will be assumed to be discrete, unless otherwise noted.

In order to derive any useful information from any variate, either discrete or continuous, it is essential that we know its distribution. The most common means of specifying a distribution is to give the admissible range and density function of the variate. The admissible range of a variate can usually be surmised from the nature of the variate. The density function can be found either experimentally or theoretically. The following example illustrates the experimental approach.

Example 1. Let X be the variate associated with the unbiased roll of a certain die (as in Example 32 of Chapter 3). Then the admissible range of X is $\{1, 2, 3, 4, 5, 6\}$. The density function of X is to be estimated experimentally. The die is rolled 1000 times and the number of occurrences of each admissible value of X is given in the following table:

Table 1

All trials	1	2	3	4	5	6
1000	150	200	100	150	175	125

Taking the relative frequency of each value of X to be its probability, we obtain the density function f for X as defined in Table 2.

Table 2

x	1	2	3	4	5	6
$f(x)$	0.15	0.2	0.1	0.15	0.175	0.125

The density function arrived at experimentally is different from what we would obtain by assuming that all admissible values of X have the same probability. We might ask: Is the die unbiased and the discrepancy due to the working of chance, or can we find in Table 2 a clear indication that the die is biased? We shall consider this type of problem later.

Although the density functions of certain random variables can be estimated experimentally, others can be found using what we have already learned about probability theory. Indeed, since a density function gives the probability of each admissible value of the (discrete) random variable, and since we have already found many such probabilities, we have in fact deduced a good many density functions even though they may, or may not, have been called density functions when first presented. We now begin the study of certain discrete distributions that are important because of their wide application. Some of these distributions will be familiar; others will be new.

Definition 1. *A variate X is said to be **rectangularly distributed**, or have the **rectangular distribution**, if X has admissible range $\{1, 2, \ldots, n\}$ and density function f defined by $f(x) = 1/n$ for each admissible value of X.*

$\left(1, \dfrac{1}{n}\right) \left(2, \dfrac{1}{n}\right) \left(3, \dfrac{1}{n}\right) \qquad\qquad \left(n, \dfrac{1}{n}\right)$

```
• • •         •  •  •  •
├──┼──┼── ··· ─┼──┼──┼──┤
1  2  3                 n
```

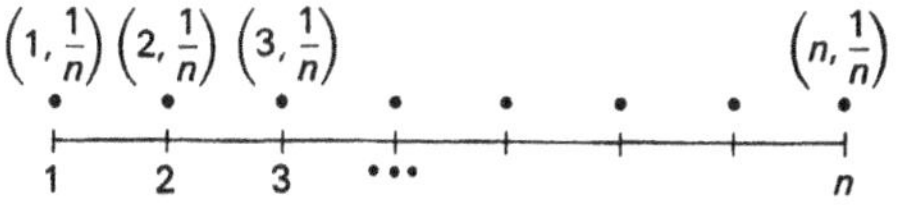

Fig. 1

The use of the term *rectangular* to describe this distribution comes naturally from the form of the graph of the density function (Fig. 1).

Some of the essential characteristics of the rectangular distribution are given in the following proposition.

Proposition 1. Let the variate X be rectangularly distributed. Then:
a) $m_x = (n+1)/2$.
b) $\sigma_x^2 = (n^2 - 1)/12$; hence $\sigma_x = ((n^2 - 1)/12)^{1/2}$.
c) Every admissible value of X is a mode. The median of X is $(n+1)/2$ if n is odd, and $n/2 + \frac{1}{2}$ if n is even.
d) $M(t; x) = \sum_{x=1}^{n} e^{tx} f(x) = \sum_{x=1}^{n} (1/n) e^{tx}$
$= (1/n)(e^t + e^{2t} + e^{3t} + \cdots + e^{nt})$.

Proof: The proofs of (a), (b), and (c) were asked for in Exercise 3 of Section 8.2. We could also have used the moment-generating function (d) of X to derive (a) and (b); we do this now. First note that

$$M'(t; x) = (1/n)(e^t + 2e^{2t} + 3e^{3t} + \cdots + ne^{nt}),$$

and

$$M''(t; x) = (1/n)(e^t + (2^2)e^{2t} + (3^2)e^{3t} + \cdots + (n^2)e^{nt}).$$

Consequently,

$$m_x = M'(0; x) = \left(\frac{1}{n}\right)(1 + 2 + \cdots + n)$$
$$= \left(\frac{1}{n}\right)(n)\left(\frac{n+1}{2}\right) = \frac{n+1}{2},$$

and

$$m_{x^2} = M''(0; x) = \left(\frac{1}{n}\right)(1^2 + 2^2 + \cdots + n^2)$$
$$= \left(\frac{1}{n}\right)\left(\frac{1}{6}\right) n(n+1)(2n+1).$$

Thus,

$$\sigma_x^2 = m_{x^2} - (m_x)^2$$
$$= \left(\frac{1}{n}\right)\left(\frac{1}{6}\right) n(n+1)(2n+1) - \frac{(n+1)^2}{4},$$

which reduces to $(n^2 - 1)/12$.

Another distribution we have previously encountered occurs in the following situation:

(1)   A collection of n objects contains p objects of Type 1 and $n - p$ objects of Type 2. A random selection s without replacement or regard to order of q of the n objects is made.

Definition 2. *The variate $X(0)$ defined as the number of Type 1 objects selected in (1) is said to be **hypergeometrically distributed**, or to have the **hypergeometric distribution**.*

The following example illustrates a variate which is hypergeometrically distributed. Despite the formidable name of the distribution, the reader will recognize the situation as one encountered many times before.

Example 2. Seven blue and nine red balls are in an urn. Six balls are drawn from the urn at random. Let x be the number of blue balls drawn. The blue balls are the Type 1 objects; the Type 2 objects are the red, or non-blue balls. In this instance, x can be any element of $\{0, 1, 2, 3, 4, 5, 6\}$. The probability that exactly x blue balls will be drawn is

$$\frac{C(7, x)C(9, 6 - x)}{C(16, 6)}$$

(see Section 2.2). The density function f is therefore defined by

$$f(x) = C(7, x)C(9, 6 - x)/C(16, 6)$$

for each admissible value of X.

More generally, a hypergeometrically distributed variate X under the conditions described in (1) has density function f defined by

$$(2) \quad f(x) = \frac{C(p, x)C(n - p, q - x)}{C(n, q)}.$$

We may also define a variate X to be hypergeometrically distributed if X has admissible range $\{0, 1, \ldots, n\}$ and density function defined by (2) for some integer p. In Example 2 we had $n = 16$, $p = 7$, and $q = 6$.

We conclude this section with another example involving a hypergeometrically distributed variate.

Example 3. A dealer receives a shipment of 200 widgets. Although it is important that none of the widgets be defective, he does not want to go to the trouble of testing more than 10 of them. How many defective widgets would there have to be in the shipment for the dealer testing ten widgets at random to have a 50% chance of finding one defective? Let p be the number of Type 1 (defective) widgets, and $X(s)$ be the number of defective widgets found in a random sample s of ten from the shipment of 200. Then the density function for X is given by

$$(3) \quad f(x) = \frac{C(p, x)C(200 - p, 10 - x)}{C(200, x)}.$$

The question to be answered is: What must p be in order for $f(1)$ to be at least $\frac{1}{2}$? To arrive at an answer, we must solve

(4) $\qquad \dfrac{1}{2} = \dfrac{C(p, 1)C(200 - p, 9)}{C(200, 10)}.$

We shall not discuss methods of solving (4).

EXERCISES

1. Find the density function of each of the variates described in the following. In each case involving a rectangular distribution, find the mean, median, and variance as well. Observe that a variate X can be defined by specifying its value x, for example, as was done in Example 2.
   a) An integer from 1 to 100 is chosen at random. Let x be the integer.
   b) Three widgets are drawn at random from a shipment of 200. There are 40 defective widgets in the shipment. Let x be the number of defectives drawn.
   c) The Bureau of Useless Statistics says that out of 200 million Americans, 25 million have red hair. An American is chosen at random. Let $x = 0$ if the person selected has red hair, and $x = 1$ if the person selected does not have red hair.
   d) Let x be the number of spades in a fairly dealt bridge hand.
   e) Let a jack, queen, king, and ace have values of 11, 12, 13, and 14, respectively, and let all other cards in a standard deck have their face values. A card is drawn at random from an ordinary deck. Let x be its value.
   f) Let a jack, queen, king, and ace have values of 1, 2, 3, and 4, respectively, and let each of all the other cards have value 0. Let x be the value of a single card drawn at random from a standard deck.
   g) Let x be the number of kings in a fairly dealt poker hand of five cards.

2. An urn contains 5 white, 14 red, and 16 blue balls. We draw 10 balls at random (without replacement) from the urn. Let X and Y be the number of red and the number of blue balls drawn, respectively.
   a) Find the density functions of X and Y.
   b) Find the density function of $Z = X + Y$.

9.2 THE BINOMIAL DISTRIBUTION

The most important discrete distribution, the *binomial distribution*, is encountered in the following situation:

(5)   Any trial of a certain experiment must have one and only one of two outcomes, A and B. (A trial which must have one and only one of two outcomes is sometimes called a *Bernoulli trial*.) Assume $P(A) = p$. The experiment is performed independently n times.

Definition 3. *The variate X whose value x is the number of times that A occurs in the n trials in (5) is said to be **binomially distributed**, or to have the **binomial distribution**.*

The groundwork for the study of the binomial distribution was laid in Section 2.3. The following example illustrates a binomially distributed variate.

Example 4. A fair coin is flipped fairly 10 times. Let x be the number of heads that occur in the 10 flips. Any trial (flip) must have either heads or tails as its outcome. $P(\text{head}) = \frac{1}{2}$. Since tails is the same as not-heads, $P(\text{tail}) = \frac{1}{2}$.

Let x be the number of times that A occurs in the n trials in (5). We shall now derive the density function of X.

Since $P(A) = p$, $P(A + B) = 1$, and $P(A, B) = 0$, we have

$$P(B) = 1 - P(A) = 1 - p.$$

The outcomes of the n trials can be presented by an ordered n-tuple

(6) $(C_1, C_2, \ldots, C_n)$,

where C_1 is the outcome of the first trial, C_2 the outcome of the second trial, etc.* Since the outcome of any one trial in no way affects the outcomes of any other trial—we have assumed that the trials are independent—we have

(7) $P(C_1, C_2, \ldots, C_n) = P(C_1)P(C_2) \ldots P(C_n).$

If A occurs as a coordinate in (6) x times, that is, if A occurs on x of the trials, then B will occur as a coordinate $n - x$ times. Therefore, in such an instance, the factor p will occur x times on the right-hand side of (7) and the factor $1 - p$ will occur $n - x$ times. Consequently, regardless of which x-coordinates of (6) are A's (on which x trials A occurs), the value of (7) will be $p^x(1 - p)^{n-x}$. We have therefore shown:

Proposition 2. *The probability of any one particular string of n trials having A as an outcome in exactly x of those trials is $p^x(1 - p)^{n-x}$.*

Proposition 2 does not give the density function for X. For, although we have calculated the probability of one particular string of n trials

---

* Each of the C_i's must be either A or B; hence there are 2^n such n-tuples (strings of n trials) in all. These strings are not as a rule equiprobable. For example, if $P(A) = 0.9$ and $P(B) = 0.1$, then a string containing all A's is more probable than a string containing all B's.

which contains x A's, there may be many such strings. If there are in fact K strings which contain exactly x A's, then since two distinct strings are mutually exclusive and each string has probability $p^x(1-p)^{n-x}$, applying Proposition 2 of Chapter 1, we find that the probability that some one of these strings will occur is $Kp^x(1-p)^{n-x}$. But the probability that some one of the strings containing x A's will occur is the probability that A will occur x times in the n trials. The density function f for x will therefore have the form $f(x) = Kp^x(1-p)^{n-x}$, where K is an expression which depends on x. We now evaluate K.

The evaluation of K reduces to the problem: In how many ways can we fill exactly x of the n coordinates of (6) with A's? This question is equivalent to: In how many ways can we select x objects from n objects without replacement? The answer is $C(n, x)$. Therefore the density function of x is defined by

(8) $\quad f(x) = C(n, x)p^x(1-p)^{n-x}$

for any admissible value of x.

We may also define a variate X to be binomially distributed if X has admissible range $\{0, 1, \ldots, n\}$ and density function defined by (8).

The right-hand side of (8) is the xth term of the *binomial expansion* of $(p + (1-p))^n$. Since $p + (1-p) = 1$, we have

$$(p + (1-p))^n = \sum_{x=0}^{n} f(x) = 1^n = 1.$$

The next proposition gives certain basic facts about a binomially distributed variate.

Proposition 3. *Let X be a binomially distributed variate. Then:*
a) *If $(n+1)p$ is an integer, then X has modes at both $(n+1)p$ and $(n+1)p - 1$. If $(n+1)p$ is not an integer, then X has a single mode at the largest integer less than $(n+1)p$.*
b) $m_x = np$.
c) $\sigma_x^2 = np(1-p)$; *hence* $\sigma_x = (np(1-p))^{1/2}$.
d) $M(t; x) = \sum_{x=0}^{n} e^{tx} C(n, x) p^x (1-p)^x = \sum_{x=0}^{n} C(n, x)(pe^t)^x(1-p)^x$
   $= (pe^t + (1-p))^n$.

The proof of (d) is its statement; (b) and (c) follow from (d). The proof of (a) is left as an exercise.

Example 5. A fair coin is flipped 12 times. Let x be the number of heads that occur in the 12 flips, A = heads, and B = tails. Then
$P(A) = P(B) = \frac{1}{2}$.

The density function for X is defined by

$$f(x) = C(12, x)(\tfrac{1}{2})^x(\tfrac{1}{2})^{12-x} = C(12, x)(\tfrac{1}{2})^{12} = C(12, x)/4096.$$

Since $(12 + 1)(\tfrac{1}{2}) = 6.5$, x has a single mode at 6. Note, however, that $f(6) = \tfrac{924}{4096}$ is only about $\tfrac{9}{40}$. The mean is 6 and the variance is 3. The standard deviation is about 1.732. Although x should "average out" to 6, the value 6 itself will occur only in about 9 trials in 40.

Example 6. A fair die is rolled fairly ten times. Let x be the number of 5's obtained in the ten rolls. If A is the event of rolling a 5 and $B = \overline{A}$, then $P(A) = \tfrac{1}{6}$ and $P(B) = \tfrac{5}{6}$. The variate X is binomially distributed with density function defined by

$$f(x) = C(10, x)(\tfrac{1}{6})^x(\tfrac{5}{6})^{10-x}$$

for each admissible value of X. Since $(10 + 1)/6$ is not an integer, there is one mode at 1. The mean and variance of X are $\tfrac{5}{3}$ and $\tfrac{25}{18}$, respectively. The distribution of X is skewed to the right.

Example 7. The probability that any one item produced by a certain stamping machine is defective is 0.001. The machine produces 1000 items per day. Let x be the number of defective items the machine will produce on a certain day. Again, X is binomially distributed; here the density function for X is given by

(9) $\quad f(x) = C(1000, x)(0.001)^x(0.999)^{1000-x}.$

The mean and mode of Y are both 1. There is, however, only about a 37% chance that the machine will make exactly one mistake on any given day. The variance is only 0.999.

Example 8. There are 12 blue and 17 red balls in an urn. A ball is drawn at random from the urn, its color is recorded, and the ball is then returned to the urn. This process is repeated 25 times. Let x be the number of red balls drawn; X is binomially distributed. (Note that if each ball drawn were not replaced, then X would be hypergeometrically distributed. The hypergeometric distribution is associated with selection without replacement; the binomial distribution with selection with replacement.) Let A be the draw of a red ball and $B = \overline{A}$. Since for any draw there will be 12 red and 17 blue balls in the urn, $P(A) = \tfrac{12}{29}$ and $P(B) = \tfrac{17}{29}$. The density function for X is given by

$$f(x) = C(25, x)(\tfrac{12}{29})^x(\tfrac{17}{29})^{25-x}.$$

Example 9. A fruitcake company believes that 5 in 12 New Yorkers eat its product. The company decides to ask 100 New Yorkers if they eat its product. The number x of people sampled who do eat the company's fruitcake is at least approximately binomially distributed with density function $f(x) = C(100, x)(\frac{5}{12})^x(\frac{7}{12})^{100-x}$, assuming (1) the company is correct in its assumption, and (2) the sample of 100 people is selected at random.

It requires a fair amount of work to evaluate numerically an expression such as (9) even for small values of x. For example, the evaluation of $f(1)$ in (9) requires raising 0.999 to the 999th power. Needless to say, such lengthy computations can consume a great deal of time and patience without adding significantly to one's understanding of probability theory. For this reason, we have rarely carried examples to their numerical conclusions. Nevertheless, when one is doing a problem for which a numerical solution is imperative, for example, in computing actuarial tables, he is faced with the following alternatives: (a) spending long hours of unpleasant computations tempered somewhat by the use of logarithms and tables, (b) getting a computer to do the computations, or (c) finding a simpler method which will give at least a reasonably good approximation to the answer. In later sections we shall encounter two approximations to the binomial distribution: the *normal* and the *Poisson approximations*.

EXERCISES

1. Find the density function. mean, mode(s), variance, and moment-generating function of each of the variates described in the following:
    a) A fair coin is flipped ten times. Let x be the number of heads that occur.
    b) Let x be the number of heads that occur in ten flips if a fair coin is flipped in such a way that heads are twice as likely as tails.
    c) Let x be the number of heads obtained in ten flips of a coin that is flipped in such a way that the probability of a head on any one flip is 0.9999.
    d) Each of 10 inspectors chooses one widget at random from a shipment of 200 widgets, records whether or not the widget is defective, and then replaces it in the shipment. (It is possible that an inspector may examine a widget which had already been inspected.) Let x be the number of times a defective widget is found, assuming that there are 10 defective widgets in the shipment.
    e) Do part (d) assuming that there are 100 defective widgets in the shipment.

2. Let X be a binomially distributed random variable [with conditions as in (5)]. Which of the following statements are true and which are false? Justify each answer you give.
    a) If n remains constant and p increases, then m_x increases.
    b) If p remains the same, but n increases, then m_x decreases.
    c) For a fixed n, σ_x increases as p increases.
    d) For a fixed p, σ_x increases as n increases.
    e) If X and Y are both binomially distributed variates, each with admissible range $\{0, 1, 2, \ldots, n\}$ and $m_x \leq m_y$, then $f(0) \leq g(0)$, where f and g are the density functions of X and Y, respectively.

3. Answer each of the following:
   a) What must p be if a variate X is to be binomially distributed with mean 5 and admissible range $\{0, 1, \ldots, 9\}$?
   b) What must n be if the variate X is to be binomially distributed with admissible range $\{0, 1, \ldots, n\}$ and $P(x = 0) = P(x = 1) = \frac{1}{2}$?

4. Prove (a) of Proposition 3. [*Hint:* Consider $f(x)/f(x-1)$. Where is this ratio ≥ 1? ≤ 1?]

9.3 DISTRIBUTIONS INVOLVING THE NUMBER OF TRIALS UNTIL SUCCESS

In Sections 9.1 and 9.2 we considered situations in which n trials were performed and the value of the random variate was the number of times a certain event occurred in the n trials. We now consider the following question: Any trial of a particular experiment has one of the events $A_1, A_2, \ldots, A_n$ as an outcome; how many trials must be performed until A_1 first occurs? This question is not well stated. For A_1 might occur on the first trial, or it might not occur until the one-millionth trial; the number of the trial on which A_1 first occurs is the value of a random variable. What we can do, and all we can do, is to compute the probability that A_1 will first occur on the first trial, or first occur on the second trial, and so on. We shall therefore find the density function of the variate X, whose value is the number of the trial on which A_1 first occurs. Evidently, this density function will depend on what type of trials we have. We illustrate now the nature of the variate we will be considering by means of two examples. We will then compute the density function of the random variable corresponding to each of several particular situations.

Example 10. A fair coin is flipped fairly. Let x be the number of the trial on which the first head is observed.

Example 11. Twenty red and 14 blue balls are in an urn. One ball at a time is drawn at random without replacement from the urn. Let x be the number of the draw on which the first red ball is drawn.

We first consider a situation involving selection without replacement as in Example 11.

(10)   There are n objects, m of which are of Type 1, and $n - m$ of which are of Type 2. One object at a time is selected at random without replacement. Let x be the number of the selection on which the first Type 1 object is drawn.

We now find the density function f for the variate X of (10).

If the first Type 1 object is to be selected on the xth draw, then Type 2 objects must be selected on the first $x - 1$ draws. The probability of drawing a Type 2 object on the first draw is $(n-m)/n$. Since there is no replacement of the object drawn, for the second draw we have $n-m-1$ Type 2 objects out of $n-1$ objects in all. Hence the probability of a Type 2 object on the second draw is $(n-m-1)/(n-1)$. The probability of a Type 2 object on the third draw assuming that Type 2 objects have been selected on the first and second draws is $(n-m-2)/(n-2)$. In general, the probability of drawing a Type 2 object on the ith draw, assuming that only Type 2 objects have been drawn on the first $i-1$ draws is

$$\frac{n-m-(i-1)}{n-(i-1)} = \frac{n-m-i+1}{n-i+1}, \qquad i=1,2,\ldots,x-1.$$

Consequently, the probability of drawing a Type 2 object on each of the first $x-1$ draws is

(11) $$\frac{n-m}{n} \cdot \frac{n-m-1}{n-1} \cdot \ldots \cdot \frac{n-m-x+2}{n-x+2}$$

(see Exercise 4, Section 1.4). After $x-1$ Type 2 objects have been drawn, there are $n-x+1$ objects remaining of which m are of Type 1. The probability of drawing a Type 1 object on the xth draw is $m/(n-x+1)$. Using this, together with (11), we find that the probability $f(x)$ that the first Type 1 object will be drawn on the xth draw is

(12) $$f(x) = \frac{n-m}{n} \cdot \frac{n-m-1}{n-1} \cdot \ldots \cdot \frac{n-m-x+2}{n-x+2} \cdot \frac{m}{n-x+1}.$$

An alternative derivation of (12) is as follows: Let A be the event of drawing $x-1$ consecutive Type 2 objects and B_x be the event of drawing the first Type 1 object on the xth draw. Then $A \cap B_x$ is the event of drawing the first Type 1 object on the xth draw, and

$$P(A \cap B_x) = P(A)P(B_x \mid A).$$

Now $P(A)$, the probability of drawing $x-1$ Type 2 objects in as many draws is $C(n-m, x-1)/C(n, x-1)$ (Sections 2.2 and 9.1), and, as we saw above, $P(B_x \mid A) = m/(n-x+1)$. Hence

(13) $$f(x) = \frac{C(n-m, x-1)}{C(n, x-1)} \cdot \frac{m}{n-x+1}.$$

Example 12. The density function of the variate X of Example 11 is

$$f(x) = \frac{C(14, x-1)}{C(34, x-1)} \cdot \frac{20}{34-x+1}.$$

The red balls are taken as the Type 1 objects, the blue, or nonred, balls as Type 2 objects. Here $m=20$. Observe that the admissible range of X is $\{1, 2, \ldots, 14, 15\}$, for the largest number of blue balls that can be drawn is 14.

Example 13. A shipment of 100 widgets contains 10 defective items. One widget at a time is drawn and inspected. Let x be the number of the draw on which the first defective widget is found. In this instance, there are 10 Type 1 objects (defective widgets), and 90 Type 2 objects (the rest of the widgets). In this instance, (13) becomes

$$f(x) = \frac{C(90, x-1)}{C(100, x-1)} \cdot \frac{10}{100-x+1}.$$

The probability that a defective widget will be found on the first draw is $\frac{10}{100} = \frac{1}{10}$. The probability that a defective widget will not be found until the 10th draw is

$$\left(\tfrac{90}{100}\right)\left(\tfrac{89}{99}\right)\left(\tfrac{88}{98}\right) \cdots \left(\tfrac{82}{92}\right)\left(\tfrac{10}{91}\right).$$

The reader will be asked to prove as an exercise that if $m = 1$, that is, if there is only one Type 1 object, then (12) reduces to $f(x) = 1/n$. Thus, when there is only one Type 1 object, the variate X of (10) is rectangularly distributed.

We now consider the situation involving selection with replacement.

(14) Any trial of a certain experiment must have one and only one of two outcomes A and B, and $P(A) = p$. [Then $P(B) = 1 - p$.] The experiment is repeated independently until A occurs. Let x be the number of the trial on which A first occurs.

We derive the density function f for the variate X of (14). If A is to occur first on the xth trial, then we must have $x - 1$ consecutive occurrences of B. Using the results of Section 9.2, we find that the probability of $x - 1$ consecutive occurrences of B is $(1-p)^{x-1}$. The probability that A will occur on the xth trial (or on any trial, for that matter) is p. Therefore the probability $f(x)$ that we will have $x - 1$ consecutive B's followed by an A is

(15) $\qquad f(x) = (1-p)^{x-1} p.$

We can derive (15) alternatively as follows: Any single string of x events which contains $x - 1$ B's and one A has probability $(1-p)^{x-1}p$ (see Proposition 2 of Section 9.2). But since a string having B's as its first $x - 1$ members and A as its last member is a string of this kind, its probability is $(1-p)^{x-1}p$.

Example 14. The density function of the variate X in Example 10 is

$$f(x) = \left(\tfrac{1}{2}\right)^{x-1}\left(\tfrac{1}{2}\right) = \left(\tfrac{1}{2}\right)^x.$$

Note that X has any positive integer as an admissible value.

Example 15. A fair die is rolled fairly. Let x be the number of the roll on which a 2 first occurs. Since $P(2) = \tfrac{1}{6}$ and $P(\overline{2}) = \tfrac{5}{6}$, (15) in this instance has the form

$$f(x) = \left(\tfrac{5}{6}\right)^{x-1}\left(\tfrac{1}{6}\right).$$

The admissible range of X here, as it will be for the variate of (14) in general, is the entire set of positive integers. Although the probability of not rolling a 2 until say the one-billionth roll may be virtually negligible, such an event is not impossible.

Example 16. The probability of a house fire on any given day in a certain town is 0.01. Let x be the number of consecutive days until a fire occurs. Let A be the event of a fire and $B = \bar{A}$; then $P(A) = 0.01$ and $P(B) = 0.99$. Therefore the probability of going x days without a fire, but having a fire on the $(x+1)$-day, is given by

$$f(x) = (0.99)^x (0.01).$$

Definition 4. *A variate X with the set of positive integers as its admissible range and density function as defined in (15) for $0 < p$ is said to be **geometrically distributed**, or to have the **geometric distribution**.*

Some basic facts about the geometric distribution are presented in the following proposition.

Proposition 4. *Let X be a geometrically distributed variate. Then:*
a) *X has a mode at 1.*
b) *$m_x = 1/p$.*
c) *$\sigma_x^2 = (1-p)/p^2$.*
d) *$M(t;x) = \sum_{x=1}^{\infty} e^{tx} f(x) = \sum_{x=1}^{\infty} e^{tx}(1-p)^{x-1} p$*

$$= pe^t \sum_{x=1}^{\infty} (e^t(1-p))^{x-1} = \frac{pe^t}{1-(1-p)e^t}.$$

Proof: The last equality in (d) follows from the fact that

$$\sum_{x=1}^{\infty} (e^t(1-p))^{x-1}$$

is a geometric series with first term 1 and ratio $e^t(1-p)$; (b) and (c) follow from (d). The proof of (a) is left as an exercise.

Example 17. In Example 14, $m_x = 2$ and $\sigma_x^2 = 2$; X has a mode at 1. In Example 15, $m_x = 6$, $\sigma_x^2 = 30$; X has a mode at 1. In Example 16, $m_x = 100$, $\sigma_x^2 = 9900$; X has a mode at 1.

Note that a geometrically distributed variate has the interesting property that its mode always occurs at 1. Thus, no matter how small the probability of A may be, it is always the first trial on which A is most likely to occur first. (We should also point out that it is also the first trial on which $\bar{A}$ is most likely to occur first.)

EXERCISES

1. Find the density function of each of the random variables described in the following. In the case of those variates which are geometrically distributed, find the mean and mode as well.
   a) A fair coin is flipped in such a way that a head is twice as probable as a tail. Let x be the number of the flip on which the first head occurs.
   b) A fair die is rolled fairly. Let x be the number of the first roll on which neither a 2 nor a 3 occurs.
   c) A bag contains slips numbered 1 through 100. The slips are drawn from the bag without replacement one at a time. Let x be the number of the draw on which slip "1" is drawn.
   d) You and 49 others are in a room waiting to be examined for possible jury duty. The prospective jurors are selected at random one at a time. Let x be the number of people questioned *before* your turn comes. Would the density function be applicable if, instead of being selected at random, the prospective jurors were questioned in alphabetic order?
   e) A bag contains five red, six blue, and one white ball. One ball at a time is drawn at random from the bag, its color is recorded, and the ball is then returned to the bag. Let x be the number of draws before the white ball is picked.

2. Show that if $p = 1$ in (10), then X is rectangularly distributed. [*Hint:* If $m = 1$, show that (12) reduces to $f(x) = 1/n$.] The mean of a rectangularly distributed variate X is $(n + 1)/2$. If m is greater than 1 in (10), will m_x be greater or less than $(n + 1)/2$?

3. Suppose that any trial of a certain experiment must have one and only one of two outcomes A and B. Let x be the number of the trial on which A occurs for the mth time if the experiment is performed independently. Prove that the density function f for X is given by

   (17) $\quad f(x) = C(x - 1, m - 1)p^{m-1}(1 - p)^{x-1-(m-1)}p$
   $\qquad\qquad = C(x - 1, m - 1)p^m(1 - p)^{x-m}.$

   Such a variate X is said to have the *negative binomial distribution* with parameters n and p. What is the admissible range of X? the mean of X? Find the density functions of the variates described in each of the following:
   a) An urn contains 100 slips labeled 1 through 100. One slip at a time is drawn at random and then replaced. Let x be the number of the draw on which the third even number is drawn.
   b) A fair coin is flipped in such a way that a head has a 0.99 chance of occurring. Let x be the number of the flip on which the second head occurs.
   c) Two fair dice are rolled together fairly. Let x be the number of the roll on which the second 7 occurs.

4. Prove that a geometrically distributed variate always has a mode at 1. Where does the variate of (11) have a mode? Where does the variate X of Exercise 3 have a mode? Does the mode seem to be a meaningful norm in these instances?

9.4 THE POISSON DISTRIBUTION

There are times when one may be interested in the number of occurrences of a particular type of event in a certain length of time. The following examples illustrate such situations in which the number of occurrences is a random variable.

Example 18. Let x be the number of phone calls placed during time t in a certain booth in New York City.

Example 19. Let x be the number of radioactive emissions in time t from a one-ounce source of pure radium.

Example 20. Let x be the number of cell reproductions which take place in a certain organism in time t.

*Definition 5. We call a situation a **Poisson process** if*

a) *independent events are occurring at random in time;*
b) *the probability of an occurrence of one of the events during an arbitrarily small interval of time dt is equal to μdt, where μ is some constant (which depends on the process);*
c) *provided the interval of time dt is small enough, then the probability of more than one occurrence in time dt is negligible.*

In Example 19, the conditions of Definition 5 are fulfilled at least approximately. We may assume that the individual emissions are independent (although this may not be strictly the case), that the probability of an emission during an interval of time dt is directly proportional to dt if dt is small enough, and that the probability of more than one emission in time dt is negligible if dt is small enough. Actually, the assumptions of Definition 5 work well only over a fairly short run, for if the radium is allowed to decay, then the constant μ will be getting smaller and smaller. That is, the smaller the quantity of radium, the smaller the probability of an emission.

In Examples 18 and 20, the assumptions of Definition 5 may be at least approximately true in the short run. In Example 18, however, the factor μ would be a function of the time of day—fewer calls would be made late at night, presumably, than during the lunch hour. In Example 20, the factor μ would tend to increase with the number of cells. In the short run, it may be impossible to consider μ as approximately constant.

Definition 5. Let x (the value of X) be the number of occurrences during time t in a Poisson process. Then X is said to have the **Poisson distribution.** The density function for the Poisson distribution is given by

(16) $$f(x) = \frac{(\mu t)^x}{x!} e^{-\mu t}.$$

Alternately, a variate X with the set of nonnegative integers as admissible range and density function (16) is said to have the Poisson distribution.

Note that the admissible range of X is the set of nonnegative integers. Observe too that t in (16) is a fixed interval of time and μ is a constant which depends on the process, but which can be approximated if it is assumed or established that a variate has the Poisson distribution. We shall make no attempt to justify (16), although its derivation is not beyond the scope of this text.

The following proposition gives some of the basic properties of the Poisson distribution.

Proposition 5. Let X be a variate with the Poisson distribution. Then:

a) $m_x = \mu t$.
b) $\sigma_x^2 = \mu t$.
c) $M(x;t) = \sum_{x=0}^{\infty} e^{tx} f(x) = \sum_{x=0}^{\infty} e^{tx} \frac{(\mu t)^x}{x!} e^{-\mu t}$
$= e^{-\mu t} \sum_{x=0}^{\infty} \frac{(e^t \mu t)^x}{x!} = e^{-\mu t}(e^{e^t \mu t})$
$= e^{-\mu t(1-e^t)}$.

Proof: The next-to-last equality in (c) follows from the fact that

$$\sum_{x=0}^{\infty} \frac{(e^t \mu t)^x}{x!}$$

is the Maclaurin series expansion for $e^{e^t \mu t}$. Parts (a) and (b) can be derived from (c).

It is left as an exercise to find the mode of a variate with the Poisson distribution.

If a variate X has the Poisson distribution, then μ can be estimated using (a) of Proposition 6. This is because we can arrive at an estimate of m_x by counting the number of occurrences in a unit interval of time. We shall discuss the estimation of means and variances in a later chapter.

One useful property of the Poisson distribution is that, under certain conditions, it is a fairly good approximation to the binomial distribution. Specifically, if in a binomial distribution p is quite small (≤ 0.1) and n is relatively large but still small enough so that $np \leq 5$, then the Poisson distribution can be used to approximate the binomial distribution; in (16), we set $\mu = np = m_x$, and $t = 1$.

Example 21. We will use (16) to approximate $f(2)$ in Example 7. In that example, $m_x = 1$. Therefore $f(2)$ is approximately equal to

$$(\tfrac{1}{2}!)e^{-1} = (\tfrac{1}{2})e^{-1},$$

which is approximately 0.2171.

Example 22. A pair of unbiased dice is rolled 10 times. What is the probability that exactly three of the rolls will be 2's? Since $P(2) = \tfrac{1}{36}$, the probability asked for is

(17) $\quad C(10, 3)(\tfrac{1}{36})^3(\tfrac{35}{36})^7.$

To estimate this probability using the Poisson distribution, we set

$\mu = np = 10(\tfrac{1}{36}) = \tfrac{5}{18}$. The approximation then is

(18) $\quad \dfrac{(\tfrac{5}{18})^3}{3!} e^{-5/18}.$

Using a table for the values of e^{-x}, one finds that (17) is approximately 0.003.

EXERCISES

1. Evaluate (17) by means of either logarithms or a table and compare its value with the value of (18).

2. Find the mode of a variate X with density function (16). [*Hint:* Compute $f(x)/f(x+1)$. Where is this ratio ≤ 1? $= 1$? ≥ 1? What does this tell us about the location of the mode?]

3. Is the Poisson distribution skewed to the right or left? Compute Pearson's measure of skewness for this distribution.

4. Prove $\sum_{x=0}^{\infty} ((\mu t)^x/x!)e^{-\mu t} = 1$. [*Hint:* Factor $e^{-\mu t}$ from the sum. What is the sum of the Maclaurin series that remains?]

5. Use the Poisson approximation to the binomial distribution to estimate each of the following probabilities. In several instances, also compute the true probability and compare it with the approximation.

    a) Suppose that the average technical text contains one misprint every 10 pages. What is the probability that a certain technical text has 50 consecutive pages without a misprint? has a misprint on each of its first 50 pages?

    b) The probability of a major fire in a certain city on any given day is 0.06. What is the probability of having no major fire for 10 consecutive days? of having three fires in a period of 100 days? of having a full year without a fire?

6. Which of the following variates would probably have the Poisson distribution? Carefully explain your decision in each case.
   a) The number of "shooting stars" that one would observe in one hour of night viewing
   b) The number of customers a certain store had between 11:00 a.m. and 12 noon of a business day
   c) The number of days of the year that a certain family will have steak for dinner
   d) The number of 6's that will be rolled in one hour by a machine which rolls a certain pair of dice once every 30 seconds.

10 Other Important Distributions

10.1 THE NORMAL DISTRIBUTION

Recall the following aspects of the binomial distribution: Any trial of a particular experiment must have one and only one of the events A and B as its outcome and $P(A) = p$. The experiment is performed independently n times; the value of the variate X is the number of times that A occurs in the n trials. Define

$$Y = \frac{X - np}{\sqrt{np(1-p)}} = \frac{X - m_x}{\sigma_x}.$$

We recognize Y as a variate with mean 0 and standard deviation 1 (see Exercise 4, Section 8.3). It can be proved that

$$P(a \leq y \leq b) \quad \text{approaches} \quad \int_a^b (1/\sqrt{2\pi}) e^{-y^2/2} \, dy$$

as $n \to \infty$. Thus the density function of Y approaches

(1) $\quad f(y) = (1/\sqrt{2\pi}) e^{-y^2/2}$

as $n \to \infty$.

If X is any variate, then $a \leq x \leq b$ if and only if
$$\frac{a - m_x}{\sigma_x} \leq \frac{x - m_x}{\sigma_x} = y \leq \frac{b - m_x}{\sigma_x}.$$
Therefore if X is binomially distributed with n rather large, we have

(2) $\quad P(a \leq x \leq b) = P\left(\frac{a - m_x}{\sigma_x} \leq y \leq \frac{b - m_x}{\sigma_x}\right)$

$$\doteq \int_{(a-m_x)/\sigma_x}^{(b-m_x)/\sigma_x} \left(\frac{1}{\sqrt{2\pi}}\right) e^{-y^2/2}\, dy.$$

We can use (2) to approximate solutions to problems involving the binomial distribution. There are, fortunately, extensive tables available to help us evaluate (2).

If we want to approximate the probability of one particular value of a binomially distributed value X by means of (2), that is, if $a = b$, then we substitute $a - \frac{1}{2}$ for a and $a + \frac{1}{2}$ for b.

Example 1. Suppose X is a binomially distributed variate with $n = 100$ and $p = 0.1$. We shall approximate $P(0 \leq x \leq 10)$ using (2). Here $m_x = (100)0.1 = 10$ and $\sigma_x = (100(0.1)(0.9))^{1/2} = 3$. Hence the limits of integration in (2) will be $(0 - 10)/3 = -\frac{10}{3}$ and $(10 - 10)/3 = 0$. In this instance, (2) becomes

(3) $\quad \int_{-10/3}^{0} (1/\sqrt{2\pi}) e^{-y^2/2}\, dy.$

Now $f(x)$ in (1) is symmetric with respect to 0 since $f(x) = f(-x)$; hence

(4) $\quad \int_{-a}^{a} f(x)\, dx = 2 \int_{-a}^{0} f(x)\, dx = 2 \int_{0}^{a} f(x)\, dx.$

Using (4), we can go to a table for the normal distribution to find that (3) is approximately 0.4996.

Example 2. Let X be binomially distributed with $n = 1600$ and $p = \frac{1}{2}$. Then $m_x = 800$ and $\sigma_x = 20$. We shall approximate the probability that $x = 800$. As limits of integration in (2), we have

$(799.5 - 800)/20 = -0.025$ and $(800.5 - 800)/20 = 0.025$.)

Thus (2) becomes

(5) $\quad \int_{-0.025}^{0.025} (1/\sqrt{2\pi}) e^{-y^2/2}\, dy.$

Using a table, we find that (5) evaluates to approximately 0.02.

Definition 1. The **general** form of the density function for a **normally distributed** variate X is

(6) $$g(x) = \left(\frac{1}{\sigma_x \sqrt{2\pi}}\right) e^{-(1/2)[(x-m_x)/\sigma_x]^2},$$

where m_x and σ_x are in fact the mean and standard deviation of X, respectively; X has the set of real numbers as its admissible range. The change of variable $Y = (X - m_x)/\sigma_x$ puts (6) in the form (1). A variate with density function (1) is said to have the **standard normal distribution.**

Any normally distributed variate must be put in standard form before we can use tables as a computational aid since the tables are almost invariably compiled for the standard normal distribution.

The general normal distribution (6) has mean, mode, and median at m_x and is symmetric with respect to m_x; its variance is σ_x^2. The proof of these facts is left as an exercise.

Proposition 1. If x is a standard normally distributed variate, then

$$M(t; x) = e^{t^2/2}.$$

Proof:

$$M(t; x) = \int_{-\infty}^{\infty} (1/\sqrt{2\pi})(e^{ty} e^{-y^2/2})$$
$$= e^{t^2/2} \int_{-\infty}^{\infty} (1/\sqrt{2\pi}) e^{-(y-t)^2/2} \, dy$$
$$= e^{t^2/2} \int_{-\infty}^{\infty} (1/\sqrt{2\pi}) e^{-x^2/2} \, dx$$

(using the change of variable $x = y - t$)

$$= e^{t^2/2}(1) = e^{t^2/2}.$$

One of the most important theorems in all probability theory is the so-called *Central Limit Theorem.* We state this theorem without proof in the next proposition.

Proposition 2. Let $X_1, X_2, \ldots, X_n$ be n independent variates which have arbitrary distributions but for which the means $m_{x_1}, \ldots, m_{x_n}$ and variances $\sigma_{x_1}^2, \ldots, \sigma_{x_n}^2$ all exist. As n goes to infinity, the variate

$$Y = X_1 + X_2 + X_3 + \cdots + X_n$$

will tend to be normally distributed with $m_y = m_{x_1} + \cdots + m_{x_n}$ and $\sigma_y^2 = \sigma_{x_1}^2 + \cdots + \sigma_{x_n}^2$.

Note that Proposition 2 makes no assumptions about the distributions of the X_i except that their means and variances exist. How good an approximation

$$f(y) = (1/\sqrt{2\pi})e^{-y^2/2}$$

is to the true density function of Y is dependent on the size of n and on the density functions of the X_i.

Proposition 2 is the Central Limit Theorem in one of its more general forms. One of the most important applications of the Central Limit Theorem involves the distribution of sample norms.

Definition 2. *The collection of all observable values of some numerically measured characteristic is called a* **population.** *A numerical quantity associated with an entire population is called a* **population parameter.**

Suppose we are given a sample of n observations $x_1, x_2, \ldots, x_n$ from a certain population. Then the **sample mean** (or **average**) $\tilde{x}$ is defined by

(7) $\quad \tilde{x} = (1/n)(x_1 + x_2 + \cdots + x_n)$.

The **sample variance** $s_{\tilde{x}}^2$ is defined by

(8) $\quad s_{\tilde{x}}^2 = ((\tilde{x} - x_1)^2 + (\tilde{x} - x_2)^2 + \cdots + (\tilde{x} - x_n)^2)/n$.

If we define the variate X by $X(x) = x$, where x is a randomly selected member of the population (remember that a population is a set of numerical quantities), then m_x and σ_x^2 are said to be the **population mean** *and* **population variance,** *respectively* (m_x and σ_x^2 are examples of population parameters).

If S is the set of all random samples of n observations from a population P, then $\tilde{X}$ defined by $\tilde{X}(s) = $ sample mean of s for each s in S is a random variable. The following proposition (the proof of which is omitted) gives the mean and variance of $\tilde{X}$.

Proposition 3. $m = m_x$, *the population mean, and* $s_{\tilde{x}}^2 = (1/n)\sigma_x^2$, *where σ_x^2 is the population variance.*

If the variate X defined in Definition 2 relative to a certain population P has a certain distribution, for example, the normal distribution, then we say that the population has the same distribution as X; for example, if X is normally distributed, we have a normally distributed population. Even if X itself is not normally distributed, the Central Limit Theorem tells us that the distribution of $\tilde{X}$ is approximately normal if n is large enough. We emphasize this fact in the following proposition.

Proposition 4. *If X is a variate defined as in Definition 2 relative to a population P (with any distribution) such that the mean and variance of X exist,* then $\bar{X}$ is approximately normally distributed with mean m_x and variance σ_x^2/n (provided that n is large enough). Moreover,

$$\frac{\bar{X} - m_x}{\sigma_x/\sqrt{n}}$$

is approximately a standard normal variate.

Example 3. A study is made of the I.Q.'s of a certain group of 5000 people. The I.Q.'s of the individuals studied form the population. Five people are chosen at random from the 5000 and found to have I.Q.'s 104, 105, 90, 96, and 110. The sample mean in this instance is

$(104 + 105 + 90 + 96 + 110)/5 = 101,$

and the sample variance is

$(104 - 101)^2 + (105 - 101)^2 + \cdots + (110 - 101)^2 = 252/5.$

Let x be the I.Q. of a randomly selected subject from the test group. The distribution of the variate X thus defined (which is the variate X of Definition 2 applied to this particular situation) can be calculated using relative frequencies. Specifically, if n is the number of people who have the admissible value x of X as an I.Q., then we set $f(x) = n/5000$; f is the density function of X. The mean and variance of X are the population mean and variance. The population mean m_x is in fact nothing more than the average I.Q. of the members of the test group (obtained by adding all the members of the population, the I.Q.'s of the 5000 people, and dividing by 5000) and σ_x^2 is the sum of the terms $(x - m_x)^2/5000$, where x is taken over the entire population.

Assume $m_x = 100$ and $\sigma_x^2 = 100$. If we take samples of 100 people, then the sample mean $\bar{X}$ will be approximately normally distributed with mean 100 and standard deviation $10/\sqrt{100} = 1$. In this situation, then, the density function for $\bar{X}$ will approximate

$f(\tilde{x}) = (1/\sqrt{2\pi}) \exp\left[-(\tfrac{1}{2})(\tilde{x} - 100)^2\right].$*

We now compute the probability that the sample mean of 100 randomly chosen representatives of the group will be between 80 and 95. This probability is

(9) $\quad \int_{80}^{95} (1/\sqrt{2\pi}) \exp\left[-(\tfrac{1}{2})(\tilde{x} - 100)^2\right] d\tilde{x}.$

Since tables can only be used to find values of integrals involving the standard normal distribution, we must make the change of variable

---

* $\exp x$ is the same as e^x.

$Y = \tilde{X} - 100$ to bring (9) into standard normal form. With the change of variable as indicated, we obtain

(10) $\quad \int_{-20}^{-5} (1/\sqrt{2\pi})e^{-(1/2)y^2} \, dy.$

The value of (10) in turn is approximately $0.5 - \int_{-5}^{0} (1/\sqrt{2\pi})e^{-y^2/2} \, dy$. From a table we find that (10), for all intents and purposes, is 0. Hence we conclude that it is very unlikely that the sample mean will lie between 80 and 95.

Example 4. The average weight of the citizens of a certain city of 10,000 is 140 pounds; the standard deviation is 16. A sample of 100 people is selected at random. What is the probability that the sample mean will be more than 141? The sample mean $\tilde{X}$ (for samples of 100) is at least approximately normally distributed with mean 140 and standard deviation $\frac{16}{10} = 1.6$.* Therefore the probability asked for is given by the integral

(11) $\quad \int_{141}^{\infty} (1/(0.4)\sqrt{2\pi})e^{-(1/2)((x-140)/0.4)^2} \, dx.$

We can transform (11) to standard form by the change of variable

$y = (x - 140)/1.6.$

With this transformation, (11) becomes

(12) $\quad \int_{0.625}^{\infty} (1/\sqrt{2\pi})e^{-y^2/2} \, dy = 0.5 - \int_{0}^{0.625} (1/\sqrt{2\pi})e^{-y^2/2} \, dy.$

We can now evaluate (12), using Table 2 of the Appendix; its value is approximately 0.27.

We cite the following fact about the normal distribution without proof.

Proposition 5. *If $X_1, X_2, \ldots, X_n$ are independent normal variates, and*

$Z = a_1 X_1 + a_2 X_2 + \cdots + a_n X_n + c,$

where the a_i and c are constants, then Z is a normally distributed variate with mean $m_z = a_1 m_{x_1} + a_2 m_{x_2} + \cdots + a_n m_{x_n} + c$ and variance

$\sigma_z^2 = (a_1 \sigma_{x_1})^2 + \cdots + (a_n \sigma_{x_n})^2.$

Example 5. If X_1 and X_2 are independent normally distributed variates, both with mean 0 and standard deviation 1, then $Z = X_1 + X_2$ is normally distributed with mean 0 and standard deviation $(1^2 + 1^2)^{1/2} = \sqrt{2}$. Since Z is normally distributed, $Z/\sqrt{2}$ is normally distributed and is a standard normal variate.

---

* If the original variate X happens to be normally distributed, then the sample mean $\tilde{X}$ will in fact be normally distributed as well. (This can be proved easily using Proposition 5.)

We saw at the beginning of this section that if X is binomially distributed variate, then if n is sufficiently large, $Y = (X - m_x)/\sigma_x$ is approximately a standard normal variate. As an example, we now assume the Central Limit Theorem and prove this fact.

Example 6. Assume that n independent trials are performed, each trial must have either A or B (but not both) as its outcome, and $P(A) = p$. If s is a string of n trials, define

$$X_i(s) = \begin{cases} 1 & \text{if } A \text{ occurs on the } i\text{th trial,} \\ 0 & \text{if } A \text{ does not occur on the } i\text{th trial,} \end{cases}$$

$i = 1, 2, \ldots, n$. Then $m_{x_i} = p$ and $\sigma_{x_i}^2 = p(1-p)$ for $i = 1, 2, \ldots, n$. The variate

$$X = X_1 + X_2 + \cdots + X_n$$

has as its value $X(s)$, the number of times A occurs in s, where s is any string of n trials. Thus X is binomially distributed. But since X is the sum of n independent variates, the distribution of X is approximately normal with mean np and variance $np(1-p)$ if n is large enough. Since $m_x = np$ and $\sigma_x^2 = np(1-p)$, then $Y = (X - m_x)/\sigma_x$ is approximately a standard normal variate.

Example 7. Suppose there are one million college professors in the United States out of a total population of 200 million. Then the probability that a randomly selected citizen of the United States is a college professor is 1/200. The number Y of college professors in a random selection of 200 citizens is approximately normally distributed with mean 1 and variance approximately 1. The probability that a random selection of 200 citizens will contain three or more college professors is approximately

$$\int_3^\infty (1/\sqrt{2\pi}) e^{-(x-1)^2/2} \, dx = \int_2^\infty (1/\sqrt{2\pi}) e^{-y^2/2} \, dy \doteq 0.0228.$$

EXERCISES

1. Use Table 2 of the Appendix to evaluate each of the following integrals.

a) $\int_0^3 \left(\dfrac{1}{\sqrt{2\pi}}\right) e^{-x^2/2} \, dx$ b) $\int_{1.5}^\infty \left(\dfrac{1}{\sqrt{2\pi}}\right) e^{-y^2/2} \, dy$

c) $\int_{-1}^1 \left(\dfrac{1}{\sqrt{2\pi}}\right) e^{-y^2/2} \, dy$ d) $\int_{-\infty}^4 \left(\dfrac{1}{\sqrt{2\pi}}\right) e^{-z^2/2} \, dz$

e) $\int_0^1 \left(\dfrac{1}{2\sqrt{2\pi}}\right) \exp\left[-\left(\dfrac{1}{2}\right)\left(\dfrac{x-3}{2}\right)^2\right] dx$

f) $\int_{0.5}^{1} \left(\frac{1}{9\sqrt{2\pi}}\right) \exp\left[-\left(\frac{1}{2}\right)\left(\frac{x-3}{9}\right)^2\right] dx$

2. Find each of the probabilities asked for in the following. Where the distribution involved is binomial or approximately binomial, use the normal approximation to the binomial distribution to obtain your answer.
   a) A variate X is normally distributed with mean 6 and variance 9. Find the probability that x will lie somewhere between 5 and 7.
   b) There are 90 million overweight people in the United States. If the population of the United States is 200 million people, what is the probability that more than half of a group of 100 randomly selected people will be overweight?
   c) What is the probability of rolling exactly sixty 2's in 2000 tosses of a pair of unbiased dice?
   d) A shipment of 100 lambchops is received by a market. The mean weight of the chops is 3 ounces with a variance of 1. A customer buys 10 chops selected at random. What is the probability that the mean weight for the customer's purchase lies between 2.5 and 3.5 ounces?
   e) A shipment of 1000 widgets contains 40 defective units. What is the probability of finding at least one defective widget in a random sample of 100 of the widgets?
   f) Suppose X is a normally distributed variate with mean 1 and variance 1. Find the probability that $3x + 7$ lies between 0 and 10.

3. Let X be normally distributed (with mean m_x and variance σ_x^2). Find b such that 95% of the area under the density function for X lies over $[m_x - b, m_x + b]$. Do this problem, substituting 99% for 95%.

4. Suppose X is normally distributed with variance 9 and the probability that x is greater than 20 is 0.05. Find m_x.

5. The I.Q.'s of 10,000 college students are found to be normally distributed with a mean of 110 and a variance of 49. What percentage of college students would one expect to have an I.Q. of 130 or more? Find b such that 90% of the college students can be expected to have an I.Q. in the interval $[m_x - b, m_x + b]$.

6. A test of a large number of lightbulbs indicates that their mean life is 750 hours with a variance of 100. What percentage of lightbulbs can be expected to last between 750 and 800 hours?

7. A test is run concerning the height of college freshmen boys. A random sample of six heights is taken; the heights recorded are 68, 72, 75, 70, 71, and 70 inches. What is the sample mean? the sample variance? What population is involved in the test? Suppose the population mean is 70 and the population variance is 25. What is the density function (approximately) for the sample means of samples of 100? What is the probability that the mean of a random sample of 100 heights lies between 70 and 78?

10.2 STUDENT'S t-DISTRIBUTION

Statisticians often try to estimate information about a large group of objects by examining a small proportion of those objects. For example, a small percentage of the voters may be polled, yet the results of an election are predicted. Medical men estimate the percentage of heavy smokers who will contract lung cancer by studying only a relatively small number of heavy smokers. In instances such as those cited, conclusions are being drawn about a population parameter from population samples.

We can, for example, estimate a population mean by using the sample mean of a random sample from the population. This procedure would not be unreasonable since, as was seen in Proposition 3 of the preceding section, the population mean equals the mean of the sample mean. If the sample used is fairly large, then the sample mean $\bar{X}$ will be approximately normally distributed with mean m_x and variance σ_x^2/n, where m_x and σ_x^2 are the population mean and variance; we can use this fact to determine how accurately we may expect the sample mean to approximate the population mean. This procedure is illustrated in the following example.

Example 8. A random sample of 10,000 adults gives an average weight of 160 pounds with a variance of 121. The sample mean $\bar{X}$ (of random samples of 10,000 weights of individual adults) is normally distributed with mean m_x, where m_x is the average weight of a randomly chosen adult; we shall assume that the population variance (the population being all adult weights) is the same as the sample variance since the sample is fairly large. Therefore the variance of $\bar{X}$ is

$$\sigma_{\bar{x}} = (121/10{,}000)^{1/2} = 11/100 = 0.11.$$

Consequently,

$$Y = (\bar{X} - m_x)/0.11$$

is a standard normal variate. The observed value of $\bar{X}$ is 160 and m_x is to be estimated. We cannot compute m_x exactly unless we examine the entire population, but we can find the probability that m_x lies within certain limits; in particular, we shall compute b such that m_x has a probability of 0.95 of being in $[\bar{x} - b, \bar{x} + b] = [160 - b, 160 + b]$. Such an interval is called a 0.95 *confidence interval* for m_x.

We find from a table that the probability that y is in $[-1.96, 1.96]$ is 0.95; that is, the probability that

(13) $-1.96 \leq (160 - m_x)/0.11 \leq 1.96$

is 0.95. From (13), then, the probability that

$$160 - 0.2156 \leq m_x \leq 160 + 0.2156$$

is 0.95; thus $b = 0.2156$. We can thus conclude that there is a 0.95 chance that $m_x = 160 \pm 0.2156$.

If we had a sample of only 100 adults, then $\sigma_{\bar{x}}$ becomes $\frac{11}{10} = 1.1$. The smaller sample size also means that $\bar{X}$ is probably less normally distributed than with a sample of 10,000. With a sample size of 100, the variate $Y = (\bar{X} - m_x)/1.1$ is approximately a standard normal variate, where the probability that

$$-1.96 \leq (160 - m_x)/1.1 \leq 1.96$$

is 0.95. In this case the probability that

$$160 - 2.156 \leq m_x \leq 160 + 2.156$$

is 0.95, or $b = 2.156$.

If we are dealing with small samples (say, less than 30), then the sample standard deviation, which we shall denote by $s_{\bar{x}}$, is not a good estimate of the population standard deviation σ_x. If we have a random sample of n elements from the population and if n is large, then $s_{\bar{x}}$ may be expected to approximate σ_x rather closely, but if n is small, then

(14) $$s = \left(\frac{n}{n-1}\right)^{1/2} s_{\bar{x}}$$

is used to estimate σ_x. Note that as n gets large $(n/(n-1))^{1/2}$ approaches 1. If s is a "better" estimate of σ_x than $s_{\bar{x}}$ is, then $s/\sqrt{n}$ is a better estimate of $\sigma_{\bar{x}}$, the standard deviation of the sample mean X than $s_{\bar{x}}/\sqrt{n}$ is (see Proposition 3 of the preceding section). We shall denote $s/\sqrt{n}$ by $\bar{s}_{\bar{x}}$; then $\bar{s}_{\bar{x}} = s_{\bar{x}}/\sqrt{n-1}$.

The variate

(15) $$t = \frac{\bar{X} - m_x}{\bar{s}_{\bar{x}}}$$

is a standard variate (that is, has mean 0 and variance 1), but t is not normally distributed since $\sigma_x/\sqrt{n}$ has been replaced by $s_{\bar{x}}$, and because n is small, the approximation to the normal distribution is poor anyway.* The following proposition, stated without proof, tells us what the distribution of t is.

Proposition 6. *If X is a population which is at least approximately normally distributed with mean m_x, then if $\bar{X}$ is the sample mean of a random sample*

---

* We break the rule here of denoting random variables by capital letters and their values by lower-case letters of the same kind because the variate t of (15) is almost universally designated by the small letter t.

of n elements of X, then the variate t of (15) has density function f defined by

(16) $$f(t) = k\left(\frac{1+t^2}{n-1}\right)^{-n/2},$$

where k is a constant such that the area under the graph of f over the entire set of real numbers is 1.

The distribution described in Proposition 6 is known as *Student's t-distribution with $n-1$ degrees of freedom*. "Student" was the pen name of a statistician, W. S. Gosset, who worked at an Irish brewery.

We illustrate the use of this distribution on the following examples.

Example 9. Twenty randomly chosen adults are found to have an average height of 68 inches with a variance of 81. Then

$$\bar{s}_{\bar{x}} = 9/\sqrt{19} \doteq 2.29.$$

The variate

$$t = (\bar{X} - m_x)/2.29$$

has Student's t-distribution with 19 degrees of freedom.

We wish to find the probability that the true mean adult height lies between 68 and 69 inches, that is $P(68 \le m_x \le 69)$. Now $68 \le m_x \le 69$ if and only if

$$\frac{\bar{X} - 69}{2.29} \le t = \frac{\bar{X} - m_x}{2.29} \le \frac{\bar{X} - 68}{2.29}.$$

Therefore

$$P(68 \le m_x \le 69) = P\left(\frac{68-69}{2.29} \doteq -0.436 \le t \le \frac{68-68}{2.29} = 0\right)$$

since $\bar{x} = 68$. This probability, in turn, is equal to

(17) $$\int_{-0.436}^{0} f(t)\, dt, \quad \text{with } f(t) \text{ as in (16).}$$

We use a table of the Student t-distribution to evaluate (17). From Table 4 in the Appendix we find, in the row for 19 degrees of freedom, that

$$\int_{-0.391}^{0.391} f(t)\, dt \doteq 1 - 0.70 = 0.30.$$

We can therefore estimate (17) to be about $0.30/2$. In so doing we make use of the fact that $\int_{-a}^{a} f(t)\, dt = 2 \int_{-a}^{0} f(t)\, dt$ because f is symmetric with respect to 0.

We now ask: Is a true mean adult height of 65 consistent with the data given? If $m_x = 65$, then since $\bar{x} = 68$, we have

$$t = (68-65)/2.29 \doteq 1.31.$$

We see from Table 4 that the probability of t being at least as large as 1.31 is about 0.10; that is, $P(t \geq 1.31) \doteq .1$; hence it does not appear too inconsistent with the data to hold that m_x is really 65.

Example 10. Over a 10-year period the mean annual rainfall in a certain area is 49 inches with a variance of 16. We shall find b such that

$$P(49 - b \leq m_x \leq 49 + b) = 0.95,$$

where m_x is the true mean annual rainfall.

Since 10 is a small sample, we use Student's t-distribution with 9 degrees of freedom. In this instance,

$$\bar{s}_{\bar{x}} = 4/\sqrt{9} = \tfrac{4}{3} \doteq 1.33.$$

Therefore

$$t = \frac{\tilde{x} - m_x}{\tfrac{4}{3}} = \frac{3(49 - m_x)}{4}.$$

From Table 4 we find that there is a 0.95 probability that

$$-2.262 \leq t \leq 2.262.$$

Therefore there is a 0.95 probability that

$-2.262 \leq 3(49 - m_x)/4 \leq 2.262$, or

$49 - 3.016 \leq t \leq 49 + 3.016$;

that is, $b = 3.016$.

EXERCISES

1. Find the probability in Example 9 that the mean adult height lies between 68 and 69 inches, assuming that $\tilde{X}$ is normally distributed in accordance with the discussion of Section 10.1. Compare this answer with the answer given by the t-distribution.

2. In Example 10, find number b, assuming that $\tilde{X}$ is normally distributed and compare your answer with that found using the t-distribution.

3. We shall denote the density function for Student's t-distribution with n degrees of freedom by $f(t; n)$. Given that t is a variate with density function $f(t; n)$ and the entire set of real numbers as admissible range, find the mean, mode, and median of t. Prove that the distribution of t is symmetric with respect to 0. Using Table 4 find the value of t asked for, and the values of b in (d) and (e).

a) $\int_0^t f(t; 5)\, dx = 0.45$ b) $\int_{-t}^t f(t; 7)\, dx = 0.90$

c) $\int_t^\infty f(t; 17)\, dx = 0.05$

d) Find b such that $P(-b \leq t \leq b) = 0.95$ if t has density function $f(t; 11)$.
e) Find b such that $P(0 \leq t \leq b) = 0.475$, where the density function of t is $f(t; 15)$.

4. Twelve randomly chosen tomatoes are weighed; it is found that the mean weight is 6 ounces with a variance of 9. Find the probability that the true mean weight of tomatoes lies between 5 and 7 ounces. Find b such that there is a 0.95 probability that the mean weight of tomatoes lies between $6 - b$ and $6 + b$.

5. Twenty-five ball bearings are found to have an average radius of 5.001 inches with a variance of 0.0025. What is the probability that the mean radius is really less than 5? Compute this probability using both the t- and normal distributions. Compare your results.

6. A random sample of 10 elements from a certain population has a mean of 15 and a variance of 16.

    a) Find b such that there is a 0.95 probability that the population mean lies between $15 - b$ and $15 + b$.
    b) Find the b asked for in (a), assuming that the sample contains 15 observations (and all other factors remain unaltered).
    c) Find the b asked for in (a), assuming that the sample contains 20 observations. As the number of elements in the sample increases, what happens to b?

10.3 MORE ABOUT THE t-DISTRIBUTION

The t-distribution can also be used in testing whether or not two populations have the same mean, provided certain conditions are satisfied and the samples are small enough to warrant the use of the t-distribution instead of the normal distribution. The details are given in the following proposition.

Proposition 7. Let X and Y be two independent normally distributed (or at least approximately normally distributed) populations. ("Independent" here means that specific observations of X and Y will not depend on, or be tied to, one another.) Let $\bar{X}$ be the mean of a sample of n_1 observations of X, and $\bar{Y}$ be the mean of a sample of n_2 observations of Y. If $m_x - m_y = d$ and

(18) $\quad U = (\bar{X} - \bar{Y}) - d,$

then if n_1 and n_2 are "large," U is approximately normally distributed with mean 0 and variance

(19) $\quad \sigma_{x-y}^2 \left(\dfrac{1}{n_1} + \dfrac{1}{n_2} \right),$

where σ_{x-y}^2 is estimated by

$$\frac{n_1 s_1^2 + n_2 s_2^2}{n_1 + n_2},$$

where s_1^2 and s_2^2 are the sample variances.

If random samples of size n_1 and n_2 of observations of X and Y, respectively, are given with sample variances s_1^2 and s_2^2, respectively, n_1 and n_2 are "small," and X and Y have at least approximately equal variances, then we estimate σ_x^2 by

(20) $\quad \bar{\sigma}_x^2 = \dfrac{n_1 s_1^2 + n_2 s_2^2}{n_1 + n_2 - 2},$

and define $\bar{\sigma}_u^2$ with U defined as in (18) by

(21) $\quad \bar{\sigma}_u^2 = \bar{\sigma}_x^2 \left(\dfrac{1}{n_1} + \dfrac{1}{n_2} \right).$

Then

(22) $\quad t = u/\bar{\sigma}_u$

has the t-distribution with $n_1 + n_2 - 2$ degrees of freedom.

The primary application of Proposition 7 is testing whether two populations have equal means. We illustrate this procedure in the following example.

Example 11. A random sample of 10 citizens of the United States gave an average lifetime of 70 years with a variance of 100, while a random sample of 12 citizens of France gave an average lifetime of 75 years with a variance of 120. We shall assume that X, the lifetimes of citizens of the United States, and Y, the lifetimes of citizens of France, are independent, normally distributed populations with at least approximately the same variance. We shall examine the plausibility of the statement that $m_x = m_y$; that is, Frenchmen do not have a longer mean lifetime than do citizens of the United States. Since we are not given σ_x, and since n_1 and n_2 are relatively small anyway, we must estimate σ_x by (20) and use the t-distribution with $n_1 + n_2 - 2 = 10 + 12 - 2 = 20$ degrees of freedom. Then

$\bar{\sigma}_x^2 = (10 \cdot 100 + 12 \cdot 120)/20 = 122;$

hence, applying (21), we find

$\bar{\sigma}_u^2 = (122)(\frac{1}{10} + \frac{1}{12}) \doteq 22;$

therefore $\bar{\sigma}_u \doteq 4.7.$

By Proposition 7, $t = (\tilde{x} - \tilde{y})/4.7$ is t-distributed with 20 degrees of freedom, for if $m_x = m_y$, then $m_x - m_y = 0 = d$. We are trying to determine whether or not it is plausible that $m_x = m_y$; we shall see whether the assumption that $m_x = m_y$ leads to something implausible, i.e., improbable. For the observed values of $\bar{X}$ and $\bar{Y}$, $U = 70 - 75 = -5$; hence $t = -5/4.7 \doteq -1.11$.

If the mean life of a Frenchman is longer than that of a citizen of the United States, then we would expect t to differ from 0 in the negative direction, that is, $\tilde{x} - \tilde{y}$ would be expected to be negative; but even if we did have $m_x = m_y$, it is still possible for $\tilde{x} - \tilde{y}$ to be negative. The question is: How probable is it that t will assume a value of -1.11 or less? The answer found from tables is approximately 0.15; hence there is a fair possibility that even if $m_x = m_y$, we will obtain a value of t equal to or less than the value actually observed. Because of this fairly large probability, our figures give us no reason to accept the assertion that Frenchmen live longer than Americans.

In doing this problem, we assumed (1) that X and Y were normally distributed, and (2) that they had approximately the same variance. In later sections we shall study tests to determine whether a variate is normally distributed and whether or not two variates have the same variance. We shall also devote an entire chapter to the question of testing hypotheses, a topic we have already touched upon lightly (in Example 11 we were testing the plausibility of the hypothesis "$m_x = m_y$").

In certain tests, one observation from one population is paired with one observation from another population. Since the observations are paired, they are no longer independent (unrelated); hence Proposition 7 does not apply. There is, however, another result which does. Before stating it, we consider a situation involving paired variates.

Example 12. Ten randomly selected right-handed children are paired with ten randomly chosen left-handed children. We let X be the number of words a right-handed child can read in one minute and Y be the number of words a left-handed child can read in one minute. We therefore have ten random observations of X and ten random observations of Y. These are paired according to the manner the children have been paired. Suppose the results obtained are as shown in Table 1.

Table 1

Pair	1	2	3	4	5	6	7	8	9	10
x	100	120	90	80	100	150	90	100	90	70
y	90	85	110	90	70	90	120	120	70	140

Let us assume that we are interested in finding out whether it is plausible that right-handed children, on the average, read faster than left-handed children. If such is the case, then the difference $x - y$ should tend to be positive, that is, m_{x-y} should be greater than 0. We shall return to this example after considering the general situation.

Proposition 8. *Suppose X and Y are two normally distributed populations for which we have n paired observations $(x_1, y_1), \ldots, (x_n, y_n)$. Define*

$$D_i = x_i - y_i, \quad i = 1, \ldots, n, \quad \text{and} \quad \overline{D} = \sum_{i=1}^{n} D_i/n.$$

Set

$$t = (\overline{D} - d)/s,$$

where

$$s = \left(\frac{\sum_{i=1}^{n} (D_i - d)^2}{n(n-1)} \right)^{1/2}$$

and $m_x - m_y = d$. Then t will be t-distributed with $n - 1$ degress of freedom.

Example 12 continued. We shall assume that X and Y are at least approximately normally distributed, and determine some measure of the tenability of $d = 0$, that is, that $m_x = m_y$. Straightforward calculation gives the following information.

i	1	2	3	4	5	6	7	8	9	10
D_i	10	35	−20	−10	30	60	−30	−20	20	−70

$\overline{D} = \frac{5}{10} = \frac{1}{2}$,

$s = \left(\dfrac{12{,}925}{90} \right)^{1/2} \doteq 12$

(We assume that $d = 0$, since we are trying to test this assumption.) Therefore

$t = \overline{D}/12$

has the t-distribution with 9 degrees of freedom. The observed value of $\overline{D}$ was $\frac{1}{2}$; hence the observed value of t is $(\frac{1}{2})/12 = \frac{1}{24}$. Reference to a table establishes that if $d = 0$, it is quite probable for t to be $\frac{1}{24}$ or more; in fact, this probability is more than 0.45. Therefore it is quite plausible that $m_x = m_y$ in the light of the information given.

EXERCISES

1. A sample of 10 items produced on Machine I is found to have an average deviation of 0.01 inch with a variance of 0.09, while a sample of 15 items produced independently on Machine II reveals an average deviation of 0.005 inch with a variance of 0.07. Assume that the deviations X and Y from Machines I and II, respectively, are normally distributed with the same variance. Does the assumption that the average deviation is the same for both machines seem plausible in the light of the information given? Measure the plausibility by the probability of getting a value for the difference of the sample means at least as large in absolute value as the one obtained.

2. The government finds the following figures pertaining to tar content in a test of six samples each of two unrelated brands of cigarettes.

   Tar content in milligrams

Brand A	10,	10.1,	12,	8,	10.5,	9.4
Brand B	7,	8,	7.4,	9.2,	7.3,	9.1

   a) Using Proposition 7, test the assumption that both brands have the same mean tar content. Assume that the tar contents X and Y in Brand A and Brand B, respectively, are at least approximately normally distributed with the same variance.
   b) Assume that the variates are paired, the ith figure for Brand A paired with the ith figure for Brand B. Use Proposition 8 to test the assumption that both brands have the same mean tar content. Here the only assumption that must be made is that X and Y are at least approximately normally distributed. Compare your results using Proposition 8 with the results obtained using Proposition 7.
   c) Assume that the difference of the sample means is approximately normally distributed and use the normal distribution to test the hypothesis that the two brands have the same mean tar content. Use Proposition 5 to compute σ_x^2.

3. Design an experiment to test the plausibility of the assertion that men have a higher mean I.Q. than women. Your experiment may use either the t-distribution in the context of Proposition 7 or Proposition 8, or the normal distribution. Assume that the I.Q.'s of both men and women are normally distributed with the same variance for both.

4. A certain company wishes to test the effectiveness of piped-in music in increasing the productivity of its workers. Each of ten randomly chosen workers is paired with a worker whose performance on the job has been about the same as his. One group of workers is then allowed to work with music, while the other group works under similar conditions but without music. The results are given below.

Pair	Group A (music) Output	Group B (no music) Output
1	15	10
2	12	11
3	14	10
4	13	12
5	12	12
6	11	13
7	14	10
8	15	12
9	13	11
10	12	10

Test the assertion that the music increased the productivity of the workers.

10.4 THE χ^2-DISTRIBUTION

At times certain theoretical assumptions about some experiment may lead us to expect certain events to occur with certain frequencies; however, when the experiment is performed a number of times, the observed frequencies and the predicted frequencies differ. The question then is whether or not the observed frequencies deviate sufficiently from the expected frequencies to discredit the hypothesis, or hypotheses, which led to the predicted values. For example, if a coin is flipped 1000 times, then the assumption that the coin is fair will lead us to expect 500 heads and 500 tails. In actual fact, even if the coin is fair, the probability that we will obtain exactly 500 heads and 500 tails in 1000 flips is very small; suppose we obtain 650 heads and 350 tails. We may then ask: Is the discrepancy between the expected frequencies and observed frequencies of heads and tails sufficient to make the assumption that the coin is fair virtually untenable?

In Example 1 of Chapter 9, we have an example, in which the observed relative frequencies of the various faces of a die differ from the relative frequencies that are predicted by assuming that the die is unbiased. It should be stressed that we cannot expect the correspondence between observations and theory to be *too good*. Had all the observed relative frequencies turned out to be exactly $\frac{1}{6}$, then either the experiment or the author's imagination would be open to suspicion. Again, we must ponder the question: Are the discrepancies serious enough to warrant rejecting the assumption (in this case, that the die is fair) which leads to the expected (or theoretical) frequencies?

Suppose that we are considering n events $E_1, \ldots, E_n$, and that each E_i has expected frequency e_i and actually observed frequency of o_i. One measure of the discrepancy between the observed and expected frequencies is given by the *statistic**

(23) $\quad \chi^2 = \dfrac{(o_1 - e_1)^2}{e_1} + \cdots + \dfrac{(o_n - e_n)^2}{e_n} = \sum_{i=1}^{n} \dfrac{(o_i - e_i)^2}{e_i}.$

The larger χ^2 is, the greater the discrepancy between the observed and theoretical frequencies; $\chi^2 = 0$ if and only if the observed and predicted frequencies are the same.

The statistic χ^2 is a random variable whose density function is described in the following proposition.

Proposition 9. *The density function of χ^2 as defined in (23) is approximately*

(24) $\quad g(\chi^2) = c(\chi^2)^{1/2(f-2)} e^{-\chi^2/2} = c\chi^{f-2} e^{-\chi^2/2},$

*where f is an integer as determined below and c is a constant which makes the area under the graph of g from 0 to ∞ (χ^2 cannot be negative) equal to 1. The approximation becomes better as the expected frequencies become larger; for reasonable accuracy, each of the observed frequencies should be at least 5. The number f, called the number of **degrees of freedom**, is given by*

i) $f = n - 1$ if we can compute the expected frequencies without having to estimate any population parameters from the sample, or

ii) $f = n - 1 - k$ if we can compute the expected frequencies only by estimating k population parameters from the sample.

Since we wish to find the probability that the measure χ^2 of discrepancy between observed and expected values will be at least as large as some number, say N, we will generally want to find

(25) $\quad \displaystyle\int_{N}^{\infty} g(x)\,dx.$

The evaluation of (25) is best carried out by means of tables.

Example 13. Consider Example 1 of Chapter 9. If we assume that the die is fair, then we can set up Table 2.

Table 2

Event	1	2	3	4	5	6
Expected ($\doteq$)	167	167	167	167	167	167
Observed	150	200	100	150	175	125

---

* A *statistic* is a numerical quantity which can be calculated from a sample. The sample mean and sample standard deviation are statistics.

In this instance

$$\chi^2 = (17)^2/167 + (-33)^2/167 + \cdots + (22)^2/167 \doteq 3.9.$$

Since no population parameters were estimated from the sample (observations) given, there are $6 - 1 = 5$ degrees of freedom. We see from Table 3 of the Appendix that there is a better than 50% chance that χ^2 will have a value at least as large as that found; hence on the basis of the observations given there are no reasonable grounds for rejecting the assumption that the die is fair.

One important use of Proposition 9 is to test the possibility that some population has a particular distribution. We illustrate this use in the following example.

Example 14. The heights of 1000 randomly selected college men can be categorized as follows:

Table 3

Height, feet	5–5.25	5.25–5.5	5.5–5.75	5.75–6	6–6.25	6.25–6.5	6.5–6.75
Observed	10	40	200	350	200	150	50

Suppose the sample mean is 5.9 and the sample variance is 0.16. We will determine the tenability of the assumption that the heights of male college students form a normally distributed population with mean 5.9 and variance 0.16. Note that m_x and σ_x have been estimated from the sample, that is, two population parameters have been estimated from the sample; hence we have two degrees of freedom less than if the population parameters were found by some other means. There are in all seven events, the seven regions into which we have divided the heights (this division was arbitrary and finer divisions could be used to get a more accurate test); hence there are $7 - 1 - 2 = 4$ degrees of freedom in all.

We standardize the variate by setting

$$Y = (X - 5.9)/0.4,$$

so that we can use tables to aid our computations. In terms of Y, we can construct Table 4. The values in Table 4 have been rounded to the nearest hundredth. A table can be used to find the area under the graph of the standard normal density function for each interval. This, in turn, gives us the proportion of observations which should fall in that interval. Using the data from Tables 3 and 4 we compute

$$\chi^2 = (30)^2/40 + (70)^2/110 + \cdots + (30)^2/120 + 0 \doteq 115.5.$$

By consulting a table, we see that the probability that χ^2 will have a value

Table 4

y	−2.25–1.625	−1.625–1.0	−1.0–0.375	−0.375–0.25
Observed: 1000	0.01	0.04	0.20	0.35
Expected ratio, assuming normal distribution	0.04	0.11	0.19	0.25
Expected observation	40	110	190	250

y	0.25–0.875	0.875–1.5	1.5–2.125
Observed: 1000	0.20	0.15	0.05
Expected ratio, assuming normal distribution	0.21	0.12	0.05
Expected observation	210	120	50

as large as the one we have obtained is negligible. Therefore we will discard the assumption that X is normally distributed with $m_x = 5.9$ and $\sigma_x = 0.4$.

Another use of the χ^2-distribution is in conjunction with *contingency tables*. A contingency table is an array in which data are arranged according to two aspects, one aspect determining the row in which a number appears and the other aspect determining the column in which it is placed. A contingency table is given in the following example.

Example 15. Table 4 shows the results that a certain drug had in the treatment of a disease. Two groups, I and II, of 100 randomly selected victims of the disease were formed. Group I was given the drug, while Group II did not get the drug.

Note that the data are classified according to both group and reaction to the disease. Table 5 is a 2 × 2 contingency table. Contingency tables in general may be said to be $h \times k$, where h is the number of rows and k is the number of columns.

Table 5

	Recovered	Not recovered
Group I	80	20
Group II	40	60

Corresponding to each cell, or position, in a contingency table there may be both an observed frequency and a theoretical, or expected, frequency. The expected frequency is computed on the basis of some hypothesis. If we have a contingency table of n cells, we define χ^2 as in (23), where now the sum is taken over all cells in the table.

Proposition 10. *If the contingency table is $h \times k$, $h > 1$ and $k > 1$, then χ^2 has the density function (24) with $f = (h - 1)(k - 1)$ if we can compute the expected frequencies without estimating any population parameters from the recorded observations, and $f = (h - 1)(k - 1) - m$ if we must estimate m population parameters from the observed frequencies in order to compute the expected frequencies.*

We now continue Example 14 and "test" the hypothesis: Those treated with the drug had a better chance of recovery than those treated without the drug. If we assume that the drug made no difference in treatment, then, since we have 200 patients in all of which 120 recover, we set the overall chance of recovery at $120/200 = 0.6$. On this basis, the expected values of the entries in the contingency table are as in Table 6.

Table 6

	Recovered	Not recovered
Group I	60	40
Group II	60	40

Therefore

$$\chi^2 = (-20)^2/60 + (20)^2/40 + (20)^2/60 + (-20)^2/40 \doteq 33.7.$$

Since we have a 2×2 table and no population parameters were estimated, there is one degree of freedom. Referring to a table for χ^2 in the row for one degree of freedom, we see that the probability is virtually negligible that χ^2 have a value as large or larger than the one found. Therefore it seems unlikely that the drug is really ineffective. We shall make the idea and techniques of hypothesis testing more precise in the next chapter.

The variate χ^2 is actually discrete, whereas the tables associated with χ^2 treat χ^2 as being continuous. As a result, some degree of accuracy is lost. For more than one degree of freedom, this loss of accuracy is generally negligible. However, where one degree of freedom occurs, it sometimes is desirable to use

$$(26) \qquad \chi^2 = \sum_{i=1}^{n} \frac{(|o_i - e_i| - \tfrac{1}{2})^2}{e_i}$$

in place of (23). The correction for χ^2 given in (26) is called *Yates' correction for continuity*. In general, one prefers (26) to (23) to measure χ^2 only when (23) leads to some doubt as to whether or not a given hypothesis should be rejected.

EXERCISES

1. A certain coin is flipped 400 times and 240 heads are obtained. Assume that the coin is fair and compute χ^2. What is the probability of getting a value of χ^2 at least as large as the one obtained? Does the hypothesis that the coin is fair seem plausible? Do this problem, assuming that the coin is flipped 1000 times and 550 heads are obtained.

2. A real estate agent claims that about 40 cars a day pass a certain house. The traffic is observed for 10 consecutive days with the following results:

Day	1	2	3	4	5	6	7	8	9	10
Cars	60	52	39	45	43	59	38	55	47	46

Does the agent's statement appear to be plausible? What is the probability of getting a value of χ^2 as large or larger than the observed value if the agent's statement is correct?

3. The manufacturer of a patent medicine claims that the medicine is effective in preventing colds. To test the claim a randomly selected sample of 1000 people is divided randomly into two groups of 500 each. One group is given the medicine, while the second group is not. The results over one month of testing are recorded in Table 7. If Group I received the medicine, is the manufacturer's claim credible? Do this problem, using χ^2 as given in (23), and then using the adjusted χ^2 of (26).

Table 7

	Had at least one cold	No colds
Group I	275	225
Group II	245	255

4. A car manufacturer claims that its cars require fewer repairs than those made by a competitor. Fifty cars from each company are selected at random and the repair record over a two-year period is observed. The results are recorded in Table 8. On the basis of Table 8, does the manufacturer's claim seem justified? Use χ^2 to investigate the claim.

Table 8

Number of repairs

	0	1	2	3	4 or more
Manufacturer	6	7	18	10	9
Competitor	4	6	24	5	11

10.5 SOME OTHER DISTRIBUTIONS

The exponential distribution. A variate X with admissible range $[0, \infty)$ and density function

(27) $\quad f(x) = \mu e^{-\mu x},$

where μ is a positive constant, is said to be *exponentially distributed*, or to have the *exponential distribution*.

Observe that the distribution function for X is $F(x) = 1 - e^{-\mu x}$.

Proposition 11. *Assume X is exponentially distributed. Then*

a) $M(x;t) = \int_0^\infty e^{xt} \mu e^{-\mu x}\, dx = \mu/(\mu - t);$

b) $m_x = 1/\mu;$

c) $\sigma_x^2 = 1/\mu^2.$

The verification of (a), (b), and (c) is left as an exercise.

Usually x is the length of time until some specified event occurs, or until some process is completed. A variate representing such a length of time but whose distribution is uncertain should be tested (for example, using χ^2) to determine whether the distribution is exponential. Given a particular set of observations, one may also test for the exponential distribution by plotting the data on semilog graph paper; if the points plotted lie on an approximately straight line, then the variate is probably exponentially distributed.

An important relationship between the Poisson distribution and the exponential distribution is stated in the following proposition.

Proposition 12. *If we are dealing with completions, or lengths of time until a specified event occurs, and the delays between successive completions are in-*

dependent, exponentially distributed variates, all of which have mean $1/\mu$, then the number of completions in a fixed time t is a Poisson-distributed variate with mean μt.

Example 16. A certain study showed that the lengths x (in minutes) of phone conversations form an exponentially distributed population with $\mu = 1/2.26$. We shall find the probability that a call lasts between 5 and 10 minutes. This probability is given by

(28) $$\int_5^{10} (1/2.26)e^{-x/2.26}\,dx = e^{-5/2.26} - e^{-10/2.26}.$$

Using a table of values of e^{-y} (Table 7), we find that (28) is approximately equal to 0.1. Thus we can expect about 10% of all phone calls to last between 5 and 10 minutes.

We note from the table that $e^{-3} \doteq 0.05$. Therefore

$$\int_0^{3(2.26)} (1/2.26)e^{-x/2.26}\,dx = 1 - e^{-3} \doteq 0.95.$$

We can therefore say that only about 5% of all phone calls will last longer than $3(2.26) = 6.78$ minutes.

The mean length of a phone call is 2.26 minutes.

Example 17. It is found that the length of the time interval between one person entering a certain store and the next person entering it is exponentially distributed with a mean of 1.5 minutes. Assuming that the intervals between one person and the next are independent of one another, then the number of people to enter the store in 1 hour, that is, 60 minutes, is a variate Y which is Poisson-distributed with mean $1/1.5(60) = 40$. The density function f for Y is given by

$$f(y) = \left(\frac{(40)^y}{y!}\right)e^{-40}.$$

F-distribution. A distribution used to test whether two normally distributed populations have the same variance is the F-distribution. Recall that in applying the t-distribution, we must sometimes know, or reasonably suppose, that two variates have the same variance. The following proposition gives information about using the F-distribution to test for equal variances.

Proposition 13. *Suppose X and Y are normally distributed populations with the same standard deviation. If n_1 and n_2 random observations*

$$O_x: x_1, \ldots, x_{n_1} \quad \text{and} \quad O_y: y_1, \ldots, y_{n_2}$$

are made from X and Y, respectively, then

(29) $$F = \frac{s_{\bar{x}}^2}{s_{\bar{y}}^2}$$

has the F-distribution with $f_1 = n_1 - 1$ and $f_2 = n_2 - 1$, where

$$\tilde{s}_{\tilde{x}}^2 = \sum_{i=1}^{n_1} \frac{(x_i - \tilde{x})^2}{n_1 - 1} \quad \text{and} \quad \tilde{s}_{\tilde{y}}^2 = \sum_{i=1}^{n_2} \frac{(y_i - \tilde{y})^2}{n_2 - 1}.$$

(Recall that $\tilde{x}$, the sample mean, and $\tilde{s}_{\tilde{x}}^2$, an estimate of the population variance, have already been encountered.) The density function g for F above is given by

(30) $$g(F) = \frac{cF^{(f_1-2)/2}}{(f_2 + f_1 F)^{(f_1+f_2)/2}}, \quad c \text{ a constant.}$$

The quantities f_1 and f_2 are called **degrees of freedom**.

We shall make no attempt whatsoever to justify Proposition 13 or to investigate the basic properties of (30), the definite integrals of which are best estimated by means of tables. In point of fact, until more is known about hypothesis testing, little can be done with Proposition 13 that would make much sense. For the moment, we shall merely compute a value of F by way of an example.

Example 18. We will assume that the temperature recorded in any given area at noon is a normally distributed variate. The temperatures in Miami and Buffalo at noon are recorded on six and seven days, respectively, with the results shown in Table 9.

Table 9

Miami	70	74	80	66	76	78	
Buffalo	60	54	36	40	56	62	70

If X denotes the Miami temperatures and Y denotes the Buffalo temperatures, then $\tilde{x} = 74$ and $\tilde{y} = 54$. Straightforward computation yields $\tilde{s}_{\tilde{x}}^2 \doteq 27.2/6$ and $\tilde{s}_{\tilde{y}}^2 \doteq 136.7/7$. Consequently,

$$F = (27.2/136.7)(7/6) \doteq 0.23.$$

A discussion of the significance of obtaining a value of F such as this with $f_1 = 5$ and $f_2 = 6$ must wait until we have studied hypothesis testing (and know more about the F-distribution). We close by noting that if F is as defined in (29), then $1/F = \tilde{s}_{\tilde{y}}^2/\tilde{s}_{\tilde{x}}^2$ also has the F-distribution, but with $f_1 = n_2 - 1$ and $f_2 = n_1 - 1$ (since the roles of x and y have been interchanged).

EXERCISES

1. Suppose that a variate X is exponentially distributed with mean 4. Find the probability that $5 \leq x \leq 6$. Find the probability that x is greater than 2. Find a value which X can be expected to exceed 95% of the time.

2. Suppose that the time between births in the United States is exponentially distributed with a mean of $\frac{3}{4}$ minute.
   a) What is the probability of there being no births for one hour?
   b) What is the probability that a birth will occur in five minutes or less?
   c) Assuming that the intervals between successive births are independent, what is the probability of having 60 births in one hour?
   d) What is the probability of having 20 or fewer births in one hour?

3. Using χ^2, test the feasibility of the hypothesis that the following random observations are from a population X which is exponentially distributed and has sample mean m_x.

   1.5, 1.7, 2, 1.4, 1, 0.6, 0.4, 1.8, 0.1, 1.9, 1.3, 2.3, 5, 0.1, 2.9

4. Assume that the following are random observations of two normally distributed populations X and Y.

x	2	6	7	11	5	9	5
y	100	101	98	95	106	110	93

   Compute F as in (29). What are the values of f_1 and f_2 in this instance?

11 Hypothesis Testing

11.1 STATISTICAL INFERENCE

We have already touched upon hypothesis testing when studying the t- and χ^2-distributions. In Chapter 10 we investigated the tenability of certain assumptions in the light of given observations. If an assumption is true, we can, in general, expect that the logical consequences of that assumption will be realized. Thus, for example, if a coin is fair, we can reasonably expect the numbers of heads and tails flipped to be about equal. If there is a substantial disparity between the number of heads and the number of tails, we will tend to reject the hypothesis that the coin is fair in favor of a hypothesis that the coin is really biased. In sum, we expect an agreement between observation and what a hypothesis says we should observe if the hypothesis is to be tenable; if the agreement is not there, then we tend to reject the hypothesis.

Not all hypotheses lend themselves to investigation by statistical means. Hypotheses which can be either proved or disproved absolutely, such as theorems of geometry, or highly subjective statements, such as, "This cigarette tastes good," are not material suitable for statistical investigation. In general, the only hypotheses which can be tested statistically are those which make some assumption about a population parameter, or value of a probability, or nature of some particular distribution. A *statistical hypothesis* makes a statement about a population which can be tested using the tools of statistics, usually by observing a random selection of objects from the population (a sample) and determining whether or not the observations gained from the sample are consistent with the hypothesis. If the observations seem to be consistent with the hypothesis, that is, if the observations are not "too improbable," the hypothesis is accepted; otherwise, we reject the hypothesis. A hypothesis which is put forth for the purpose of being rejected is called a *null hypothesis*; the hypothesis to be accepted if the null hypothesis is rejected is called the *alternative hypothesis*. The following example illustrates the use of a null hypothesis to test an alternative hypothesis.

Example 1. It is not known whether a certain coin is fair, but it is suspected that the coin is weighted in favor of heads. We wish to test the hypothesis:

(1)   The probability of flipping a head with the given coin is greater than $\frac{1}{2}$.

Statement (1) does not indicate what the true probability of flipping a head might be, only that it is greater than $\frac{1}{2}$. Since (1) is imprecise in this regard, it will not lead to a precise probability distribution. Suppose that it is decided to test the coin by flipping it 10 times and observing the number x of heads that occur. If the true probability of flipping a head is p, then the density function for X will be

(2) $f(x) = C(10, x)p^x(1 - p)^{10-x}$.

Instead of testing (1) directly, which is impossible, since we cannot compute specific probabilities from (1), we shall formulate a hypothesis which, if rejected, will lead to the acceptance of (1). We choose as null hypothesis:

(3)   The probability of flipping a head with the given coin is $\frac{1}{2}$.

Therefore (3) gives the density function

(4) $g(x) = C(10, x)(\frac{1}{2})^x(\frac{1}{2})^{10-x} = C(10, x)(\frac{1}{2})^{10}$

for X. The events most inconsistent with (3) are getting all heads or no heads. If we got 10 tails, we might feel justified in rejecting (3), but, at the same time, we would hardly be justified in rejecting (3) *in favor of the alternative hypothesis (1)*. The event most inconsistent with (3) and consistent with (1) is to get 10 heads.

Table 1 gives the values of $g(x)$ for each admissible value of x.

Table 1

x	0	1	2	3	4	5
$g(x)$	$\frac{1}{1024}$	$\frac{10}{1024}$	$\frac{45}{1024}$	$\frac{120}{1024}$	$\frac{210}{1024}$	$\frac{252}{1024}$

x	6	7	8	9	10
$g(x)$	$\frac{210}{1024}$	$\frac{120}{1024}$	$\frac{45}{1024}$	$\frac{10}{1024}$	$\frac{1}{1024}$

The event most inconsistent with (3) and consistent with (1) after 10 heads is to get at least 9 heads, that is, either 9 or 10 heads. The probability of getting 10 heads is about 0.001; that of getting 9 or 10 heads is about 0.011. If we decide that we will reject (3) in favor of (1) if and only if 10 heads are obtained, then we will have 0.001 as the *level of significance*; if we are satisfied with "9 *or* 10" heads, then we have a 0.011 level of significance.

Why did we not say that the next most inconsistent event was *9 heads* rather than *at least 9 heads*? The reason is that we are trying to decide for what values of X we will reject (3) in favor of (1). If we say that we will reject (3) in favor of (1) if and only if $x = 9$, then this implies that we would accept (3) if x were 10, which is quite an inconsistency. Thus, if we are willing to reject a hypothesis if a discrepancy of a certain magnitude occurs, we ought to be even more willing to reject the hypothesis if an even greater discrepancy occurs. If we are willing to settle for a level of significance of 0.05 in rejecting (3) in favor of (1), we will do so if x is at least 9, since there is a probability of more than 0.05 of x being at least 8.

A statistical hypothesis implies an experiment involving one or more random variables. The nature of the experiment, the variates, and the hypothesis should enable us to find the distribution of the variates. We shall assume that only one variate X is involved and the admissible range of X is A. We decide first how improbable an event must occur before we are willing to reject the hypothesis. If we decide to reject the hypothesis if an event of probability less than or equal to α occurs, then α is said to be the *level of significance* of the test. A subset C of A in which X has probability less than α of assuming a value is said to be a *critical region* at the level of significance α. There may be many critical regions for the same level of significance. Which critical region is used depends on the nature of the hypothesis and what we wish to establish either directly or indirectly. If the critical region is to include extreme values of X on both sides of the mean of X, then the test is said to be *two-tailed* (see Fig. 1). If the critical region lies entirely to one side of m_x, then the test is said to be *one-tailed* (see Fig. 2).

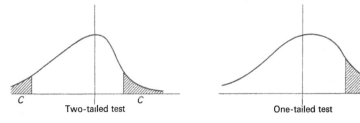

Fig. 1   Fig. 2

Example 2. If we decide upon a level of significance of 0.001 for the test of Example 1, then, since we wish to reject (3) in favor of (1), we shall choose {10} as the critical region. A suitable critical region for a level of significance of 0.05 would be {9, 10}, while {8, 9, 10} would suffice as a critical region for a 0.06 level of significance. The test in Example 1 is one-tailed.

Had we wished to test the hypothesis

(5) The coin is not fair

in Example 1, we would have had to use a two-tailed test since we would then have no preference for tails or heads, but would be looking for large deviations from the expected value of X predicted by (3). Thus, if we were testing (5) indirectly by means of (3) and wished a 0.01 level of significance, the critical region {0, 10} would be appropriate; for a 0.05 level of significance, the critical region {0, 1, 9, 10} would serve.

EXERCISES

1. Examine each of the following hypotheses for the possibility of testing it using statistical inference. If a hypothesis cannot, or should not, be tested using probability theory, indicate why not. If the hypothesis can be tested using probability theory, indicate an experiment which the hypothesis suggests which can be used to test it. What is the random variable of the experiment? Let the variate of the experiment be X. Do the hypothesis and experiment give a well-defined probability distribution for X? If not, can this distribution be found experimentally? If the distribution of X is not well-defined, can a null hypothesis be formulated which does lead to a well-defined distribution and which can be rejected in such a way as to lead to the acceptance of the hypothesis we wish to test? Find the density function of X, if you can, or at least indicate what density function is appropriate. If you cannot find a probability distribution for X, then it is possible that either the experiment cannot be used to test the hypothesis, or the hypothesis cannot be tested statistically.

    a) We are given a certain die. The hypothesis to be tested is: The die is biased toward 1's.

b) A certain woman claims that she can predict rain with a greater probability than mere chance would allow because before it rains, her corns hurt. *Test:* The woman is telling the truth.
c) *Test:* Cooper was a better author than Dickens.
d) *Test:* The majority of English majors in the class of '65 at Harvard think that Cooper was a better author than Dickens.
e) A certain man claims to have the power to draw aces from an ordinary deck of cards without looking at the cards. *Test:* The man's power to draw aces is no better than chance would allow.
f) *Test:* A given die is unbiased. [*Hint:* Consider the χ^2-distribution.]
g) *Test:* The height of a male citizen of the United States is a normally distributed variate with mean 68 and variance 25.
h) *Test:* A weight will take one second to fall 16 feet if dropped from a point 100 feet above sea level.

2. In Example 1, suppose that the true probability p of flipping a head is 0.6. If we are satisfied with a 0.05 level of significance, what is the probability that we will obtain a value of X in the critical region $\{9, 10\}$, and hence reject (3) in favor of (1)? What is this probability if p is really 0.9? Observe that even if (1) is true, the closer p is to 0.5, the less chance we have of detecting the bias in the coin by the test we have chosen. How large must p be before we have a 0.5 chance of rejecting (3) in favor of (1) at a 0.05 level of significance?

11.2 MORE ABOUT CRITICAL REGIONS

It would, of course, be ideal never to make a mistake, but such is not the destiny of man. Even the methods of statistical inference properly applied will sometimes lead to mistakes. The improbable might, in fact, occur and we will reject a hypothesis which is true. For example, in Example 1 we might get 10 heads even though the coin is unbiased. In such an instance, we would reject (3) even though it is true. On the other hand, we might get fewer than 9 heads even though the coin is biased in favor of heads; thus, we would accept (3) (at the 0.05 level of significance) even though (3) is false.

Definition 1. *The rejection of a true hypothesis is called a* **Type 1 error.** *The acceptance of a false hypothesis is called a* **Type 2 error.**

In any test we would like to minimize the chance of both types of errors. However, lowering the chance of one type of error usually raises the chance of the other type of error. For example, raising the level of significance say from 0.05 to 0.01 makes it less probable that we will reject a true hypothesis, but, at the same time, increases our chance that the hypothesis being tested will be accepted even if it is false.

A 0.05 level of significance is acceptable in most instances. However, what level of significance is used depends on the judgment of the person doing the testing, or, in some cases, legal standards. A 0.05 level of significance may be adequate for a hypothesis involving a coin, but totally inadequate in testing the efficacy of a drug which affects the lives of human beings. Where the hypothesis may only have to be "approximately" true, for example, in the case of testing whether two variates have the same variances, a low level of significance may suffice.

Once the level of significance is decided upon and the specific form of the test established, a suitable critical region must be chosen. Whether the region is one-sided or two-sided (a one-tailed or two-tailed test) depends on whether we are testing for a particular kind of deviation in the *test variate* (the random variable of the test) or are looking for extreme values of the test variate on both sides of the mean. Were we using x^2, for example, then we would be interested in large values of x^2 for rejecting a null hypothesis, since small values would be in agreement with the null hypothesis, that is, would indicate an agreement between observed and predicted frequencies. Hence in such an instance the test would be one-tailed and the critical region would consist of all values of x^2 larger than some value Z, where Z would depend on the level of significance desired.

We have seen that tables are of great help in evaluating definite integrals involving certain density functions. Certain distributions, for example, the t- and x^2-distributions, are used almost exclusively for testing hypotheses. In such instances, the tables for the density functions of those distributions are set up to give the critical regions for certain levels of significance rather than for evaluating specific definite integrals.

Example 3. Table 2 gives the row for the t-distribution with 9 degrees of freedom. The entries in the row are t_p, where

$$\int_{-t_p}^{t_p} f(t) \, dt = 2 \int_0^{t_p} f(t) \, dt = 1 - p;$$

$f(t)$ is the density function for t with 9 degrees of freedom. The area which the entries in the table represent under the graph of f is indicated in Fig. 3.

Table 2

p	0.30	0.20	0.10	0.05	0.02	0.01	0.001
t_p	1.100	1.383	1.833	2.262	2.821	3.250	4.781

For a two-sided test, then, with a 0.05 level of significance, the critical region would be $(-\infty, -2.262] \cup [2.262, \infty)$. Because of the symmetry of the t-distribution, the area on each side of 0 is exactly $\frac{1}{2}$ of the total area; in this case, the area on each side of 0 would be 0.025. For a one-sided

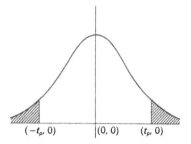

Fig. 3

test and a 0.05 level of significance, we would use the value of t_p under $p = 0.10$ (since the area on one side would be 0.05).

We have to have $|t| \geq 4.781$ to reject a hypothesis at the 0.001 level of significance in a two-tailed test where the test variate t has the t-distribution with 9 degrees of freedom.

In the case of the F-distribution (Section 10.5), since this is customarily used to test the hypothesis that two variances σ_x^2 and σ_y^2 are equal and $F = s_x^2/s_y^2$, we will tend to reject the hypothesis if F is quite small or quite large; hence we have a two-tailed test. If the hypothesis that $\sigma_x^2 = \sigma_y^2$ is true, then F should have a mean value of 1. Unfortunately, the F-distribution is not symmetric with respect to 1. As a rule, the tables given for the F-distribution are for one or two levels of significance (see Table 5 of the Appendix) and a one-tailed test. We present the following method for determining the critical region for a two-tailed test for a level of significance α; however, we make no attempt to justify the procedure. Assume that F is distributed with f_1 and f_2 degrees of freedom. Find U by looking in a table for the $\alpha/2$ level of significance for the F-distribution under f_1 and f_2 degrees of freedom. Find L as follows: Look in the table for the $\alpha/2$ level of significance under f_2 and f_1 degrees of freedom; let L' be the entry found; set $L = 1/L'$. Then the critical region will be all values of F outside the interval (L, U).

Example 4. We now go back and evaluate the significance of the value of F found in Example 18 of Chapter 10. Since Table 5 of the Appendix gives values for a one-tailed test at the 0.01 level of significance, we can use it to obtain a 0.02 level of significance for a two-tailed test. In Example 18, we have $f_1 = 5$ and $f_2 = 6$. Looking in the 5-column and 6-row, we find the value 8.75; this will be U. Looking in the 6-column and 5-row, we see 10.67; this is L'. Hence $L = 1/L' \doteq 0.09$. We will tend to reject the hypothesis that the variances are equal at the 0.02 level of significance if F lies outside (0.09, 8.75). Since the computed value of F lies inside the interval, we accept the hypothesis that the variances are equal.

It is also always possible to make the F-test for equal variances into a one-tailed test by labeling the larger sample variance s_x^2.

EXERCISES

1. a) Find a critical region for a one-tailed test involving a t-distribution with 13 degrees of freedom at the 0.05 level of significance; the 0.01 level of significance. (Assume that the critical region lies to the right of the mean.)
   b) Find a two-sided critical region for a test involving an F-distribution with $f_1 = 7$ and $f_2 = 9$ at the 0.02 level of significance.
   c) Do (b) with $f_1 = 4$ and $f_2 = 8$.

2. Random observations from two normally distributed populations X and Y are given below:

x	5	3	2	4.2	4.8	5.8
y	1	5	5.3	4.1	3.2	6

   a) Test the hypothesis that $m_x = m_y$.
   b) Test the hypothesis that the variates have equal variances.

3. The manufacturer of a breakfast cereal CRUNCHIES wants to determine whether CRUNCHIES are more popular than a competitor's MUNCHIES. A group of 100 adults are selected at random and divided randomly into two groups of 50 each. Both groups are given both cereals to taste. The results are given below:

	Group I	Group II
Prefers C	30	35
Prefers M	20	15

   Test the hypothesis that the cereals are equally popular. Does the experiment as it is outlined seem to be a genuine test of the popularity of the cereals? What criticisms do you have of the experiment? Design an experiment of your own to test the relative popularity of the cereals.

11.3 SOME REMARKS ON THE DESIGN OF EXPERIMENTS

There is of course far more to hypothesis testing than we can discuss in this book. Even a modestly developed exposition of sampling theory would require a text nearly half as long as this one. Nevertheless, the basic principles are here; these should enable the reader to do a fair amount of statistical analysis and hypothesis testing as well as serve as the foundation for further study.

As we have seen, testing a hypothesis statistically does not enable us to prove or disprove the hypothesis. Neither does it give us a probability that the hypothesis is true; the hypothesis is either true or false and will not be true part of the time and false part of the time. The test rather gives a measure of the tenability of the hypothesis in the light of observed events.

We have also seen that if a hypothesis H is to be directly testable by statistical methods, it is essential that H suggest an experiment E involving a random variable, say X, such that H and E specify the distribution of X. Very often a hypothesis can be tested indirectly by testing a hypothesis whose rejection will lead to acceptance of the given hypothesis. In this instance, the test must be designed in such a way that discrediting the null hypothesis will lead to accepting the alternative hypothesis we wish to test.

One hypothesis may suggest a number of different experiments to test it. Some of these experiments may be more practical than others, or preferable for other reasons. The choice of the test is left to the tester. We shall shortly examine several tests of one hypothesis.

The design of the test of a hypothesis must take into account the level of significance desired. For example, even though a coin may be tested for bias by flipping it n times, n must be large enough to accommodate the level of significance we want. It would be impossible to test a coin for bias at the 0.05 level of significance using only one or two flips. The critical region for a test of a hypothesis H will depend on the distribution of the test variate arrived at in accordance with H—as opposed to the true distribution of the test variate which would be different if H were false— the desired level of significance, and what we would accept if H were rejected. The probability that the variate will have a value in the critical region must not exceed the acceptable level of significance. Generally, one would choose as large a critical region as is consistent with the nature of the test and the level of significance desired.

It is of essential importance that the test variate not be influenced in any way that is inconsistent with the test. In other words, care must be taken that we have a true random variable; otherwise statistical methods are invalid. For example, in testing a coin for bias by flipping it, we must be sure that the coin is not influenced toward one face or the other by the way we flip it. If we happen to be testing a gambler's claim that he can throw more 7's with a pair of unbiased dice than chance would normally allow, we must be certain that the dice are fair and that only the gambler's way of rolling the dice is a significant factor in determining how the dice turn up.

It is, of course, not at all an easy task to keep extraneous factors from influencing the outcome of a test. Certainly a fair percentage of tests will be invalid due to improper control of the test variate; and, if an experimenter happens to have an unusually strong desire to validate a certain hypothesis, he may subconsciously or consciously perform the experiment in such a way as to obtain the result he wants. The following example illustrates how we may take safeguards to keep a variable random.

Example 5. A company has two types of fruitcake, A and B, that it can produce, but it wants to know which of the types will have the greater market appeal. The hypothesis which the company tests is:

(6)   Fruitcake A has the same appeal as Fruitcake B.

It is decided that the hypothesis will be tested by asking the opinions of 100 adults, and examining the variate X, whose value is the number of these people who prefer fruitcake A. One hundred adults are selected at random from the population of the city where the company is located. The company could have restricted its selection of adults to known consumers of fruitcake. This would have "biased" the test, perhaps in a justifiable manner. The company may in fact wish to restrict its testing to those who are known to like some type of fruitcake. As a matter of fact, by restricting the choice of subjects to a particular locale, the test already has a certain bias. This latter bias might also be desirable if the company's market were primarily in the city where its plant was located.

Each subject is to be allowed to taste both fruitcakes and express his opinion as to which he prefers. Since there may be a tendency toward choosing the fruitcake tasted last, the 100 subjects are divided randomly into two groups, I and II, of 50 each. Group I will taste Fruitcake A first, while Group II will taste Fruitcake B first. One method of random division would be as follows: Two subjects at a time are called forward in alphabetical order. A fair coin is then flipped fairly. If the coin turns up heads, the first person in alphabetical order is placed in Group I, and the second person in Group II. If the coin comes up tails, the first person goes to Group II and the second person to Group I. One method of nonrandom division would be to put the 50 oldest subjects in Group I and the rest in Group II.

Observe that those conducting the test decide how many people should be tested, into how many groups these people should be divided, and how many people should be placed in each group. Chance should decide which people are actually tested and the group into which each one is placed.

EXERCISES

1. Suppose we have a table of integers from 1 to 1000 arranged in random order.
    a) Design a way in which the table might be used to select 100 people at random from a population of 1000 people.
    b) Design a way in which the table might be used to divide 1000 people randomly into two groups of 500 each.
    c) Design a way in which such a table might be used to select 10 professors at random from a population of 100 professors.

2. Suppose in Example 4 that 54 people of Group I and 53 people of Group II prefer Fruitcake A. Can we reject (6) at the 0.05 level of significance?

3. In each of the following a hypothesis is to be tested. In each case, design a test for the hypothesis. In drawing up your test, indicate how the integrity of the test will be safeguarded; in particular, explain the procedures to be used to keep the results of the test from being warped by extraneous factors.

   a) A cigarette company wants to know how many doctors smoke their brand. The hypothesis to be tested is: More doctors smoke Brand X than all other brands combined.

   b) A floor wax manufacturer must choose to produce one of two new floor waxes, A and B. The hypothesis to be tested is: Wax A wears better under heavy traffic than Wax B.

   c) A manufacturer of toothpaste wonders whether sending free samples to people is an effective means of getting them to change brands. The hypothesis to be tested is: At least 50% of those who receive a free sample of a new toothpaste will buy a tube of that toothpaste within a month after receiving the sample.

11.4 AN EXAMPLE

This section will be devoted to discussing various ways in which the fruitcake company of *Example 5* can test its hypothesis that Fruitcake A has the same appeal as Fruitcake B. We shall consider three distinct tests of this hypothesis.

Test I (*the Sign Test*). One hundred adults are selected at random; the 100 subjects selected are then divided randomly into 10 groups of 10 each. We will designate the groups by $1, 2, \ldots, 10$. All groups are allowed to taste each fruitcake; half of the subjects in each group taste Fruitcake A first, the other half taste Fruitcake B first. The preferences are recorded as shown in Table 3.

Table 3

Group	1	2	3	4	5	6	7	8	9	10
A	5	7	6	4	8	5	6	6	8	4
B	5	3	4	6	2	5	4	4	2	6

It appears from Table 3 that A is more popular than B. For each group we consider the quantity

(7)   Number of subjects who prefer A — number of subjects who prefer B.

We define x to be the number of groups for which (7) is positive; that is, x will be the number of groups in which A is preferred by a majority of members. If A and B are equally appealing, then X will have the density function f defined by

(8) $\quad f(x) = C(10, x)(\frac{5}{11})^x(\frac{6}{11})^{10-x}$.

The probability $\frac{5}{11}$ that X will assume a positive value in any one group comes from the fact that there are 11 possibilities for *number of subjects who prefer A* for any one group and exactly 5 of these possibilities make (7) positive.

Since the data of Table 3 seem to indicate that A is preferred to B, we shall use a one-tailed test to see whether (6) can be rejected in favor of accepting

(9) Fruitcake A is preferable to Fruitcake B.

It is decided that a 0.05 level of significance is sufficient. Since the test is one-tailed, the critical region is found to be $\{8, 9, 10\}$; this is the largest set of values of X consistent with (9) such that X has less than a 0.05 probability of assuming value in that set. The observed value of X, however, is 6. Therefore using this test we cannot reject (6) in favor of (9) at the 0.05 level of significance. (We would not have been able to reject (6) using a two-tailed test at the 0.05 level of significance either.)

Test II (*the Magnitude Test*). Test I was very crude. If Fruitcake A were not decisively preferred to Fruitcake B, then Test I would not tend to lead us to reject (6) in favor of (9). The situation resembles that of Example 1 when p is quite close to $\frac{1}{2}$; in that example, the larger p is, the better chance the test has of working. But if p is quite close to $\frac{1}{2}$, we have a very good chance of making a Type 2 error. Likewise, if Fruitcake A is only slightly preferable to Fruitcake B, we are not likely to detect the preference using Test I. We now design a test which uses the magnitude of (7) instead of merely its sign. Clearly this should be more sensitive than Test I.

Table 4

Group	1	2	3	4	5	6	7	8	9	10
(7)	0	4	2	−2	6	0	2	2	6	−2

If we assume that (6) is true, then the sign of each entry in Table 4 [the entries being the observed values of (7)] is just as likely to be the opposite of what is observed. In other words, we assume that we are dealing with 10 pairs of numbers,

(10) $\quad$ (0, 0), (4, −4), (2, −2), (−2, 2), (6, −6),
$\qquad$ (0, 0), (2, −2), (2, −2), (6, −6), (−2, 2),

with the first number in each pair being as likely as the second number. We let

$$x = y_1 + y_2 + \cdots + y_{10},$$

where y_i is one of the numbers from the ith pair of numbers. If the y_i occur randomly, as during the experiment, then x defines a random variable X. The largest value x can assume is

$$0 + 4 + 2 + 2 + 6 + \cdots + 2 = 26$$

and the smallest value it can assume is -26. The closer the actual value of x obtained in the test is to 26, the larger the inconsistency of (6). There are 2^8 ten-tuples (there are only 2^8 instead of 2^{10}, since $(0, 0)$ occurs twice) which can be formed using one number from each of the 10 pairs of numbers in (10); only the ten-tuple

$$(0, 4, 2, 2, 6, 0, 2, 2, 6, 2)$$

gives $x = 26$. If f denotes the density function of X, then under hypothesis (6), which implies that all ten-tuples have the same probability, we have

$$f(26) = \tfrac{1}{2}^8 = \tfrac{1}{256} \doteq 0.004.$$

The next smallest admissible value of X is 22 (formed by changing any one of the 2's to a -2). There are 5 ten-tuples in all which sum to 22; hence

$$f(22) = \tfrac{5}{256} \doteq 0.02.$$

If we wanted a two-tailed test at the 0.05 level of significance, since $f(x) = f(-x)$, we would use a critical region of $\{-26, -22, 22, 26\}$. The probability that x will assume a value in this critical region is approximately 0.048.

If we wish to reject (6) in favor of (9), we use a one-tailed test. The next smallest admissible value of X (after 22) is 18, obtained by changing two of the 2's to -2's, or keeping everything positive except the 4; there are 11 ten-tuples, which give 18. Therefore

$$f(18) = \tfrac{11}{256} \doteq 0.044.$$

Thus the largest critical region that can be used for a one-tailed test [with the idea of accepting (9) if (6) is rejected] is $\{22, 26\}$. The value of x actually obtained is 18. Consequently, with either a one-tailed or a two-tailed test, we cannot reject (6).

If we had been willing to settle for a 0.10 level of significance, then we could have used the critical region $\{18, 22, 26\}$ for a one-tailed test. We could then reject (6) in favor of (9).

Test III (*using the normal distribution*). Let the 100 subjects be denoted by $1, 2, \ldots, 100$. We define $y_i = 1$ if Person i prefers Fruitcake A and

$y_i = 0$ if Person i prefers Fruitcake B. Then Y_i is a random variable with admissible range $\{0, 1\}$ and density function f_i defined by

$$f_i(1) = f_i(0) = \tfrac{1}{2}$$

[since, according to (6) it is just as likely that the subject will prefer Fruitcake A as Fruitcake B]. Define

$$X = \sum_{i=1}^{100} Y_i.$$

According to the Central Limit Theorem (Section 10.1), X is approximately normally distributed with mean $(100 \times \tfrac{1}{2})$ and variance $(100)(\tfrac{1}{2})^2$, since each Y_i is binomially distributed with $n = 1$ and $p = \tfrac{1}{2}$. Therefore

$$Z = \frac{X - 50}{5}$$

can be considered a standard normal variate.

If we are to reject (6) in favor of (9), then we should have a critical region containing values of Z which deviate from 0 in the positive direction (since we want the value of X to be larger than 50). If we want a 0.05 level of significance with a one-tailed test, then we see from Table 2 of the Appendix that $[1.64, \infty)$ is a suitable critical region. Since the observed value of Z is $(59 - 50)/5 = 1.8$, we *can* reject (6) in favor of (9).

The three tests presented here are not exhaustive as far as testing (6) on the basis of the information presented in Table 3 is concerned. Tests might also be designed, for example, using the χ^2-distribution. The design of such a test is left as an exercise.

EXERCISES

1. Find a critical region for Test I for a two-tailed test at the 0.05 level of significance.

·2. Would there be any substantial difference in Test I between having half of the groups taste Fruitcake A first rather than having half the members of each group taste Fruitcake A first? What if there was a tendency to pick the fruitcake tasted last?

3. Prove that the number of ten-tuples in Test II, which give $x = 18$, is 11; that is, prove $f(18) = \tfrac{11}{256}$. What is $f(-18)$? What is the largest admissible value of X less than 18? Find $f(w)$, where w is that value. Prove that $f(x) = f(-x)$ for each admissible value of X.

4. Design a test for (6) using χ^2. There is more than one such test.

5. Design several tests to determine whether a certain new drug is effective in treating a certain ailment. After you have formulated several tests, discuss

each test from the viewpoint of its sensitivity to the data to be gathered and its practicality from the viewpoint of implementation.

6. Assume that there are 300,000 medical doctors in the United States. A cough syrup company claims that 9 out of 10 doctors recommend its product. Design a test using a random sample of 500 doctors to test the company's claim.

7. What would the probability of a person choosing Fruitcake A have to be in order for us to have a 0.50 chance of rejecting (6) at the 0.05 level of significance according to the procedure of Test I?

8. Which of the three tests given seems the most sensitive to the data? which is the least sensitive? Since all three tests are based on essentially the same data, there is little difference in their practicality.

12 Functions of Several Variables

12.1 MULTIVARIATE FUNCTIONS

Thus far we have restricted our attention to functions of one variable. A function f of one variable pairs each element x of its domain with a unique element $f(x)$ of its range; given x, we can find $f(x)$. In such a case, $f(x)$ depends only on x (and the function). There are many instances, however, in which some one factor depends on many other factors or variables, and not just one. If we set

(1) $\quad z = x^2 + xy + y^2, \quad 0 \leq x \leq 1, \quad 0 \leq y \leq 1,$

then z depends not only on x, but on y as well. In order to determine z, we must have values for x and y. We can say that z is a function of both x and y. If we wanted to do so, it would be appropriate to write $z(x, y)$ to show the dependence of z on x and y. Then $z(x, y)$ associates each ordered pair (x, y) of real numbers (such that $0 \leq x \leq 1$ and $0 \leq y \leq 1$ with the real number $x^2 + xy + y^2$. The domain of z is represented in

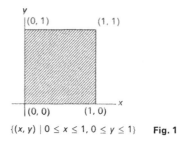

$\{(x, y) \mid 0 \leq x \leq 1, 0 \leq y \leq 1\}$ **Fig. 1**

Fig. 1. We see that z is a function from the portion of the plane depicted in Fig. 1 into R, the set of real numbers.

The reader has already encountered functions of several variables in the area formulas of elementary geometry; for example, the area A of a rectangle is a function of both the length x and width y of the rectangle, $A(x, y) = xy$. The volume of a rectangular parallelepiped (box) is a function of three variables, the length, width, and height.

Given two random variables X and Y defined on the same sample space (which may or may not be independent), we may wish to know probabilities associated with (x, y), the simultaneous or *joint* occurrence of the values x and y.

Example 1. Coin A is flipped in such a way that a head is twice as probable as a tail, while coin B is a fair coin flipped fairly. Define variates X and Y by letting x and y be 0 or 1, depending on whether a head or a tail appears on coin A and coin B, respectively. The density functions f and g of X and Y, respectively, are given by

(2) $\quad f(0) = \tfrac{2}{3}, \quad f(1) = \tfrac{1}{3};$
$\quad\quad g(0) = g(1) = \tfrac{1}{2}.$

The variates X and Y are independent in the sense that the value of one does not affect the probability distribution of the other. Therefore

$P(x, y)$ (the probability that x and y will occur jointly) $= f(x)g(y)$

(see Section 1.4). We give $P(x, y)$, the *joint density function* of X and Y, in Table 1.

Table 1

y \ x	0	1	Sum $= g(y)$
0	$\tfrac{1}{3}$	$\tfrac{1}{6}$	$\tfrac{1}{2}$
1	$\tfrac{1}{3}$	$\tfrac{1}{6}$	$\tfrac{1}{2}$
Sum $= f(x)$	$\tfrac{2}{3}$	$\tfrac{1}{3}$	

The density functions of X and Y can be recovered from Table 1. The sum of the entries in the first column gives us $f(0)$. This is because $x = 0$ is equivalent to $(x = 0, y = 0) \cup (x = 0, y = 1)$. Hence

$P(x = 0) = f(0) = P(0, 0) + P(0, 1) = \frac{1}{3} + \frac{1}{3}$.

We could also use Table 1 to find the density function h of the variate $Z = X + Y$. It is left as an exercise to verify that h is defined by

(3) $h(0) = P(0, 0) = \frac{1}{3}, \quad h(1) = \frac{1}{2}, \quad h(2) = \frac{1}{6}.$

We shall generalize the notion of a probability distribution function to the case involving several variates.

Definition 1. *Given random variables* $X_1, X_2, \ldots, X_n$, *all defined on the same sample space, the **joint distribution function** F of the n variates is defined by*

$F(y_1, \ldots, y_n) = P(x_1 \leq y_1, x_2 \leq y_2, \ldots, x_n \leq y_n).$

The variates $X_1, \ldots, X_n$ *are said to be **independent** if*

$F(x_1, x_2, \ldots, x_n) = F_1(x_1) F_2(x_2) \cdots F_n(x_n),$

where F_i is the distribution function of x_i, $i = 1, \ldots, n$.

Observe that a joint distribution function is a straightforward generalization of the notion of a distribution function for a single random variable. We define the *joint density function* for several discrete variates as follows:

Definition 2. *Given discrete random variables* $x_1, \ldots, x_n$, *the **joint density function** f of the n variates is defined by*

$f(x_1, \ldots, x_n) = P(x_1, x_2, \ldots, x_n).$

*Assume that the admissible range of X_i is A_i, $i = 1, \ldots, n$. Then the **marginal density function** f_i of X_i is defined by*

(4) $f_i(x_i) = \sum_{\substack{\text{all } A_j \\ i \neq j}} f(x_1, \ldots, x_i, \ldots, x_n).$

We shall define the density function for the case of continuous variates after we have introduced partial differentiation. The joint density function of Definition 2 has many properties analogous to the properties of a density function of a single variate. We state some of these properties in the following proposition; the proof is left to the reader.

Proposition 1. *Suppose the variates* $X_1, \ldots, X_n$ *are all discrete with joint distribution function F and joint density function f, and each variate X_i*

has admissible range A_i, $i = 1, \ldots, n$. Then:

a) If $a_1, \ldots, a_n$ are admissible values of $X_1, \ldots, X_n$, respectively, it follows that

$$F(a_1, a_2, \ldots, a_n) = \sum_{\substack{\text{all } x_i \text{ in } A_i \\ x_i \le a_i \\ i=1,2,\ldots,n}} f(x_1, \ldots, x_n).$$

c) The marginal density function f_i of X_i as defined in (4) is the density function of X_i.

d) The $X_1, \ldots, X_n$ are mutually independent if and only if

$$f(x_1, x_2, \ldots, x_n) = f_1(x_1) f_2(x_2) \cdots f_n(x_n),$$

where the f_i are the (marginal) density functions of x_i.

Example 2. Consider the bivariate distribution whose density function is defined by Table 2.

Table 2

x_2 \ x_1	0	1	2	3
0	0.1	0.03	0.17	0.09
1	0.05	0.2	0.1	0.08
2	0.04	0.01	0.03	0.1

The admissible ranges of X_1 and X_2 are

$A_1 = \{0, 1, 2, 3\}$ and $A_2 = \{0, 1, 2\}$,

respectively. The admissible range of (X_1, X_2) is $A_1 \times A_2$. Let F be the joint distribution function of X_1 and X_2. By way of example, we will compute $F(1, 2)$:

$$F(1, 2) = f(0, 0) + f(0, 1) + f(0, 2) + f(1, 0) + f(1, 1) + f(1, 2)$$
$$= 0.1 + 0.05 + 0.04 + 0.03 + 0.2 + 0.01 = 0.43.$$

Observe, too, that the sum of all of entries of Table 2 is 1. Why must this be the case?

The density functions f_1 and f_2 for X_1 and X_2, respectively, can be computed from Table 2. For example,

$$f_1(0) = f(0, 1) + f(0, 1) + f(0, 2) = 0.1 + 0.05 + 0.04 = 0.19.$$

Complete definitions of f_1 and f_2 are as follows:

(5) $\quad f_1(0) = 0.19, \quad f_1(1) = 0.24, \quad f_1(2) = 0.30, \quad f_1(3) = 0.27,$
$\quad\quad\; f_2(0) = 0.39, \quad f_2(1) = 0.43, \quad f_2(2) = 0.18.$

Evidently, X_1 and X_2 cannot be independent since $f(x_1, x_2)$ is not always equal to $f_1(x_1)f_2(x_2)$. For example,

$$f(0, 0) = 0.1 \neq f_1(0)f_2(0) = (0.19)(0.39).$$

We can, using Table 2 and the probability axiom

$$P(A \cap B) = P(A)P(B \mid A),$$

find a *conditional density function* of $X_1 \mid X_2$ (or $X_2 \mid X_1$). For example, if g is this function,

$g(y_1 \mid y_2)$ (the probability that $x_1 = y_1$ if $x_2 = y_2$) $= f(y_1, y_2)/f_2(y_2)$.

Thus

$$g(0 \mid 1) = f(0, 1)/f_2(1) = 0.05/0.43 = \tfrac{5}{43}.$$

Note that although the density functions of X_1 and X_2 can be deduced from Table 2, the joint density function f of X_1 and X_2 cannot be deduced from these individual density functions unless X_1 and X_2 are independent, in which case $f(x_1, x_2) = f_1(x_1)f_2(x_2)$. If X_1 and X_2 are not independent, then both the density functions for X_1 and X_2 and the conditional density function for $X_1 \mid X_2$ or $X_2 \mid X_1$ would have to be known before we could find the entries for Table 2.

Before proceeding to joint density and distribution functions for several continuous variates, we must learn something about differentiating and integrating functions of several variables. We begin this study in the next section.

EXERCISES

1. Consider Table 3 (X_1 and X_2 are discrete variates)

Table 3

x_2 \ x_1	1	2	3	4
10	0.06	0.12	0.18	0.24
11	0.04	0.08	0.12	0.16

 a) Find the density functions f_1 and f_2 of X_1 and X_2, respectively.
 b) Let F be the joint distribution function of X_1 and X_2. Find $F(2, 11)$, $F(10, 4)$, and $F(11, 4)$.
 c) Let g be the conditional density function of $X_1 \mid X_2$. Find $g(x_1 \mid x_2 = 10)$ for each admissible value of X_1.
 d) Determine whether X_1 and X_2 are independent.

2. Prove each statement of Proposition 1.

3. a) Define $Z = X_1 + X_2$, where the joint distribution of X_1 and X_2 is determined by Table 3. Find the density function for Z.
   b) Define $W = X_1 X_2$, where the joint distribution of X_1 and X_2 is determined by Table 2. Find the density function for W.

4. Determine what the entries of Table 2 would have to be if the variates X_1 and X_2 of Example 2 were to be independent. In Example 2, find the values of $g(x_1 \mid x_2 = 2)$ for each admissible value of X_1.

5. Two fair dice are rolled together ten times. Let x be the number of 2's obtained with the first die and y be the number of 6's obtained with the second die. Find the joint density function of X and Y, assuming that X and Y are independent.

6. A fair die is rolled ten times. Let x be the number of times that a 6 occurs and y be the number of times that a 5 occurs in the ten rolls. Find the joint density function of X and Y. Can X and Y be independent?

12.2 PARTIAL DIFFERENTIATION

Suppose f is a function of two variables x and y. There are several ways in which we could generalize the notion of a derivative of a function at a point a (Definition 3 of Chapter 4) to f. For example, we could define $f'(a, b)$ to be

(6) $\quad \lim_{h \to 0} \dfrac{f(a+h, b+h) - f(a, b)}{h}.$

The point (a, b) is a point of the plane. If f were a function only of x, then $x + h$ could "approach" a (as h approaches 0) only along a line (Fig. 2).

```
         a + h₁      a + hₙ
  ─────────●─●─●●●●─────────  Fig. 2
              a + h₂  a
```

Since (a, b) is a point of the plane, if we wish to consider a sequence of points which approach (a, b), the sequence could approach (a, b) along any one of infinitely many lines, or not along a line at all (Figs 3 and 4). In (6), we are only considering points which approach (a, b) along the line through (a, b) with slope 1 (Fig. 5). We may consider (6) to give the derivative of f in the direction of slope $+1$. Reasoning in a similar fashion, we see that

(7) $\quad \lim_{h \to 0} \dfrac{f(a + 2h, b - h) - f(a, b)}{h}$

would be a derivative of f in the direction of slope $-\frac{1}{2}$. Expression (7), if the limit exists, gives the rate of change of $f(x, y)$ at (a, b) along the line through (a, b) with slope $-\frac{1}{2}$.

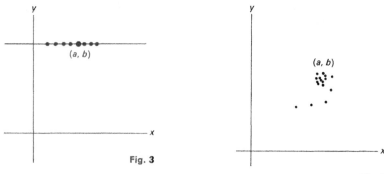

Fig. 3

Fig. 4

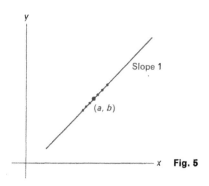

Fig. 5

The two "most important" lines through (a, b) are those parallel to the x- and y-axes. The rates of change of $f(x, y)$ at (a, b) in the direction of these lines have special significance. The derivatives of f along these lines are given by

(8) $\quad \lim_{h \to 0} \dfrac{f(a + h, b) - f(a, b)}{h}$

and

(9) $\quad \lim_{h \to 0} \dfrac{f(a, b + h) - f(a, b)}{h}.$

In (8), y is constant while x varies; in (9), x is constant, while y varies. In (8), the effect is the same as that we would obtain by considering f only as a function of x and taking the derivative with respect to x at a (with $y = b$). In (9), f is considered as a function of y alone with $x = a$. We call (8), if it exists, the *first partial derivative of f with respect to x at* (a, b); (9) would be the *first partial derivative of f with respect to y at* (a, b). We generalize (8) and (9) now to the case of a function of two or more variables.

Definition 3. *Suppose f is a function of the variables $x_1, \ldots, x_n$. Then the first partial derivative of f at $(a_1, a_2, \ldots, a_n)$ **with respect to** x_i is defined to be*

(10) $$\lim_{h \to 0} \frac{f(a_1, \ldots, a_i + h, \ldots, a_n) - f(a_1, \ldots, a_n)}{h},$$

if this limit exists. We denote (10) by

$$\left.\frac{\partial f}{\partial x_i}\right|_{(a_1, \ldots, a_n)}.$$

We may also take the formal derivative of f with respect to x_i, considering the variables other than x_i as constants (that is, we would be finding $f'(x_i)$ as if f were a function of x_i alone). This formal derivative is called the **first partial derivative of f with respect to** x_i and is denoted by $\partial f/\partial x_i$.

If $\partial f/\partial x_i$ is defined for $(a_1, \ldots, a_n)$, then its value at $(a_1, \ldots, a_n)$ will be

$$\left.\frac{\partial f}{\partial x_i}\right|_{(a_1, \ldots, a_n)}.$$

Example 3. Suppose

$$f(x, y) = y^2 + xe^y + 3.$$

Then $\partial f/\partial x = e^y$. Note that wherever y appears, it is treated as a constant in computing $\partial f/\partial x$. If we wish to compute

$$\left.\frac{\partial f}{\partial x}\right|_{(3,4)},$$

we substitute 3 for x and 4 for y in $\partial f/\partial x$; thus,

$$\left.\frac{\partial f}{\partial x}\right|_{(3,4)} = e^4.$$

We also have

$$\frac{\partial f}{\partial y} = 2y + xe^y \quad \text{and} \quad \left.\frac{\partial f}{\partial y}\right|_{(3,4)} = 2(4) + 3(e^4) = 8 + 3e^4.$$

It is also possible to take the partial derivative with respect to either x or y of either $\partial f/\partial x$ or $\partial f/\partial y$. Thus,

$$\frac{\partial(\partial f/\partial x)}{\partial x} = 0,$$

but

$$\frac{\partial(\partial f/\partial x)}{\partial y} = e^y.$$

If f is a function of $x_1, \ldots, x_n$, then

$$\frac{(\partial f/\partial x_i)}{\partial x_j}$$

is usually denoted by

$$\frac{\partial^2 f}{\partial x_i\, \partial x_j}.$$

This function, if it is defined, is one of the second derivatives of f. In certain pathological instances (with which we shall not concern ourselves), $\partial^2 f/\partial x_i\, \partial x_j$ is not the same as $\partial^2 f/\partial x_j\, \partial x_i$. Thus, it is possible for a function of only two variables x_1 and x_2 to have as many as four distinct second derivatives:

$$\frac{\partial^2 f}{\partial x_1\, \partial x_1},\quad \frac{\partial^2 f}{\partial x_2\, \partial x_2},\quad \frac{\partial^2 f}{\partial x_1\, \partial x_2},\quad \frac{\partial^2 f}{\partial x_2\, \partial x_1}.$$

The definition and notation for derivatives higher than the second are analogous to those for the second derivative. Hence

$$\frac{\partial^3 f}{\partial x_1\, \partial x_2\, \partial x_2}$$

denotes

$$\frac{\partial(\partial f^2/\partial x_1\, \partial x_2)}{\partial x_2}.$$

Example 4. Consider the function

$$f(x_1, x_2, x_3) = x_1 x_2^2 + e^{x_1 x_2 x_3} + (x_3)^4.$$

Then

$$\frac{\partial f}{\partial x_3} = (x_1 x_2) e^{x_1 x_2 x_3} + 4(x_3)^4$$

and

$$\frac{\partial^2 f}{\partial x_3\, \partial x_2} = (x_1 x_3)(x_1 x_2)^{x_1 x_2 x_3}.$$

Suppose now that f is a function of the variables x and y. If $f(a, b)$ is to be a relative maximum of $f(x, y)$, then $f(a, b)$ will also be a relative maximum of the functions $f(x, b)$ and $f(a, y)$ of one variable. For $f(a, b)$ will be a relative maximum of $f(x, y)$ as we approach (a, b) along any line through (a, b); hence it will be a relative maximum along the lines parallel to the x- and y-axes. The geometrical situation is illustrated in Fig. 6.

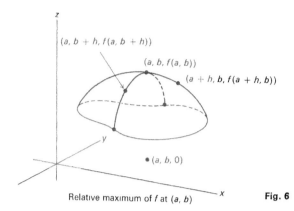

Relative maximum of f at (a, b) Fig. 6

If $f(x, b)$ has a relative maximum at $x = a$, and

$$f'(a, b) = \left.\frac{\partial f}{\partial x}\right|_{(a,b)}$$

exists, then $f'(a, b) = 0$. Similarly, if $f(a, y)$ has a maximum at $y = b$, then if

$$f'(a, b) = \left.\frac{\partial f}{\partial y}\right|_{(a,b)}$$

exists, it will equal 0. We therefore conclude that if f is suitably differentiable and $f(a, b)$ is a relative maximum of f, then

$$(11) \quad \left.\frac{\partial f}{\partial x}\right|_{(a,b)} = \left.\frac{\partial f}{\partial y}\right|_{(a,b)} = 0.$$

It also follows that if $f(a, b)$ is a relative minimum of $f(x, y)$ and f is suitably differentiable, then (11) will hold.

Unfortunately, however, (11) can be satisfied even when we have no relative maximum or minimum of f. (Recall that when f was a function of x alone, $f'(a) = 0$ did not ensure that f had either a relative maximum, or minimum, at a.) Condition (11) may be satisfied where there is only a point of inflection with respect to either x or y, or we may even have a maximum with respect to x and a minimum with respect to y. This latter state of affairs, which gives a so-called *saddle point*, is illustrated in Fig. 7.

Definition 4. *A function of f of the variables $x_1, \ldots, x_n$ is said to have a* **critical point** *at $(a_1, \ldots, a_n)$ if*

$$\left.\frac{\partial f}{\partial x_i}\right|_{(a_1,\ldots,a_n)} = 0, \quad i = 1, \ldots, n.$$

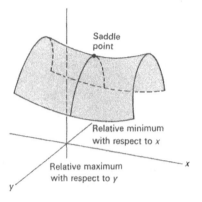

Fig. 7

Therefore if the function f in Definition 4 has a relative maximum or minimum at $(a_1, \ldots, a_n)$ and if $\partial f/\partial x_i$ exists for each x_i at $(a_1, \ldots, a_n)$, then $(a_1, \ldots, a_n)$ will be a critical point of f. But a critical point of f need not yield a relative maximum or minimum of f. The next proposition gives some information about testing a critical point for a function of two variables.

Proposition 2. *Suppose f is a function of x and y and (a, b) is a critical point of f. Evaluate*

(12) $$\left(\frac{\partial^2 f}{\partial x^2}\right)\left(\frac{\partial^2 f}{\partial y^2}\right) - \left(\frac{\partial^2 f}{\partial x\, \partial y}\right).$$

at (a, b). Then if

a) *(12) is positive and $\partial^2 f/\partial x^2 + \partial^2 f/\partial y^2$ is negative, $f(a, b)$ is a relative maximum;*

b) *(12) is positive and $\partial^2 f/\partial x^2 + \partial^2 f/\partial y^2$ is positive, $f(a, b)$ is a relative minimum;*

c) *(12) is negative, then $f(a, b)$ is neither a relative maximum nor a relative minimum;*

d) *(12) is 0, then the test yields no information about the nature of $f(a, b)$.*

Example 5. Suppose $f(x, y) = x^2 + xy - y^2 + x$, for all real numbers x and y. Then $\partial f/\partial x = 2x + y + 1$ and $\partial f/\partial y = x + 2y$. Solving

$2x + y + 1 = 0,$
$x + 2y = 0$

simultaneously, we find that f has a critical point at $(\tfrac{1}{3}, \tfrac{1}{3})$. Now

$$\frac{\partial^2 f}{\partial x^2} = 2, \qquad \frac{\partial^2 f}{\partial y^2} = 2, \qquad \frac{\partial^2 f}{\partial x\, \partial y} = 1.$$

Therefore, in this instance,

$(12) = (2)(2) - (1) = -4 - 1 = -5.$

We also find that $\partial^2 f/\partial x^2 + \partial^2 f/\partial y^2 = 4$. Therefore by (b) of Proposition 2, $f(\frac{1}{3}, \frac{1}{3})$ is a relative minimum of $f(x, y)$.

Example 6. Suppose $f(x, y) = x^2 - y^2$. Then

$$\frac{\partial f}{\partial x} = 2x \quad \text{and} \quad \frac{\partial f}{\partial y} = -2y.$$

Also

$$\frac{\partial^2 f}{\partial x^2} = 2, \quad \frac{\partial^2 f}{\partial y^2} = -2, \quad \frac{\partial^2 f}{\partial x\, \partial y} = 0.$$

Therefore f has a critical point at $(0, 0)$. The value of (12) for this example is $(2)(-2) - (0)^2 = -4$. Therefore by (c) of Proposition 2, f has neither a relative maximum nor minimum at $(0, 0)$. Note that at $(0, 0)$, f has a relative maximum with respect to y, but a relative minimum with respect to x; we have the geometric situation of Fig. 7.

Suppose now that F is the joint distribution function of the continuous variates $X_1, \ldots, X_n$. If F were a function only of the single variate X_1, then the density function for X_1 would be $F'(x_1)$. Since F is a function of n variates, to obtain a density function, we shall take the derivative with respect to each of the variates. Specifically, we make the following definition.

Definition 5. *If F is the joint distribution function of the variates $X_1, \ldots, X_n$, then the **joint density function** f of the variates $X_1, \ldots, X_n$ is defined to be*

$$f(x_1, \ldots, x_n) = \frac{\partial^n F}{\partial x_1\, \partial x_2 \cdots \partial x_n},$$

if this function exists.

As with distributions of one variate, the joint distribution of several variables $X_1, \ldots, X_n$ can be specified by giving the admissible range of $(X_1, \ldots, X_n)$ together with either the joint distribution function or the joint density function.

Example 7. Suppose the variates X and Y have the joint distribution function F defined by

$$F(x, y) = 1 - e^{-x} - e^{-y} + e^{-(x+y)},$$

where the admissible range of (X, Y) consists of all points of the first quadrant of the plane, that is $\{(x, y) \mid x \geq 0, y \geq 0\}$. Then the joint density function f of X and Y is given by

$$f(x, y) = \frac{\partial^2 F}{\partial x\, \partial y} = e^{-(x+y)}.$$

X and Y are both exponentially distributed variates, and have distribution functions $F_1(x) = 1 - e^{-x}$ and $F_2(y) = 1 - e^{-y}$, respectively; thus they would be independent since $F(x, y) = F_1(x)F_2(y)$.

We further note that since X and Y are exponentially distributed, their density functions are $f_1(x) = e^{-x}$ and $f_2(y) = e^{-y}$; hence we would also have $f(x, y) = f_1(x)f_2(y)$. In general, the following is true:

Proposition 3. *Continuous variates $X_1, \ldots, X_n$ with joint density function $f(x_1, \ldots, x_n)$ are mutually independent if and only if*

$$f(x_1, \ldots, x_n) = f_1(x_1)f_2(x_2) \cdots f_n(x_n),$$

where $f_i(x_i)$ is the density function of X_i.

EXERCISES

1. Find $\partial f/\partial x_1$, $\partial f/\partial x_2$, $\partial^2 f/\partial x_1 \partial x_2$, and $\partial^2 f/\partial x_2^2|_{(1,1,\ldots,1)}$ for each of the following functions of several variables.
   a) $f(x_1, x_2) = x_1 x_2$
   b) $f(x_1, x_2) = x_1^2 + x_2 + 3$
   c) $f(x_1, x_2) = e^{x_1 x_2} + e^{-x_1 x_2}$
   d) $f(x_1, x_2, x_3) = x_1 x_2 x_3 + x_3^2$
   e) $f(x_1, x_2, x_3) = x_1 - \dfrac{x_3}{(x_1 + x_2)}$

2. Find all the critical points of the following functions. Where possible, use Proposition 2 to test the nature of these critical points.
   a) $f(x, y) = x^2 + y^2$
   b) $f(x, y) = xe^{-x(1+y)}$
   c) $f(x, y) = x^3 + y^3 + 6xy$
   d) $f(x, y) = x^2 - y^2$
   e) $f(x, y) = x^2 - 2y^3 + 3xy + 7$

3. Find the value of the derivative of $f(x, y) = xy$ at $(0, 0)$ along the line $y = 2x$.

4. Prove that the rectangular parallelepiped of maximum volume with surface area 54 is a cube with side 3. [*Hint:* If x, y, and z are the dimensions of the parallelepiped, then its volume is $V(x, y, z) = xyz$. We wish to find a maximum of $V(x, y, z)$, while satisfying the additional condition
$$54 = 2(xy + yz + xz).$$

5. Find three positive numbers x, y, and z whose sum is 300 and such that $f(x, y, z) = xy + yz + xz$ is maximal.

12.3 MULTIPLE INTEGRATION

If we can differentiate a function of several variables with respect to one variable at a time, that is, take the derivative of such a function with respect to one variable while considering the other variables as constants, then we can also integrate a function of several variables with respect to one variable at a time. This process is illustrated in the following example.

Example 8. Consider

(13) $\quad \int_0^1 \int_x^{x^2} xy\, dy\, dx.$

To evaluate (13), we integrate first with respect to y, considering x as a constant. Since x is considered constant during the first integration, the limits of integration for the first integration can involve x. We shall give a geometric interpretation of using limits of integration involving expressions in variables shortly. We use the notation $g(x)\big|_{L_2}^{L_1}$ to stand for $g(L_1) - g(L_2)$. Carrying out the first integration called for in (13), we obtain

(14) $\quad \int_0^1 \frac{(xy^2)}{2}\bigg|_x^{x^2} dx = \int_0^1 \left(\frac{x^5}{2} - \frac{x^3}{2}\right) dx.$

The second integration gives

$$\frac{x^6}{12} - \frac{x^4}{8}\bigg|_0^1 = \frac{1}{12} - \frac{1}{8} = -\frac{1}{24}.$$

Recall from our discussion of single integration, that integration is essentially a process of summation (or the limit of a process of summation). Geometrically, $\int_a^b f(x)\, dx$ gives us the area under the graph of f from a to b. Then

(15) $\quad \int_{a(y)}^{b(y)} f(x, y)\, dx$

gives the area under the graph of f from $a(y)$ to $b(y)$. This area is represented in Fig. 8. Note that the graph of f is a surface since f is a function of two variables. During the integration y is held constant. If we then take

(16) $\quad \int_c^d \left(\int_{a(y)}^{b(y)} f(x, y)\, dx \right) dy,$

we are "summing" all areas of the type given by (15), thus obtaining the *volume* of the figure in Fig. 9.

Although double integrals can be interpreted geometrically as volumes, double integration (the integration with respect to one variable at a time with two variables) can also be used to find areas.

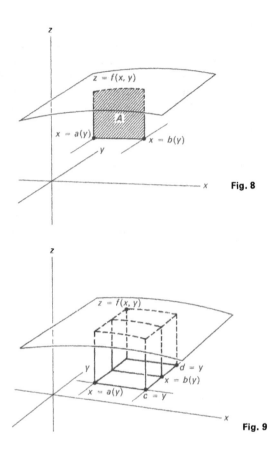

Fig. 8

Fig. 9

Example 9. The graphs of the functions $f(x) = x^2$ and $g(x) = x^3$ cross at the points $(0, 0)$ and $(1, 1)$. We wish to find the area between the two graphs from 0 to 1 (Fig. 10). The integral

(17) $$\int_{x^3}^{x^2} (1)\, dy$$

gives the area under $h(x, y) = 1$ from x^3 to x^2. Then, since y varies from 0 to 1,

(18) $$\int_0^1 \left(\int_{x^3}^{x^2} dy \right) dx$$

gives the volume (sum of the areas) of the cylinder with base area equal to the area we are trying to find and height 1. The volume of this cylinder

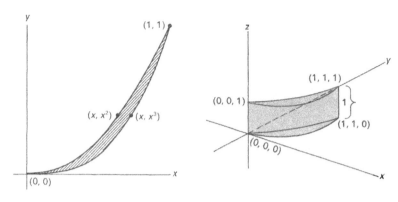

Fig. 10   Fig. 11

then is the same as the area we want (Fig. 11). Evaluating (18), we obtain

$$\int_0^1 (x^2 - x^3)\, dx = x^3/3 - x^4/4 = \tfrac{1}{3} - \tfrac{1}{4} = \tfrac{1}{12}.$$

We can obtain the area between the two graphs found in Example 9, using single integration by determining the area under both graphs and then subtracting the smaller from the greater. We should determine the area in this manner to verify that the result obtained is the same as that obtained in Example 9.

If we have a function of more than two variables, then we may have triple, quadruple, or integrals iterated as many times as there are variables; we may even integrate more than once with respect to the same variable. Geometrically, the value of the integral would represent a higher-dimensional "volume."

Example 10. Consider

$$\int_{-1}^{1} \int_{z}^{z^2} \int_{y-z}^{y+z} x\, dx\, dy\, dz.$$

Evaluating $\int_{y-z}^{y+z} x\, dx$, we obtain

$$\left.\frac{x^2}{2}\right|_{y-z}^{y+z} = \frac{(y+z)^2}{2} - \frac{(y-z)^2}{2} = 2yz.$$

We then evaluate $\int_z^{z^2} 2yz\, dy$ to be $zy^2\big|_z^{z^2} = z^5 - z^3$. We complete the triple integration by finding

$$\int_{-1}^{1} (z^5 - z^3)\, dz = \left(\frac{z^6}{6} - \frac{z^4}{4}\right)\bigg|_{-1}^{1} = 0.$$

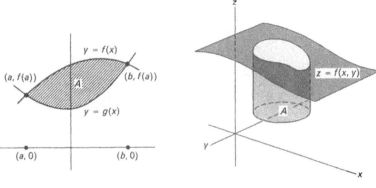

Fig. 12   Fig. 13

Suppose we are given a portion A of the xy-plane bounded by the graphs of functions f and g as shown in Fig. 12, and suppose h is a function of x and y defined for each point of A. We may wish to integrate h over A, thus, geometrically, obtaining the volume of the solid bounded by the graph of h and the xy-plane over A (Fig. 13). The integral to be evaluated is

(19) $\quad \int_a^b \int_{f(x)}^{g(x)} h(x, y)\, dy\, dx.$

The graphs of f and g are determined by the relationships $y = f(x)$ and $y = g(x)$, respectively. We may solve for x in terms of y so that (19) might be expressed as a double integral, with the integration with respect to x carried out first. Suppose it is found that $x = f_1(y)$ and $x = g_1(y)$ and that as x varies from a to b, y varies from c to d. Then with integration with respect to x first, (19) becomes

(20) $\quad \int_c^d \int_{f_1(y)}^{g_1(y)} h(x, y)\, dy\, dx.$

Exanple 11. We shall transform the integral

(21) $\quad \int_0^1 \int_{x^2}^{x} dy\, dx$

into an equivalent integral, with the integration to be carried out first with respect to x. Before we transform any integral, it is best, if possible, to sketch the area over which the integral is to be taken. Observe that y varies from x^2 to x, that is, the area over which the integration is taking place lies between the curves $y = x^2$ and $y = x$. Furthermore, y varies between 0 and 1. Hence the area is bounded by the curves $y = x^2$, $y = x$, $x = 0$, and $x = 1$. This area is shown in Fig. 14.

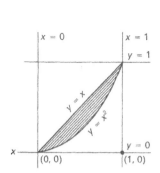

Fig. 14

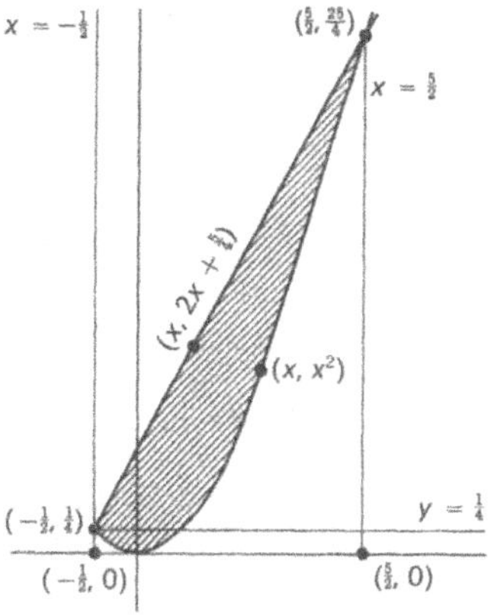

Fig. 15

Since $y = x$ and $y = x^2$, we also have $x = y$ and $x = y^{1/2}$, if x is positive, as is the case for the area here. We also see from Fig. 14 that x varies from 0 to 1; thus the area over which we are integrating is bounded by $x = y, x = y^{1/2}, y = 0$, and $y = 1$. Consequently, (21) is equivalent to

(22) $\quad \int_0^1 \int_y^{y} \tfrac{1}{2} \, dx \, dy.$

At times a single multiple integral will have to be expressed as the sum of two or more multiple integrals if the order of integration is changed. This point is illustrated in the following example.

Example 12. Consider

(23) $\quad \int_{-1/2}^{5/2} \int_{x^2}^{2x+5/4} xy \, dy \, dx.$

The area of the plane over which we are integrating is bounded by the curves $y = x^2, y = 2x + \tfrac{5}{4}, x = -\tfrac{1}{2}$, and $x = \tfrac{5}{2}$. The area is depicted in Fig. 15. If $y = x^2$, then $x = y^{1/2}$ if x is nonnegative, and $x = -y^{1/2}$ if x is negative. Part of the area in Fig. 15 is bounded by $x = y^{1/2}$, $x = -y^{1/2}, y = 0$, and $y = \tfrac{1}{4}$. The rest of the area of Fig. 15 is bounded by $x = y^{1/2}, x = y/2 - \tfrac{5}{8}, y = \tfrac{1}{4}$, and $y = 6\tfrac{1}{4}$. Thus (23), in order to have integration with respect to x first, must be expressed as two integrals, one for each part of the area of Fig. 15. Specifically, (23) is equivalent to

(24) $\quad \int_0^{1/4} \int_{-y^{1/2}}^{y^{1/2}} xy \, dx \, dy + \int_{1/4}^{25/4} \int_{y/2-5/8}^{y^{1/2}} xy \, dx \, dy.$

The following proposition (presented without proof) provides a relationship between the joint density function of several continuous variates and the probabilities associated with the joint variate. The form of this proposition should not be unexpected since it is nothing but the higher-dimensional analog of Proposition 2 of Chapter 7, which gives the situation for density functions of a single variate.

Proposition 4. *Suppose f is the joint density function of the continuous variates $X_1, \ldots, X_n$ and $(X_1, \ldots, X_n)$ has admissible range A. If B is a subset of A, then the probability that $(X_1, \ldots, X_n)$ will assume a value in B is equal to*

(25) $\quad \int_B f(x_1, \ldots, x_n) \, dx_1 \cdots dx_n.$

In particular, if B is bounded by the graphs of $x_1 = g_1(x_2, \ldots, x_n)$, $x_1 = h_1(x_2, \ldots, x_n)$, $x_2 = g_2(x_3, \ldots, x_n)$, $x_2 = h_2(x_3, \ldots, x_n)$, \ldots, $x_n = a$, $x_n = b$, then (25) is equal to

(26) $\quad \int_a^b \int_{g_{n-1}(x_n)}^{h_{n-1}(x_n)} \cdots \int_{g_1(x_2,\ldots,x_n)}^{h_1(x_2,\ldots,x_n)} f(x_1, \ldots, x_n) \, dx_1 \cdots dx_n.$

Example 13. Consider the bivariate density function $f(x, y) = e^{-(x+y)}$ of Example 8. Here the admissible range $A = \{(x, y) \mid 0 \leq x, 0 \leq y\}$. We first show that $\int_A f(x, y) \, dx \, dy = 1$, which is merely a statement that one of the admissible values of (X, Y) is certain to occur. Now x varies from 0 to ∞ and y varies from 0 to ∞; hence the integral over the entire admissible range of (x, y) is

$$\int_0^\infty \int_0^\infty e^{-(x+y)} \, dx \, dy = \int_0^\infty \int_0^\infty e^{-y} e^{-x} \, dx \, dy$$

$$= \int_0^\infty e^{-y} \left(\int_0^\infty e^{-x} \, dx \right) dy$$

(since e^{-y} is considered constant with respect to the integration with respect to x)

$$= (1) \int_0^\infty e^{-y} \, dy = (1)(1) = 1.$$

We now find the probability that both x and y will be greater than 1, that is, the probability that (x, y) will lie in the portion of A shown in Fig. 16. According to Proposition 4 this probability will be given by

$$\int_1^\infty \int_1^\infty e^{-(x+y)} \, dx \, dy = e^{-2}.$$

The density functions of X and Y can both be found from A and the joint density function of X and Y. In this example, we observe that

(27) $\quad e^{-x} = f(x)$ (the density function of X) $= \int_0^\infty e^{-(x+y)} dy,$

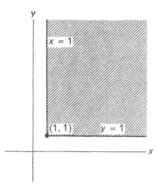

Fig. 16

and

(28) $\quad e^{-y} = g(y)$ (the density function of Y) $= \int_0^\infty e^{-(x+y)}\, dx.$

Recall that if $f(x, y)$ is the joint density function of X and Y, where x and y are discrete variates with admissible ranges A_1 and A_2, then the density function f of X is given by

$$f(x) = \sum_{y \text{ in } A_2} f(x, y).$$

In the continuous case, summation changes to integration, hence the following:

Proposition 5. *Suppose that $X_1, \ldots, X_n$ are continuous variates with admissible ranges $A_1, A_2, \ldots, A_n$, respectively, and joint density function f. Then the (marginal) density function f_1 of X_1 is given by $f_1(x_1) =$*

$$\int_{A_2} \int_{A_3} \cdots \int_{A_n} f(x_1, \ldots, x_n)\, dx_2\, dx_3 \cdots dx_n.$$

In Example 13 both A_1 and A_2 are $[0, \infty)$. The integrals called for in Proposition 5 are (27) and (28). Even though, given the joint density function of several variates, it is possible to find their individual density functions, if we are given the density functions of the variates, we cannot find the joint density function unless the variates are independent (see Proposition 3).

EXERCISES

1. Evaluate each of the following multiple integrals.

a) $\displaystyle\int_0^1 \int_0^1 xy\, dy\, dx$ b) $\displaystyle\int_7^8 \int_x^{2x} y e^x\, dy\, dx$

c) $\displaystyle\int_0^{10} \int_{1-y}^{y^2} y\, dx\, dy$ d) $\displaystyle\int_0^1 \int_0^1 \int_0^1 xz\, dz\, dx\, dy$

e) $\int_0^{12} \int_{x_1}^{x_1+1} \int_{x_1 x_2}^{x_1+x_2} x_1 x_2^2 \, dx_3 \, dx_2 \, dx_1$

2. In each of the following a double integral is given with integration with respect to y to be performed first. Change the integral to an equivalent integral with integration with respect to x to be performed first. See, e.g., Example 11.

a) $\int_0^1 \int_0^1 xy \, dy \, dx$ b) $\int_0^1 \int_{x^3}^{x^4} y \, dy \, dx$ c) $\int_0^\infty \int_{e^{-2x}}^{e^{-x}} e^{-y} \, dy \, dx$

d) $\int_0^1 \int_0^{(1-x^2)^{1/2}} dy \, dx$ e) $\int_{-1}^1 \int_x^{x^3} xe^y \, dy \, dx$ f) $\int_{-2}^4 \int_{x^2}^{x^3} xy^2 \, dy \, dx$

3. Integrate the bivariate functions given in each of the following over the area given with each function. Sketch the area in each case.

    a) $f(x, y) = xy$; the area bounded by $x = 2$, $x = 4$, $y = 4$, $y = -1$.
    b) $f(x, y) = e^{-2x-y}$; the entire first quadrant of the plane.
    c) $f(x, y) = 3$; the area bounded by $y = x^2$, $y = x^{1/2}$, $x = 0$, $x = 1$.
    d) $f(x, y) = 1$; the area inside the circle of radius 1 with center $(0, 0)$.
    e) $f(x, y) = x$; the area bounded by $y = x^4$ below and above by the line which passes through the point $(0, 4)$ with slope 2.

4. Find the density functions of the variates X and Y, where the joint density function of X and Y is given by $f(x, y) = xe^{-x(1+y)}$ and the admissible range of both X and Y is $[0, \infty)$. If A is the admissible range of (X, Y), confirm that $\int_A xe^{-x(1+y)} \, dx \, dy = 1$.

13 Regression and Correlation

13.1 LINEAR REGRESSION

Occasions arise where one variate X is paired with another variate Y.

Example 1. A college professor gives a midyear examination, as well as a final examination. If s is one of the professor's students, let $X(s)$ and $Y(s)$ be the marks that s obtains on the mid year and final examinations, respectively.

Example 2. A company gives an aptitude test to each of its employees. The company also rates each employee on the basis of his actual performance on the job. For each employee s, let $X(s)$ and $Y(s)$ be the score on the aptitude test and the performance rating, respectively.

Definition 1. *If for each value of a variate X, we have a corresponding value of a variate Y, then the set of pairs of corresponding values of X and Y is said to be a **bivariate population;** we denote the bivariate population of X and Y by (X, Y).*

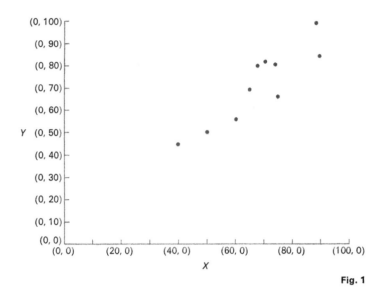

Fig. 1

If S is a finite subset of a bivariate population (X, Y), then a graphical representation of S is called a **scatter diagram**.

Example 3. The variates X and Y of Examples 1 and 2 give bivariate populations. We now consider the bivariate population (X, Y) of Example 2. The company compiles a random selection of 10 elements of (X, Y) as shown in Table 1.

Table 1

X	70	40	75	74	65	90	76	91	52	63
Y	80	45	65	80	68	96	80	85	50	55

A scatter diagram for the data of Table 1 is given in Fig. 1.

It appears from Fig. 1 that the randomly selected points of the bivariate population (X, Y) are "approximately" collinear. It is therefore reasonable to seek that straight line which gives the "best approximation" to the data of Table 1.

Suppose we have n points $(x_1, y_1), \ldots, (x_n, y_n)$ from a bivariate population (X, Y) and we wish to "fit" a straight line to these points. One method of doing this is to visually estimate a straight line on the scatter diagram determined by the points—the line which gives the best fit. Such a procedure has the obvious disadvantage that the line chosen will vary with the person who chooses it. Nevertheless, if Y and X are related to

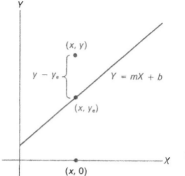

Fig. 2

each other in a linear fashion, one should be able to estimate approximately where the line of "best fit" should be on the scatter diagram.

If X and Y are related linearly, then we ought to be able to estimate Y from X by a relationship of the form

(1) $Y_e = mX + b,$

where m and b are constants, and Y_e indicates the estimate of Y. For a particular value (x, y) of (X, Y), $(y - y_e)^2$ is a measure of how much the estimated value of Y deviates from the actual value of Y (Fig. 2). Given n elements of (X, Y), we choose as line of "best fit" that line for which

(2) $\sum_{i=1}^{n} (y_i - y_{e_i})^2$

will be a minimum. We therefore want to find m and b such that with $y_{e_i} = mx_i + b$, $i = 1, \ldots, n$, (2) will be a minimum.

With the substitution $y_{e_i} = mx_i + b$, (2) becomes

(3) $\sum_{i=1}^{n} (y_i - mx_i - b)^2.$

We can find the values of m and b for which (3) is a minimum by taking the partial derivatives of (3) with respect to m and b and applying the methods of Section 12.2. In this way, it is found that m and b are the simultaneous solutions to the equations

(4) $$bn + m \sum_{i=1}^{n} x_i = \sum_{i=1}^{n} y_i,$$
$$b \sum_{i=1}^{n} x_i + m \sum_{i=1}^{n} x_i^2 = \sum_{i=1}^{n} x_i y_i.$$

The pair of equations in (4) can be solved simultaneously for m and b using standard algebraic techniques. The solutions can be simplified by setting

(5) $\bar{x} = \dfrac{\sum_{i=1}^{n} x_i}{n}$ and $\bar{y} = \dfrac{\sum_{i=1}^{n} y_i}{n}$.

In terms of the x_i, y_i, $\bar{x}$, and $\bar{y}$, the solution to (4) can be expressed as

(6) $m = \dfrac{\sum_{i=1}^{n} x_i y_i - n\bar{x}\bar{y}}{\sum_{i=1}^{n} x_i^2 - n\bar{x}^2}$, $b = \bar{y} - m\bar{x}$.

Definition 2. The line $Y_e = mX + b$ fitted to the points $(x_1, y_1), \ldots, (x_n, y_n)$ with m and b as in (6) is called the **line of regression of Y on X**.

Example 4. We use the data of Table 1 to compute the line of regression of Y on X in Example 3.

Table 2

X	Y	XY	X^2
70	80	5600	4900
40	45	1800	1600
75	65	4875	5625
74	80	5920	5476
65	68	4420	4225
90	96	8640	8100
76	80	6080	5776
91	85	7735	8281
52	50	2600	2704
63	55	3465	3969
Sum 696	704	51,135	50,658

Consequently, $\bar{x} = 69.6$ and $\bar{y} = 70.4$. Therefore

$$m = \dfrac{51{,}135 - (10)(69.6)(70.4)}{50{,}658 - (10)(69.6)^2} \doteq 0.96,$$

and

$b = 70.4 - (0.96)(69.6) \doteq 3.6$.

Thus the equation of the line of regression of Y on X in this instance is

(7) $Y_e = 0.96X + 3.6$.

EXERCISES

1. Take the partial derivatives of (3) with respect to m and b and find the critical points; that is, derive the set of equations in (4). The simultaneous solution to these equations always gives a minimum.

2. In a certain test, paired values of X and Y are found as shown in Table 3.

   Table 3

X	1	2	3	4	5
Y	5	1	−2	−6	−10

   a) Draw a scatter diagram for the information presented in Table 3.
   b) Find the equation of the line of regression of Y on X. Graph this line on the scatter diagram.
   c) Use the equation of the line of regression to predict the value of Y for $x = 6$; for $x = -3$.

3. Each student in a calculus class is asked to estimate the mark he will obtain on the final examination; this mark is then compared with his actual performance on the finals. The data are presented in Table 4; the values of X and Y are the predicted and actual marks on the final examination, respectively.

   Table 4

X	60	55	80	85	40	65	80
Y	80	65	70	95	45	75	90

   a) Draw a scatter diagram for the information presented in Table 4.
   b) Find the equation of the line of regression of Y on X.
   c) Does there seem to be "as strong" a linear relationship between Y and X in this exercise as there is between Y and X in Examples 3 and 4?

13.2 MEASURES OF CORRELATION

Given a bivariate population (X, Y) such that the value of Y can always be determined from the value of X, for example, X and Y satisfy an equation such as $Y = X^2 + 4$, we say that X and Y are *perfectly correlated*. If there is no relationship between X and Y, then we say that X and Y are *uncorrelated*.

Example 5. If X is a random variable on a sample space S and Y is defined by $Y = (X - 4)^2$, then X and Y are perfectly correlated. If X and Y were independent variates on S, however, then X and Y would be uncorrelated.

Example 6. Two dice are rolled together. Let the values of X and Y be the numbers obtained on the first and second dice, respectively. Since X and Y can be assumed to be independent variates, they will be uncorrelated.

Evidently, the variates X and Y from a bivariate population (X, Y) may be correlated to some degree even though the correlation is not perfect. For example, we see that the variates X and Y of Examples 3 and 4 appear to be rather closely related in a linear fashion even though the correlation is not perfect.

Definition 3. *If the points of a bivariate population (X, Y) lie on or near a straight line of nonzero slope, then there is said to be a **linear correlation** between X and Y. If the points of (X, Y) lie on or near a nonlinear curve, then there is said to be a **nonlinear correlation** between X and Y.*

We shall restrict the discussion of this chapter to linear correlation. One can also speak about the correlation among many variates; we shall also restrict our discussion to bivariate populations. The transition from linear correlation of two variates to nonlinear correlation and multivariate populations consists mainly of extending the principles and methods that we shall consider in this chapter.

If we have n elements $(x_1, y_1), \ldots, (x_n, y_n)$ of a bivariate population (X, Y), we can calculate the equation $Y_e = mX + b$ of the line of regression of Y on X by the procedure indicated in Section 13.1. Equation (1) is an attempt to find a linear correlation between X and Y; we now ask: How can we measure in a quantitative manner the "degree" of linear correlation between X and Y?

Both X and Y are random variables; thus $Y_e = mX + b$ and $Y - Y_e$ are random variables. From (1), we obtain

(8) $\quad Y = mX + b + (Y - Y_e).$

From (8) we see that $Y - Y_e$ is a random variable which measures by how much X and Y miss being perfectly linearly correlated. It is reasonable to assume that $Y - Y_e$ will have a mean value of 0, in fact, that its distribution will be symmetric with respect to 0. What we will assume, in other words, is that the mean value of Y will be linearly related to X (according to the equation $y = mx + b$) and the observed values of Y will be distributed evenly on both sides of the line of regression of Y on X.

Since $Y - Y_e$ has mean 0, the estimate of its standard deviation obtained from the sample of n points of (X, Y) is

(9) $\quad \hat{s}_e = \left(\dfrac{\sum_{i=1}^{n} (y_i - y_{e_i})^2}{n} \right)^{1/2},$

where $y_{e_i} = mx_i + b$.

Definition 4. *The statistic $\hat{s}_e$ of (9) is called the **standard error of estimate**. (The notation used to indicate the standard error of estimate varies widely from text to text.)*

*For rather large n, $\hat{s}_e$ provides a fairly good estimate for σ_{y-y_e}. If n is small (say, less than 30), most statisticians prefer a **modified standard error of estimate** defined by*

(10) $\quad s = \hat{s}_e \left(\dfrac{n}{n-2}\right)^{1/2}$

as an estimate of σ_{y-y_e}.

Letting $\bar{y}$ be defined as in (5), we also make the following definitions.

*The **sum of squares for regression** is defined to be*

$$\text{SSR} = \sum_{i=1}^{n} (y_{e_i} - \bar{y})^2.$$

*The **sum of squares for error** is defined to be*

$$\text{SSE} = \sum_{i=1}^{n} (y_i - y_{e_i})^2.$$

*The **total variation** is defined by*

$$\text{TV} = \sum_{i=1}^{n} (y_i - \bar{y})^2.$$

Note that $\hat{s}_e^2 = \text{SSE}/n$, while $s^2 = \text{SSE}/(n-2)$.

The SSR measures the dispersion of the predicted values of Y about $\bar{y}$, the sample mean of the observed values of Y. The SSE measures the dispersion of the observed values of Y about the predicted values of Y (the corresponding values on the line of regression). A basic relationship among the TV, SSE and SSR is given in the following proposition.

Proposition 1. $\text{TV} = \text{SSE} + \text{SSR}$.

We omit the proof of this proposition since it consists of lengthy, although fairly straightforward, algebraic manipulations.

If n is fairly large, then $Y - Y_e$ is at least approximately normally distributed with mean 0 and variance $\hat{s}_e^2$ (we present no argument here to show that this is in fact the case). Thus, if we draw the two lines parallel to the line of regression of Y on X and at a vertical distance $\hat{s}_e$ from it (Fig. 3), then about 68% of the values of $Y - Y_e$ should lie between the two lines. Were we to draw the two lines parallel to the line of regression and at a vertical distance $2\hat{s}_e$ from it, we could expect about 95% of the values of $Y - Y_e$ to fall between these lines.

The following proposition (left unproved) gives a formula which is sometimes useful in computing $\hat{s}_e^2$.

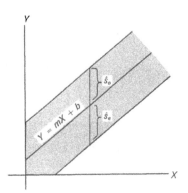

Fig. 3

Proposition 2. If the points $(x_1, y_1), \ldots, (x_n, y_n)$ of (X, Y) are used to compute the equation $Y_e = mX + b$ of the line of regression of Y on X, then

$$\hat{s}_e^2 = \frac{\sum_{i=1}^n y_i^2 - b \sum_{i=1}^n y_i - m \sum_{i=1}^n x_i y_i}{n}.$$

We close this section by computing some of the quantities defined in this section for the situation presented in Example 3.

Example 7. We return again to Example 3. Table 5 gives certain values related to X and Y not already given in Tables 1 and 2.

Table 5

X	Y	Y^2	Y_e	$Y_e - \bar{y}$	$(Y_e - \bar{y})^2 = (Y_e - 70.4)^2$
			(All values rounded off)		
70	80	6400	70.7	0.3	0.1
40	45	2025	42	−28.4	804
75	65	4225	75.6	4.2	17.4
74	80	6400	74.5	4.1	16.8
65	68	4624	66	−4.4	18.5
90	96	9216	90	19.6	382
76	80	6400	76.6	6.2	38.4
91	85	7225	91	20.6	423
52	50	2500	53.5	−16.9	285
63	55	3025	64.1	−6.3	39.5
		Sum 52,040			Sum 2024.7

From Proposition 2 we have then that

(11) $$\hat{s}_e^2 = \frac{52{,}040 - 704(3.6) - (0.96)(51{,}135)}{10} = \frac{416}{10} = 41.6.$$

From (11), we also have

SSE = 416.

From Table 5, we see that

SSR = 2024.7.

Proposition 1 then enables us to say

TV = SSE + SSR = 416 + 2024.7 = 2440.7.

The modified standard error of estimate in this instance is

$s = (\frac{416}{8})^{1/2} = 52^{1/2} \doteq 7.23$,

as compared with

$\hat{s}_e \doteq 6.46$.

EXERCISES

1. Using the information given in Table 3 (Exercise 2 of Section 13.1), find the standard error of estimate, the modified standard error of estimate, the sum of squares for regression, the sum of squares for error, and the total variation.
2. Do Exercise 1 above with Table 4 (Exercise 3 of Section 13.1) substituted for Table 3.

13.3 THE COEFFICIENT OF CORRELATION

Once more we assume that n elements $(x_1, y_1), \ldots, (x_n, y_n)$ of a bivariate population (X, Y) have been used to find a line of regression of Y on X with equation $Y = mX + b$. Consider the

$$\text{SSR} = \sum_{i=1}^{n} (y_{e_i} - \bar{y})^2.$$

This quantity is sometimes called the *explained variation* because there is a definite order to it; in other words, $Y_e - \bar{y}$ is not a random variable, but can be computed precisely given any value of X (and the equation of the line of regression of Y on X). The

$$\text{SSE} = \sum_{i=1}^{n} (y_i - y_{e_i})^2,$$

on the other hand, is sometimes called the *unexplained variation*. This term is used because $Y - Y_e$ is a random variable; thus the SSE is itself a random variable for random samples of n points from (X, Y).

If X and Y are perfectly linearly correlated (by $Y = mX + b$), then $Y - Y_e$ will always be 0; hence the TV will entirely be explained variation. But if X and Y are uncorrelated, then Y_e will always be $\bar{y}$ (or m_y, the expected value of Y); hence the TV will entirely be unexplained variation. We therefore see that the quantity

(12)   SSR/TV = explained variation/total variation

is a measure of the degree of linear correlation between X and Y.

Definition 5. *The quantity (12), which we denote by r^2, is called the **coefficient of determination**.*

*The **coefficient of correlation** r is defined to be $+(r^2)^{1/2}$ if the slope m of the line of regression of Y on X is positive and $-(r^2)^{1/2}$ if m is negative; if $r^2 = 0$, we set $r = 0$.*

Example 8. In Example 7, we have TV = 2440.7 and SSR = 2024.7; hence in this instance $r^2 = 2024.7/2440.7 \doteq 0.832$. Since $m = 0.96$ is positive, r is the positive square root of 0.832, which is approximately 0.915.

The following proposition gives properties of the coefficients of determination and correlation which follow immediately from their definitions.

Proposition 3

a) *Since the* SSR *and* TV *are always nonnegative and* SSR $\leq$ TV *(by Proposition 1), we have* $0 \leq r^2 \leq 1$.
b) *Since r^2 lies between 0 and 1, r will always lie between -1 and 1.*
c) *If all of the sample points from (X, Y) lie on a straight line, then $r^2 = 1$; hence r will be 1 if the line has positive slope, or -1 if the line has negative slope. If there is no linear correlation between X and Y, $r^2 = 0$; hence r will be 0.*

In Example 8, we found $r = 0.915$; hence we may expect a high degree of linear correlation between X and Y. It may be possible, however, to randomly select n points of (X, Y) which give a nonzero value of r^2, even if there is no, or very little, linear correlation between X and Y. Indeed, since r^2 can assume any value between 0 and 1 if X and Y are continuous variates, it would be all but impossible in such an instance for r^2 to be 0 even if X and Y were uncorrelated. Thus a nonzero coefficient of correlation does not mean that there is necessarily some linear correlation between X and Y, that is, that the population coefficient of correlation is not 0. We shall investigate the significance of the coefficient of correlation in the next section.

We should also keep in mind that a low coefficient of correlation does not necessarily mean that X and Y are uncorrelated. As we have seen, the workings of chance may give a low coefficient of correlation even when the population coefficient of correlation is close to 1. It may also be that

X and Y are correlated, but not linearly. For example, if $Y = X^2$, then X and Y are perfectly correlated, but the coefficient of correlation obtained from a fairly large sample of elements of (X, Y) is likely to be rather low, since it measures only the degree of linear correlation.

We shall devote the rest of this section to the computation and basic algebraic properties of the coefficients of correlation and determination.

We continue to assume that we have a random sample of n elements $(x_1, y_1), \ldots, (x_n, y_n)$ from the bivariate population (X, Y) and that the equation of the line of regression of Y on X is $Y_e = mX + b$. We set

$$\hat{s}_x^2 = \frac{\sum_{i=1}^n (x_i - \bar{x})^2}{n} \quad \text{and} \quad \hat{s}_y^2 = \frac{\sum_{i=1}^n (y_i - \bar{y})^2}{n}.$$

From the definition of the SSR and the TV, we have immediately that

(13) $$r^2 = \frac{\sum_{i=1}^n (y_{e_i} - \bar{y})^2}{\sum_{i=1}^n (y_i - \bar{y})^2}.$$

Using various algebraic manipulations (which we omit), we can derive the various expressions for r^2 presented in the following proposition.

Proposition 4. *The following expressions are all equal to r^2:*

(14) $$\frac{(\sum_{i=1}^n (x_i - \bar{x})(y_i - \bar{y}))^2}{(\sum_{i=1}^n (x_i - \bar{x})^2 \sum_{i=1}^n (y_i - \bar{y})^2)};$$

(15) $$\frac{(\sum_{i=1}^n (x_i - \bar{x})(y_i - \bar{y}))^2}{n^2 \hat{s}_x^2 \hat{s}_y^2};$$

(16) $$\frac{(\sum_{i=1}^n x_i y_i - (\sum_{i=1}^n x_i \cdot \sum_{i=1}^n y_i)/n)^2}{(\sum_{i=1}^n x_i^2 - (\sum_{i=1}^n x_i)^2/n)(\sum_{i=1}^n y_i^2 - (\sum_{i=1}^n y_i)^2/n)};$$

(17) $$\frac{(\sum_{i=1}^n x_i y_i - n\bar{x}\bar{y})^2}{(\sum_{i=1}^n x_i^2 - n\bar{x}^2)(\sum_{i=1}^n y_i^2 - n\bar{y}^2)};$$

(18) $(\hat{s}_x^2/\hat{s}_y^2) m^2;$

(19) $1 - (\hat{s}_e^2/\hat{s}_y^2);$

(20) $(\text{TV} - \text{SSE})/\text{TV} = \text{SSR}/\text{TV}.$

Expression (20), which follows at once from the definition of r^2 and Proposition 1, can be interpreted as follows: r^2 measures the fraction of the total variation that is explained by taking X into account, that is, by using the equation of the line of regression to estimate Y.

It can be shown that the value of r^2 is independent of the type of unit used to express X and Y (for example, r^2 will remain the same regardless of whether X and Y are expressed in inches, feet, or centimeters).

EXERCISES

1. Prove that if $X' = kX$ and $Y' = kY$, where k is a nonzero constant, then r^2 relative to X' and Y' is the same as r^2 relative to X and Y.

2. Prove that if $X' = X - k$ and $Y' = Y - k'$, where k and k' are both constants, then r^2 relative to X' and Y' is the same as r^2 relative to X and Y.

3. Compute r^2 relative to the data presented in Exercise 2 of Section 13.1.

4. Compute r^2 relative to the data presented in Exercise 3 of Section 13.1.

5. Derive as many of the expressions for r^2 given in Proposition 4 as you can.

13.4 THE SIGNIFICANCE OF r AND r^2

Throughout this section we shall consider a bivariate population (X, Y) with a population coefficient of correlation ρ and with the property that for any fixed value of one variate, the values of the other variate are at least approximately normally distributed. For example, for a value x of X,

$Y_x =$ the second coordinate of (x, Y)

is approximately normally distributed.

We consider now the population $n(X, Y)$ of samples of n members of (X, Y). Each member of $n(X, Y)$ gives a coefficient of correlation r. Therefore r defines a random variable R on $n(X, Y)$. If we were able to determine the distribution of R, then we would be able to answer such important questions as: What is the probability of getting r at least as large as 0.1 if in fact X and Y are uncorrelated, that is, if $\rho = 0$? The distribution of R will depend not only on n, the size of the sample, but also on ρ. The actual density function for R in terms of ρ and n is a very involved expression whose reproduction here would serve little purpose. We shall concern ourselves with propositions about the distribution of R which involve distributions we have encountered previously.

Proposition 5. *If (X, Y) satisfies the conditions indicated at the beginning of this section and $\rho = 0$, then for samples of size n*

$$(21) \qquad t = \frac{R}{((1 - R^2)/(n - 2))^{1/2}}$$

satisfies Student's t-distribution with f, the number of degrees of freedom, equal to $n - 2$.

Example 9. We shall assume that the variates X and Y of Example 3 give a bivariate *normal* population (X, Y) (that is, a bivariate population which is normal with respect to each variate) and test the hypothesis $\rho = 0$ at the 0.05 level of significance. We first ask: If $\rho = 0$, in what interval about 0 can we expect t to lie with a probability of 0.95? The number of degrees of freedom here is $10 - 2$; using Table 4 of the Appendix, we see that t has a probability of 0.95 of lying in the interval $(-2.3, 2.3)$. Therefore, if the observed value of t lies outside this interval, we can reject the hypothesis that $\rho = 0$ at the 0.05 level of significance. Now the observed value of t (using the results of Example 8) is

$$\frac{0.915}{((1 - 0.83)/8)^{1/2}} \geq 6.$$

Since 8 is well outside $(-2.3, 2.3)$, we can easily reject the hypothesis $\rho = 0$.

Example 10. A sample of 62 elements of the (normal) bivariate population (X, Y) yields $r = 0.12$. We shall test the hypothesis that $\rho = 0$ at the 0.05 level of significance. For 60 degrees of freedom, t has a probability of 0.95 of lying in the interval $(-2, 2)$. The observed value of t is

$$\frac{0.12}{((1 - 0.014)/60)^{1/2}} \doteq 12;$$

therefore we can reject the hypothesis $\rho = 0$ at the 0.05 level of significance.

Proposition 5 is useful in determining whether an observed value of R is significantly different from 0, that is, in testing the hypothesis $\rho = 0$. The next proposition enables us to test hypotheses involving $\rho \neq 0$.

Proposition 6. *Define*

$$Z = (\tfrac{1}{2}) \ln [(1 + R)/(1 - R)],$$

where the value of R for each element of $n(X, Y)$ is the coefficient of correlation. Set

$$m_z = (\tfrac{1}{2}) \ln [(1 + \rho)/(1 - \rho)],$$

where ρ is the population coefficient of correlation, and

$$\sigma_z = \frac{1}{(n - 3)^{1/2}}.$$

Then Z is approximately normally distributed with mean m_z and standard deviation σ_z.

Example 11. We shall test the hypothesis that the population coefficient of correlation ρ for the situation of Examples 2 and 3 is really 0.5. For $\rho = 0.5$,

$$m_z = \left(\frac{1}{2}\right) \ln \left(\frac{1.5}{0.5}\right) = \left(\frac{1}{2}\right) \ln 3 \doteq 0.55,$$

and
$$\sigma_z = \frac{1}{(10-3)^{1/2}} = \frac{1}{7} \doteq 0.377.$$

Therefore $(Z - 0.55)/0.377$ is approximately normally distributed. It follows, then, that $(Z - 0.55)/0.377$ has a 0.95 probability of lying in the interval $(-1.96, 1.96)$. Consequently, we can reject the hypothesis $\rho = 0.5$ at the 0.05 level of significance if the observed value of

$(Z - 0.55)/0.377$

lies outside $(-1.96, 1.96)$. The observed value of Z is

$(\frac{1}{2}) \ln [(1 + 0.92)/(1 - 0.92)] \doteq 1.16$;

hence, in this instance, $(Z - 0.55)/0.377$ is approximately 2.8. Since 2.8 lies outside $(-1.96, 1.96)$, we can reject the hypothesis $\rho = 0.5$ at the 0.05 level of significance.

Given an observed value r of R, we can also use Proposition 6 to solve the following problem: Find an interval $[a, b]$ which contains r such that ρ, the population coefficient of correlation, has a probability p of lying in the interval $[a, b]$. We call $[a, b]$ a *p-confidence interval* for ρ. (In Example 8 of Chapter 10 a 0.95-confidence interval for the population mean was calculated using the normal distribution.) The procedure for finding a confidence interval for ρ using Proposition 6 is illustrated in the following example.

Example 12. For the situation presented in Examples 2 and 3, we shall find a 0.95 confidence interval for ρ. Let m_z be the true mean of

$Z = (\frac{1}{2}) \ln ((1 + R)/(1 - R))$.

Since $(Z - m_z)/0.37$ is a standard normal variate (the 0.37 was computed in Example 11), there is a 0.95 probability of its value lying in the interval $[-1.96, 1.96]$. We have

$-1.96 \leq (Z - m_z)/0.37 \leq 1.96$

if and only if

(22) $Z - (1.96)(0.37) \leq m_z \leq Z + (1.96)(0.37)$.

The observed value of Z is approximately 1.19 (Example 11); hence (22) becomes

(23) $0.86 \leq m_z \leq 2.34$.

Now

$$m_z = \left(\frac{1}{2}\right) \ln \left(\frac{1+\rho}{1-\rho}\right).$$

Straightforward calculation or the use of Table 7 of the Appendix indicates that (23) is equivalent to

$0.70 \leq \rho \leq 0.80$.

Therefore [0.70, 0.80] is the 0.95-confidence interval for ρ.

We close this section with some words of caution about the use of the coefficient of correlation. First of all, you have already learned that the coefficient of correlation measures only linear correlation. Two variates may be perfectly correlated in a nonlinear way, and yet yield a coefficient of correlation close to 0. Techniques to investigate nonlinear correlations are fairly straightforward generalizations of the methods used to investigate linear correlation, although, as would be expected, the computations involved are more cumbersome. In some instances, the investigation of a nonlinear correlation can be transformed into one involving a linear correlation. In particular, if X and Y are exponentially or normally related, as for example, in $Y = e^X$, then there will be a linear correlation between X and $\ln Y$.

Second, one should not expect a variate Y to be a function of just one variate X. Although Y and X may be correlated to some degree, there will, in general, be many more factors determining Y than X alone. We may find, for example, a correlation between annual rainfall and the mean annual temperature, but clearly both of these variates are related to many other factors as well. One can generalize bivariate correlation techniques to multivariate situations.

We must also keep in mind that the larger the sample from (X, Y) that we have to work with, the more meaningful the coefficient of correlation will be. A coefficient of correlation of 0.5 based on a sample of 100 is far more meaningful than the same value obtained from a sample of only 12. The coefficient of determination r^2 is a more meaningful indicator of the correlation than is r. For example, with a large sample, $r = 0.1$ may indicate a correlation between two variates X and Y. But since $r^2 = 0.01$, we see that the use of the equation of the regression line to estimate Y given X will only explain about 1% of the total variation, certainly a weak correlation even if significantly different from zero.

EXERCISES

1. Is the r found in Exercise 2 of Section 13.3 significantly different from 0 at the 0.05 level of significance?

2. Test the hypothesis (Exercise 3 of Section 13.3) that $\rho = 0$. Use the 0.05 level of significance.

3. Find a 0.95-confidence interval for the coefficient of correlation of a population if a random sample of 22 elements from the population yields $r = 0.4$.

4. Find a 0.98-confidence interval for the coefficient of correlation of a population if a random sample of 42 elements from the population yields $r = 0.85$.

5. A sample of 22 elements from a population yields $r = 0.4$. Test the hypothesis $\rho = 0.7$ at the 0.05 level of significance.

6. How large would the coefficient of correlation obtained from a sample of 16 elements from a population have to be to reject the hypothesis $\rho = 0$ at the 0.05 level of significance?

7. A statistician claims that the value $r = 0.4$ obtained from a sample of n elements of a certain population enables him to reject the hypothesis $\rho = 0$ at the 0.05 level of significance. What is the minimum number of elements the statistician's sample could contain?

8. If the population coefficient of determination is 0.25 and the slope of the line of regression is positive, what is the probability of obtaining a value of R between 0.1 and 0.3 if a sample of size 18 is used? if a sample of size 66 is used?

Appendix

Table 1 Common Logarithms

N	0	1	2	3	4	5	6	7	8	9
10	.0000	.0043	.0086	.0128	.0170	.0212	.0253	.0294	.0334	.0374
11	.0414	.0453	.0492	.0531	.0569	.0607	.0645	.0682	.0719	.0755
12	.0792	.0828	.0864	.0899	.0934	.0969	.1004	.1038	.1072	.1106
13	.1139	.1173	.1206	.1239	.1271	.1303	.1335	.1367	.1399	.1430
14	.1461	.1492	.1523	.1553	.1584	.1614	.1644	.1673	.1703	.1732
15	.1761	.1790	.1818	.1847	.1875	.1903	.1931	.1959	.1987	.2014
16	.2041	.2068	.2095	.2122	.2148	.2175	.2201	.2227	.2253	.2279
17	.2304	.2330	.2355	.2380	.2405	.2430	.2455	.2480	.2504	.2529
18	.2553	.2577	.2601	.2625	.2648	.2672	.2695	.2718	.2742	.2765
19	.2788	.2810	.2833	.2856	.2878	.2900	.2923	.2945	.2967	.2989
20	.3010	.3032	.3054	.3075	.3096	.3118	.3139	.3160	.3181	.3201
21	.3222	.3243	.3263	.3284	.3304	.3324	.3345	.3365	.3385	.3404
22	.3424	.3444	.3464	.3483	.3502	.3522	.3541	.3560	.3579	.3598
23	.3617	.3636	.3655	.3674	.3692	.3711	.3729	.3747	.3766	.3784
24	.3802	.3820	.3838	.3856	.3874	.3892	.3909	.3927	.3945	.3962
25	.3979	.3997	.4014	.4031	.4048	.4065	.4082	.4099	.4116	.4133
26	.4150	.4166	.4183	.4200	.4216	.4232	.4249	.4265	.4281	.4298
27	.4314	.4330	.4346	.4362	.4378	.4393	.4409	.4425	.4440	.4456
28	.4472	.4487	.4502	.4518	.4533	.4548	.4564	.4579	.4594	.4609
29	.4624	.4639	.4654	.4669	.4683	.4698	.4713	.4728	.4742	.4757
30	.4771	.4786	.4800	.4814	.4829	.4843	.4857	.4871	.4886	.4900
31	.4914	.4928	.4942	.4955	.4969	.4983	.4997	.5011	.5024	.5038
32	.5051	.5065	.5079	.5092	.5105	.5119	.5132	.5145	.5159	.5172
33	.5185	.5198	.5211	.5224	.5237	.5250	.5263	.5276	.5289	.5302
34	.5315	.5328	.5340	.5353	.5366	.5378	.5391	.5403	.5416	.5428
35	.5441	.5453	.5465	.5478	.5490	.5502	.5514	.5527	.5539	.5551
36	.5563	.5575	.5587	.5599	.5611	.5623	.5635	.5647	.5658	.5670
37	.5682	.5694	.5705	.5717	.5729	.5740	.5752	.5763	.5775	.5786
38	.5798	.5809	.5821	.5832	.5843	.5855	.5866	.5877	.5888	.5899
39	.5911	.5922	.5933	.5944	.5955	.5966	.5977	.5988	.5999	.6010
40	.6021	.6031	.6042	.6053	.6064	.6075	.6085	.6096	.6107	.6117
41	.6128	.6138	.6149	.6160	.6170	.6180	.6191	.6201	.6212	.6222
42	.6232	.6243	.6253	.6263	.6274	.6284	.6294	.6304	.6314	.6325
43	.6335	.6345	.6355	.6365	.6375	.6385	.6395	.6405	.6415	.6425
44	.6435	.6444	.6454	.6464	.6474	.6484	.6493	.6503	.6513	.6522
45	.6532	.6542	.6551	.6561	.6571	.6580	.6590	.6599	.6609	.6618
46	.6628	.6637	.6646	.6656	.6665	.6675	.6684	.6693	.6702	.6712
47	.6721	.6730	.6739	.6749	.6758	.6767	.6776	.6785	.6794	.6803
48	.6812	.6821	.6830	.6839	.6848	.6857	.6866	.6875	.6884	.6893
49	.6902	.6911	.6920	.6928	.6937	.6946	.6955	.6964	.6972	.6981
50	.6990	.6998	.7007	.7016	.7024	.7033	.7042	.7050	.7059	.7067
51	.7076	.7084	.7093	.7101	.7110	.7118	.7126	.7135	.7143	.7152
52	.7160	.7168	.7177	.7185	.7193	.7202	.7210	.7218	.7226	.7235
53	.7243	.7251	.7259	.7267	.7275	.7284	.7292	.7300	.7308	.7316
54	.7324	.7332	.7340	.7348	.7356	.7364	.7372	.7380	.7388	.7396

(cont)

Table 1 continued

N	0	1	2	3	4	5	6	7	8	9
55	.7404	.7412	.7419	.7427	.7435	.7443	.7451	.7459	.7466	.7474
56	.7482	.7490	.7497	.7505	.7513	.7520	.7528	.7536	.7543	.7551
57	.7559	.7566	.7574	.7582	.7589	.7597	.7604	.7612	.7619	.7627
58	.7634	.7642	.7649	.7657	.7664	.7672	.7679	.7686	.7694	.7701
59	.7709	.7716	.7723	.7731	.7738	.7745	.7752	.7760	.7767	.7774
60	.7782	.7789	.7796	.7803	.7810	.7818	.7825	.7832	.7839	.7846
61	.7853	.7860	.7868	.7875	.7882	.7889	.7896	.7903	.7910	.7917
62	.7924	.7931	.7938	.7945	.7952	.7959	.7966	.7973	.7980	.7987
63	.7993	.8000	.8007	.8014	.8021	.8028	.8035	.8041	.8048	.8055
64	.8062	.8069	.8075	.8082	.8089	.8096	.8102	.8109	.8116	.8122
65	.8129	.8136	.8142	.8149	.8156	.8162	.8169	.8176	.8182	.8189
66	.8195	.8202	.8209	.8215	.8222	.8228	.8235	.8241	.8248	.8254
67	.8261	.8267	.8274	.8280	.8287	.8293	.8299	.8306	.8312	.8319
68	.8325	.8331	.8338	.8344	.8351	.8357	.8363	.8370	.8376	.8382
69	.8388	.8395	.8401	.8407	.8414	.8420	.8426	.8432	.8439	.8445
70	.8451	.8457	.8463	.8470	.8476	.8482	.8488	.8494	.8500	.8506
71	.8513	.8519	.8525	.8531	.8537	.8543	.8549	.8555	.8561	.8567
72	.8573	.8579	.8585	.8591	.8597	.8603	.8609	.8615	.8621	.8627
73	.8633	.8639	.8645	.8651	.8657	.8663	.8669	.8675	.8681	.8686
74	.8692	.8698	.8704	.8710	.8716	.8722	.8727	.8733	.8739	.8745
75	.8751	.8756	.8762	.8768	.8774	.8779	.8785	.8791	.8797	.8802
76	.8808	.8814	.8820	.8825	.8831	.8837	.8842	.8848	.8854	.8859
77	.8865	.8871	.8876	.8882	.8887	.8893	.8899	.8904	.8910	.8915
78	.8921	.8927	.8932	.8938	.8943	.8949	.8954	.8960	.8965	.8971
79	.8976	.8982	.8987	.8993	.8998	.9004	.9009	.9015	.9020	.9025
80	.9031	.9036	.9042	.9047	.9053	.9058	.9063	.9069	.9074	.9079
81	.9085	.9090	.9096	.9101	.9106	.9112	.9117	.9122	.9128	.9133
82	.9138	.9143	.9149	.9154	.9159	.9165	.9170	.9175	.9180	.9186
83	.9191	.9196	.9201	.9206	.9212	.9217	.9222	.9227	.9232	.9238
84	.9243	.9248	.9253	.9258	.9263	.9269	.9274	.9279	.9284	.9289
85	.9294	.9299	.9304	.9309	.9315	.9320	.9325	.9330	.9335	.9340
86	.9345	.9350	.9355	.9360	.9365	.9370	.9375	.9380	.9385	.9390
87	.9395	.9400	.9405	.9410	.9415	.9420	.9425	.9430	.9435	.9440
88	.9445	.9450	.9455	.9460	.9465	.9469	.9474	.9479	.9484	.9489
89	.9494	.9499	.9504	.9509	.9513	.9518	.9523	.9528	.9533	.9538
90	.9542	.9547	.9552	.9557	.9562	.9566	.9571	.9576	.9581	.9586
91	.9590	.9595	.9600	.9605	.9609	.9614	.9619	.9624	.9628	.9633
92	.9638	.9643	.9647	.9652	.9657	.9661	.9666	.9671	.9675	.9680
93	.9685	.9689	.9694	.9699	.9703	.9708	.9713	.9717	.9722	.9727
94	.9731	.9736	.9741	.9745	.9750	.9754	.9759	.9763	.9768	.9773
95	.9777	.9782	.9786	.9791	.9795	.9800	.9805	.9809	.9814	.9818
96	.9823	.9827	.9832	.9836	.9841	.9845	.9850	.9854	.9859	.9863
97	.9868	.9872	.9877	.9881	.9886	.9890	.9894	.9899	.9903	.9908
98	.9912	.9917	.9921	.9926	.9930	.9934	.9939	.9943	.9948	.9952
99	.9956	.9961	.9965	.9969	.9974	.9978	.9983	.9987	.9991	.9996

THE USE OF TABLE 2

The use of Table 2 requires that the raw score be transformed into a z-score and that the variable be normally distributed.

The values in Table 2 represent the proportion of area in the standard normal curve which has a mean of 0, a standard deviation of 1.00, and a total area also equal to 1.00.

Since the normal curve is symmetrical, it is sufficient to indicate only the areas corresponding to positive z-values. Negative z-values will have precisely the same proportions of area as their positive counterparts.

Column B represents the proportion of area between the mean and a given z.

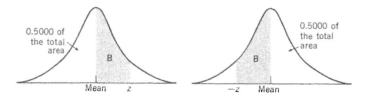

Column C represents the proportion of area beyond a given z.

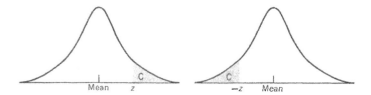

Table 2 Areas under the Standard Normal Curve

(A) z	(B) area between mean and z	(C) area beyond z	(A) z	(B) area between mean and z	(C) area beyond z	(A) z	(B) area between mean and z	(C) area beyond z
0.00	.0000	.5000	0.55	.2088	.2912	1.10	.3643	.1357
0.01	.0040	.4960	0.56	.2123	.2877	1.11	.3665	.1335
0.02	.0080	.4920	0.57	.2157	.2843	1.12	.3686	.1314
0.03	.0120	.4880	0.58	.2190	.2810	1.13	.3708	.1292
0.04	.0160	.4840	0.59	.2224	.2776	1.14	.3729	.1271
0.05	.0199	.4801	0.60	.2257	.2743	1.15	.3749	.1251
0.06	.0239	.4761	0.61	.2291	.2709	1.16	.3770	.1230
0.07	.0279	.4721	0.62	.2324	.2676	1.17	.3790	.1210
0.08	.0319	.4681	0.63	.2357	.2643	1.18	.3810	.1190
0.09	.0359	.4641	0.64	.2389	.2611	1.19	.3830	.1170
0.10	.0398	.4602	0.65	.2422	.2578	1.20	.3849	.1151
0.11	.0438	.4562	0.66	.2454	.2546	1.21	.3869	.1131
0.12	.0478	.4522	0.67	.2486	.2514	1.22	.3888	.1112
0.13	.0517	.4483	0.68	.2517	.2483	1.23	.3907	.1093
0.14	.0557	.4443	0.69	.2549	.2451	1.24	.3925	.1075
0.15	.0596	.4404	0.70	.2580	.2420	1.25	.3944	.1056
0.16	.0636	.4364	0.71	.2611	.2389	1.26	.3962	.1038
0.17	.0675	.4325	0.72	.2642	.2358	1.27	.3980	.1020
0.18	.0714	.4286	0.73	.2673	.2327	1.28	.3997	.1003
0.19	.0753	.4247	0.74	.2704	.2296	1.29	.4015	.0985
0.20	.0793	.4207	0.75	.2734	.2266	1.30	.4032	.0968
0.21	.0832	.4168	0.76	.2764	.2236	1.31	.4049	.0951
0.22	.0871	.4129	0.77	.2794	.2206	1.32	.4066	.0934
0.23	.0910	.4090	0.78	.2823	.2177	1.33	.4082	.0918
0.24	.0948	.4052	0.79	.2852	.2148	1.34	.4099	.0901
0.25	.0987	.4013	0.80	.2881	.2119	1.35	.4115	.0885
0.26	.1026	.3974	0.81	.2910	.2090	1.36	.4131	.0869
0.27	.1064	.3936	0.82	.2939	.2061	1.37	.4147	.0853
0.28	.1103	.3897	0.83	.2967	.2033	1.38	.4162	.0838
0.29	.1141	.3859	0.84	.2995	.2005	1.39	.4177	.0823
0.30	.1179	.3821	0.85	.3023	.1977	1.40	.4192	.0808
0.31	.1217	.3783	0.86	.3051	.1949	1.41	.4207	.0793
0.32	.1255	.3745	0.87	.3078	.1922	1.42	.4222	.0778
0.33	.1293	.3707	0.88	.3106	.1894	1.43	.4236	.0764
0.34	.1331	.3669	0.89	.3133	.1867	1.44	.4251	.0749
0.35	.1368	.3632	0.90	.3159	.1841	1.45	.4265	.0735
0.36	.1406	.3594	0.91	.3186	.1814	1.46	.4279	.0721
0.37	.1443	.3557	0.92	.3212	.1788	1.47	.4292	.0708
0.38	.1480	.3520	0.93	.3238	.1762	1.48	.4306	.0694
0.39	.1517	.3483	0.94	.3264	.1736	1.49	.4319	.0681
0.40	.1554	.3446	0.95	.3289	.1711	1.50	.4332	.0668
0.41	.1591	.3409	0.96	.3315	.1685	1.51	.4345	.0655
0.42	.1628	.3372	0.97	.3340	.1660	1.52	.4357	.0643
0.43	.1664	.3336	0.98	.3365	.1635	1.53	.4370	.0630
0.44	.1700	.3300	0.99	.3389	.1611	1.54	.4382	.0618
0.45	.1736	.3264	1.00	.3413	.1587	1.55	.4394	.0606
0.46	.1772	.3228	1.01	.3438	.1562	1.56	.4406	.0594
0.47	.1808	.3192	1.02	.3461	.1539	1.57	.4418	.0582
0.48	.1844	.3156	1.03	.3485	.1515	1.58	.4429	.0571
0.49	.1879	.3121	1.04	.3508	.1492	1.59	.4441	.0559
0.50	.1915	.3085	1.05	.3531	.1469	1.60	.4452	.0548
0.51	.1950	.3050	1.06	.3554	.1446	1.61	.4463	.0537
0.52	.1985	.3015	1.07	.3577	.1423	1.62	.4474	.0526
0.53	.2019	.2981	1.08	.3599	.1401	1.63	.4484	.0516
0.54	.2054	.2946	1.09	.3621	.1379	1.64	.4495	.0505

Table 2 continued

(A) z	(B) area between mean and z	(C) area beyond z	(A) z	(B) area between mean and z	(C) area beyond z	(A) z	(B) area between mean and z	(C) area beyond z
1.65	.4505	.0495	2.22	.4868	.0132	2.79	.4974	.0026
1.66	.4515	.0485	2.23	.4871	.0129	2.80	.4974	.0026
1.67	.4525	.0475	2.24	.4875	.0125	2.81	.4975	.0025
1.68	.4535	.0465	2.25	.4878	.0122	2.82	.4976	.0024
1.69	.4545	.0455	2.26	.4881	.0119	2.83	.4977	.0023
1.70	.4554	.0446	2.27	.4884	.0116	2.84	.4977	.0023
1.71	.4564	.0436	2.28	.4887	.0113	2.85	.4978	.0022
1.72	.4573	.0427	2.29	.4890	.0110	2.86	.4979	.0021
1.73	.4582	.0418	2.30	.4893	.0107	2.87	.4979	.0021
1.74	.4591	.0409	2.31	.4896	.0104	2.88	.4980	.0020
1.75	.4599	.0401	2.32	.4898	.0102	2.89	.4981	.0019
1.76	.4608	.0392	2.33	.4901	.0099	2.90	.4981	.0019
1.77	.4616	.0384	2.34	.4904	.0096	2.91	.4982	.0018
1.78	.4625	.0375	2.35	.4906	.0094	2.92	.4982	.0018
1.79	.4633	.0367	2.36	.4909	.0091	2.93	.4983	.0017
1.80	.4641	.0359	2.37	.4911	.0089	2.94	.4984	.0016
1.81	.4649	.0351	2.38	.4913	.0087	2.95	.4984	.0016
1.82	.4656	.0344	2.39	.4916	.0084	2.96	.4985	.0015
1.83	.4664	.0336	2.40	.4918	.0082	2.97	.4985	.0015
1.84	.4671	.0329	2.41	.4920	.0080	2.98	.4986	.0014
1.85	.4678	.0322	2.42	.4922	.0078	2.99	.4986	.0014
1.86	.4686	.0314	2.43	.4925	.0075	3.00	.4987	.0013
1.87	.4693	.0307	2.44	.4927	.0073	3.01	.4987	.0013
1.88	.4699	.0301	2.45	.4929	.0071	3.02	.4987	.0013
1.89	.4706	.0294	2.46	.4931	.0069	3.03	.4988	.0012
1.90	.4713	.0287	2.47	.4932	.0068	3.04	.4988	.0012
1.91	.4719	.0281	2.48	.4934	.0066	3.05	.4989	.0011
1.92	.4726	.0274	2.49	.4936	.0064	3.06	.4989	.0011
1.93	.4732	.0268	2.50	.4938	.0062	3.07	.4989	.0011
1.94	.4738	.0262	2.51	.4940	.0060	3.08	.4990	.0010
1.95	.4744	.0256	2.52	.4941	.0059	3.09	.4990	.0010
1.96	.4750	.0250	2.53	.4943	.0057	3.10	.4990	.0010
1.97	.4756	.0244	2.54	.4945	.0055	3.11	.4991	.0009
1.98	.4761	.0239	2.55	.4946	.0054	3.12	.4991	.0009
1.99	.4767	.0233	2.56	.4948	.0052	3.13	.4991	.0009
2.00	.4772	.0228	2.57	.4949	.0051	3.14	.4992	.0008
2.01	.4778	.0222	2.58	.4951	.0049	3.15	.4992	.0008
2.02	.4783	.0217	2.59	.4952	.0048	3.16	.4992	.0008
2.03	.4788	.0212	2.60	.4953	.0047	3.17	.4992	.0008
2.04	.4793	.0207	2.61	.4955	.0045	3.18	.4993	.0007
2.05	.4798	.0202	2.62	.4956	.0044	3.19	.4993	.0007
2.06	.4803	.0197	2.63	.4957	.0043	3.20	.4993	.0007
2.07	.4808	.0192	2.64	.4959	.0041	3.21	.4993	.0007
2.08	.4812	.0188	2.65	.4960	.0040	3.22	.4994	.0006
2.09	.4817	.0183	2.66	.4961	.0039	3.23	.4994	.0006
2.10	.4821	.0179	2.67	.4962	.0038	3.24	.4994	.0006
2.11	.4826	.0174	2.68	.4963	.0037	3.25	.4994	.0006
2.12	.4830	.0170	2.69	.4964	.0036	3.30	.4995	.0005
2.13	.4834	.0166	2.70	.4965	.0035	3.35	.4996	.0004
2.14	.4838	.0162	2.71	.4966	.0034	3.40	.4997	.0003
2.15	.4842	.0158	2.72	.4967	.0033	3.45	.4997	.0003
2.16	.4846	.0154	2.73	.4968	.0032	3.50	.4998	.0002
2.17	.4850	.0150	2.74	.4969	.0031	3.60	.4998	.0002
2.18	.4854	.0146	2.75	.4970	.0030	3.70	.4999	.0001
2.19	.4857	.0143	2.76	.4971	.0029	3.80	.4999	.0001
2.20	.4861	.0139	2.77	.4972	.0028	3.90	.49995	.00005
2.21	.4864	.0136	2.78	.4973	.0027	4.00	.49997	.00003

Table 3 Critical Values of Chi Square

	Probability under null hypothesis that $x^2 \geq$ Chi Square.					
	Level of significance for one-tailed test					
	.10	.05	.025	.01	.005	.0005
	Level of significance for two-tailed test					
df	.20	.10	.05	.02	.01	.001
1	1.64	2.71	3.84	5.41	6.64	10.83
2	3.22	4.60	5.99	7.82	9.21	13.82
3	4.64	6.25	7.82	9.84	11.34	16.27
4	5.99	7.78	9.49	11.67	13.28	18.46
5	7.29	9.24	11.07	13.39	15.09	20.52
6	8.56	10.64	12.59	15.03	16.81	22.46
7	9.80	12.02	14.07	16.62	18.48	24.32
8	11.03	13.36	15.51	18.17	20.09	26.12
9	12.24	14.68	16.92	19.68	21.67	27.88
10	13.44	15.99	18.31	21.16	23.21	29.59
11	14.63	17.28	19.68	22.62	24.72	31.26
12	15.81	18.55	21.03	24.05	26.22	32.91
13	16.98	19.81	22.36	25.47	27.69	34.53
14	18.15	21.06	23.68	26.87	29.14	36.12
15	19.31	22.31	25.00	28.26	30.58	37.70
16	20.46	23.54	26.30	29.63	32.00	39.29
17	21.62	24.77	27.59	31.00	33.41	40.75
18	22.76	25.99	28.87	32.35	34.80	42.31
19	23.90	27.20	30.14	33.69	36.19	43.82
20	25.04	28.41	31.41	35.02	37.57	45.32
21	26.17	29.62	32.67	36.34	38.93	46.80
22	27.30	30.81	33.92	37.66	40.29	48.27
23	28.43	32.01	35.17	38.97	41.64	49.73
24	29.55	33.20	36.42	40.27	42.98	51.18
25	30.68	34.38	37.65	41.57	44.31	52.62
26	31.80	35.56	38.88	42.86	45.64	54.05
27	32.91	36.74	40.11	44.14	46.96	55.48
28	34.03	37.92	41.34	45.42	48.28	56.89
29	35.14	39.09	42.69	46.69	49.59	58.30
30	36.25	40.26	43.77	47.96	50.89	59.70

Table 4 Critical Values of *t*

For any given m, the table shows the values of t corresponding to various levels of probability. Obtained t is significant at a given level if it is equal to or greater than the value shown in the table.

m	Level of significance for one-tailed test					
	.10	.05	.025	.01	.005	.0005
	Level of significance for two-tailed test					
	.20	.10	.05	.02	.01	.001
1	3.078	6.314	12.706	31.821	63.657	636.619
2	1.886	2.920	4.303	6.965	9.925	31.598
3	1.638	2.353	3.182	4.541	5.841	12.941
4	1.533	2.132	2.776	3.747	4.604	8.610
5	1.476	2.015	2.571	3.365	4.032	6.859
6	1.440	1.943	2.447	3.143	3.707	5.959
7	1.415	1.895	2.365	2.998	3.499	5.405
8	1.397	1.860	2.306	2.896	3.355	5.041
9	1.383	1.833	2.262	2.821	3.250	4.781
10	1.372	1.812	2.228	2.764	3.169	4.587
11	1.363	1.796	2.201	2.718	3.106	4.437
12	1.356	1.782	2.179	2.681	3.055	4.318
13	1.350	1.771	2.160	2.650	3.012	4.221
14	1.345	1.761	2.145	2.624	2.977	4.140
15	1.341	1.753	2.131	2.602	2.947	4.073
16	1.337	1.746	2.120	2.583	2.921	4.015
17	1.333	1.740	2.110	2.567	2.898	3.965
18	1.330	1.734	2.101	2.552	2.878	3.922
19	1.328	1.729	2.093	2.539	2.861	3.883
20	1.325	1.725	2.086	2.528	2.845	3.850
21	1.323	1.721	2.080	2.518	2.831	3.819
22	1.321	1.717	2.074	2.508	2.819	3.792
23	1.319	1.714	2.069	2.500	2.807	3.767
24	1.318	1.711	2.064	2.492	2.797	3.745
25	1.316	1.708	2.060	2.485	2.787	3.725
26	1.315	1.706	2.056	2.479	2.779	3.707
27	1.314	1.703	2.052	2.473	2.771	3.690
28	1.313	1.701	2.048	2.467	2.763	3.674
29	1.311	1.699	2.045	2.462	2.756	3.659
30	1.310	1.697	2.042	2.457	2.750	3.646
40	1.303	1.684	2.021	2.423	2.704	3.551
60	1.296	1.671	2.000	2.390	2.660	3.460
120	1.289	1.658	1.980	2.358	2.617	3.373
∞	1.282	1.645	1.960	2.326	2.576	3.291

Table 5 Critical Values of F

Level of significance 0.05 (light row) and 0.01 (dark row) points for the distribution of F

f_2 \ f_1	1	2	3	4	5	6	7	8	9	10	11	12	14	16	20	24	30	40	50	75	100	200	500	∞
1	161 4052	200 4999	216 5403	225 5625	230 5764	234 5859	237 5928	239 5981	241 6022	242 6056	243 6082	244 6106	245 6142	246 6169	248 6208	249 6234	250 6258	251 6286	252 6302	253 6323	253 6334	254 6352	254 6361	254 6366
2	18.51 98.49	19.00 99.01	19.16 99.17	19.25 99.25	19.30 99.30	19.33 99.33	19.36 99.34	19.37 99.36	19.38 99.38	19.39 99.40	19.40 99.41	19.41 99.42	19.42 99.43	19.43 99.44	19.44 99.45	19.45 99.46	19.46 99.47	19.47 99.48	19.47 99.48	19.48 99.49	19.49 99.49	19.49 99.49	19.50 99.50	19.50 99.50
3	10.13 34.12	9.55 30.81	9.28 29.46	9.12 28.71	9.01 28.24	8.94 27.91	8.88 27.67	8.84 27.49	8.81 27.34	8.78 27.23	8.76 27.13	8.74 27.05	8.71 26.92	8.69 26.83	8.66 26.69	8.64 26.60	8.62 26.50	8.60 26.41	8.58 26.30	8.57 26.27	8.56 26.23	8.54 26.18	8.54 26.14	8.53 26.12
4	7.71 21.20	6.94 18.00	6.59 16.69	6.39 15.98	6.26 15.52	6.16 15.21	6.09 14.98	6.04 14.80	6.00 14.66	5.96 14.54	5.93 14.45	5.91 14.37	5.87 14.24	5.84 14.15	5.80 14.02	5.77 13.93	5.74 13.83	5.71 13.74	5.70 13.69	5.68 13.61	5.66 13.57	5.65 13.52	5.64 13.48	5.63 13.46
5	6.61 16.26	5.79 13.27	5.41 12.06	5.19 11.39	5.05 10.97	4.95 10.67	4.88 10.45	4.82 10.27	4.78 10.15	4.74 10.05	4.70 9.96	4.68 9.89	4.64 9.77	4.60 9.68	4.56 9.55	4.53 9.47	4.50 9.38	4.46 9.29	4.44 9.24	4.42 9.17	4.40 9.13	4.38 9.07	4.37 9.04	4.36 9.02
6	5.99 13.74	5.14 10.92	4.76 9.78	4.53 9.15	4.39 8.75	4.28 8.47	4.21 8.26	4.15 8.10	4.10 7.98	4.06 7.87	4.03 7.79	4.00 7.72	3.96 7.60	3.92 7.52	3.87 7.39	3.84 7.31	3.81 7.23	3.77 7.14	3.75 7.09	3.72 7.02	3.71 6.99	3.69 6.94	3.68 6.90	3.67 6.88
7	5.59 12.25	4.74 9.55	4.35 8.45	4.12 7.85	3.97 7.46	3.87 7.19	3.79 7.00	3.73 6.84	3.68 6.71	3.63 6.62	3.60 6.54	3.57 6.47	3.52 6.35	3.49 6.27	3.44 6.15	3.41 6.07	3.38 5.98	3.34 5.90	3.32 5.85	3.29 5.78	3.28 5.75	3.25 5.70	3.24 5.67	3.23 5.65
8	5.32 11.26	4.46 8.65	4.07 7.59	3.84 7.01	3.69 6.63	3.58 6.37	3.50 6.19	3.44 6.03	3.39 5.91	3.34 5.82	3.31 5.74	3.28 5.67	3.23 5.56	3.20 5.48	3.15 5.36	3.12 5.28	3.08 5.20	3.05 5.11	3.03 5.06	3.00 5.00	2.98 4.96	2.96 4.91	2.94 4.88	2.93 4.86
9	5.12 10.56	4.26 8.02	3.86 6.99	3.63 6.42	3.48 6.06	3.37 5.80	3.29 5.62	3.23 5.47	3.18 5.35	3.13 5.26	3.10 5.18	3.07 5.11	3.02 5.00	2.98 4.92	2.93 4.80	2.90 4.73	2.86 4.64	2.82 4.56	2.80 4.51	2.77 4.45	2.76 4.41	2.73 4.36	2.72 4.33	2.71 4.31
10	4.96 10.04	4.10 7.56	3.71 6.55	3.48 5.99	3.33 5.64	3.22 5.39	3.14 5.21	3.07 5.06	3.02 4.95	2.97 4.85	2.94 4.78	2.91 4.71	2.86 4.60	2.82 4.52	2.77 4.41	2.74 4.33	2.70 4.25	2.67 4.17	2.64 4.12	2.61 4.05	2.59 4.01	2.56 3.96	2.55 3.93	2.54 3.91
11	4.84 9.65	3.98 7.20	3.59 6.22	3.36 5.67	3.20 5.32	3.09 5.07	3.01 4.88	2.95 4.74	2.90 4.63	2.86 4.54	2.82 4.46	2.79 4.40	2.74 4.29	2.70 4.21	2.65 4.10	2.61 4.02	2.57 3.94	2.53 3.86	2.50 3.80	2.47 3.74	2.45 3.70	2.42 3.66	2.41 3.62	2.40 3.60
12	4.75 9.33	3.88 6.93	3.49 5.95	3.26 5.41	3.11 5.06	3.00 4.82	2.92 4.65	2.85 4.50	2.80 4.39	2.76 4.30	2.72 4.22	2.69 4.16	2.64 4.05	2.60 3.98	2.54 3.86	2.50 3.78	2.46 3.70	2.42 3.61	2.40 3.56	2.36 3.49	2.35 3.46	2.32 3.41	2.31 3.38	2.30 3.36
13	4.67 9.07	3.80 6.70	3.41 5.74	3.18 5.20	3.02 4.86	2.92 4.62	2.84 4.44	2.77 4.30	2.72 4.19	2.67 4.10	2.63 4.02	2.60 3.96	2.55 3.85	2.51 3.78	2.46 3.67	2.42 3.59	2.38 3.51	2.34 3.42	2.32 3.37	2.28 3.30	2.26 3.27	2.24 3.21	2.22 3.18	2.21 3.16
14	4.60 8.86	3.74 6.51	3.34 5.56	3.11 5.03	2.96 4.69	2.85 4.46	2.77 4.28	2.70 4.14	2.65 4.03	2.60 3.94	2.56 3.86	2.53 3.80	2.48 3.70	2.44 3.62	2.39 3.51	2.35 3.43	2.31 3.34	2.27 3.26	2.24 3.21	2.21 3.14	2.19 3.11	2.16 3.06	2.14 3.02	2.13 3.00
15	4.54 8.68	3.68 6.36	3.29 5.42	3.06 4.89	2.90 4.56	2.79 4.32	2.70 4.14	2.64 4.00	2.59 3.89	2.55 3.80	2.51 3.73	2.48 3.67	2.43 3.56	2.39 3.48	2.33 3.36	2.29 3.29	2.25 3.20	2.21 3.12	2.18 3.07	2.15 3.00	2.12 2.97	2.10 2.92	2.08 2.89	2.07 2.87

f_2																								
16	4.49 8.53	3.63 6.23	3.24 5.29	3.01 4.77	2.85 4.44	2.74 4.20	2.66 4.03	2.59 3.89	2.54 3.78	2.49 3.69	2.45 3.61	2.42 3.55	2.37 3.45	2.33 3.37	2.28 3.25	2.24 3.18	2.20 3.10	2.16 3.01	2.13 2.96	2.09 2.89	2.07 2.86	2.04 2.80	2.02 2.77	2.01 2.75
17	4.45 8.40	3.59 6.11	3.20 5.18	2.96 4.67	2.81 4.34	2.70 4.10	2.62 3.93	2.55 3.79	2.50 3.68	2.45 3.59	2.41 3.52	2.38 3.45	2.33 3.35	2.29 3.27	2.23 3.16	2.19 3.08	2.15 3.00	2.11 2.92	2.08 2.86	2.04 2.79	2.02 2.76	1.99 2.70	1.97 2.67	1.96 2.65
18	4.41 8.28	3.55 6.01	3.16 5.09	2.93 4.58	2.77 4.25	2.66 4.01	2.58 3.85	2.51 3.71	2.46 3.60	2.41 3.51	2.37 3.44	2.34 3.37	2.29 3.27	2.25 3.19	2.19 3.07	2.15 3.00	2.11 2.91	2.07 2.83	2.04 2.78	2.00 2.71	1.98 2.68	1.95 2.62	1.93 2.59	1.92 2.57
19	4.38 8.18	3.52 5.93	3.13 5.01	2.90 4.50	2.74 4.17	2.63 3.94	2.55 3.77	2.48 3.63	2.43 3.52	2.38 3.43	2.34 3.36	2.31 3.30	2.26 3.19	2.21 3.12	2.15 3.00	2.11 2.92	2.07 2.84	2.02 2.76	2.00 2.70	1.96 2.63	1.94 2.60	1.91 2.54	1.90 2.51	1.88 2.49
20	4.35 8.10	3.49 5.85	3.10 4.94	2.87 4.43	2.71 4.10	2.60 3.87	2.52 3.71	2.45 3.56	2.40 3.45	2.35 3.37	2.31 3.30	2.28 3.23	2.23 3.13	2.18 3.05	2.12 2.94	2.08 2.86	2.04 2.77	1.99 2.69	1.96 2.63	1.92 2.56	1.90 2.53	1.87 2.47	1.85 2.44	1.84 2.42
21	4.32 8.02	3.47 5.78	3.07 4.87	2.84 4.37	2.68 4.04	2.57 3.81	2.49 3.65	2.42 3.51	2.37 3.40	2.32 3.31	2.28 3.24	2.25 3.17	2.20 3.07	2.15 2.99	2.09 2.88	2.05 2.80	2.00 2.72	1.96 2.63	1.93 2.58	1.89 2.51	1.87 2.47	1.84 2.42	1.82 2.38	1.81 2.36
22	4.30 7.94	3.44 5.72	3.05 4.82	2.82 4.31	2.66 3.99	2.55 3.76	2.47 3.59	2.40 3.45	2.35 3.35	2.30 3.26	2.26 3.18	2.23 3.12	2.18 3.02	2.13 2.94	2.07 2.83	2.03 2.75	1.98 2.67	1.93 2.58	1.91 2.53	1.87 2.46	1.84 2.42	1.81 2.37	1.80 2.33	1.78 2.31
23	4.28 7.88	3.42 5.66	3.03 4.76	2.80 4.26	2.64 3.94	2.53 3.71	2.45 3.54	2.38 3.41	2.32 3.30	2.28 3.21	2.24 3.14	2.20 3.07	2.14 2.97	2.10 2.89	2.04 2.78	2.00 2.70	1.96 2.62	1.91 2.53	1.88 2.48	1.84 2.41	1.82 2.37	1.79 2.32	1.77 2.28	1.76 2.26
24	4.26 7.82	3.40 5.61	3.01 4.72	2.78 4.22	2.62 3.90	2.51 3.67	2.43 3.50	2.36 3.36	2.30 3.25	2.26 3.17	2.22 3.09	2.18 3.03	2.13 2.93	2.09 2.85	2.02 2.74	1.98 2.66	1.94 2.58	1.89 2.49	1.86 2.44	1.82 2.36	1.80 2.33	1.76 2.27	1.74 2.23	1.73 2.21
25	4.24 7.77	3.38 5.57	2.99 4.68	2.76 4.18	2.60 3.86	2.49 3.63	2.41 3.46	2.34 3.32	2.28 3.21	2.24 3.13	2.20 3.05	2.16 2.99	2.11 2.89	2.06 2.81	2.00 2.70	1.96 2.62	1.92 2.54	1.87 2.45	1.84 2.40	1.80 2.32	1.77 2.29	1.74 2.23	1.72 2.19	1.71 2.17
26	4.22 7.72	3.37 5.53	2.98 4.64	2.74 4.14	2.59 3.82	2.47 3.59	2.39 3.42	2.32 3.29	2.27 3.17	2.22 3.09	2.18 3.02	2.15 2.96	2.10 2.86	2.05 2.77	1.99 2.66	1.95 2.58	1.90 2.50	1.85 2.41	1.82 2.36	1.78 2.28	1.76 2.25	1.72 2.19	1.70 2.15	1.69 2.13
27	4.21 7.68	3.35 5.49	2.96 4.60	2.73 4.11	2.57 3.79	2.46 3.56	2.37 3.39	2.30 3.26	2.25 3.14	2.20 3.06	2.16 2.98	2.13 2.93	2.08 2.83	2.03 2.74	1.97 2.63	1.93 2.55	1.88 2.47	1.84 2.38	1.80 2.33	1.76 2.25	1.74 2.21	1.71 2.16	1.68 2.12	1.67 2.10
28	4.20 7.64	3.34 5.45	2.95 4.57	2.71 4.07	2.56 3.76	2.44 3.53	2.36 3.36	2.29 3.23	2.24 3.11	2.19 3.03	2.15 2.95	2.12 2.90	2.06 2.80	2.02 2.71	1.96 2.60	1.91 2.52	1.87 2.44	1.81 2.35	1.78 2.30	1.75 2.22	1.72 2.18	1.69 2.13	1.67 2.09	1.65 2.06
29	4.18 7.60	3.33 5.42	2.93 4.54	2.70 4.04	2.54 3.73	2.43 3.50	2.35 3.33	2.28 3.20	2.22 3.08	2.18 3.00	2.14 2.92	2.10 2.87	2.05 2.77	2.00 2.68	1.94 2.57	1.90 2.49	1.85 2.41	1.80 2.32	1.77 2.27	1.73 2.19	1.71 2.15	1.68 2.10	1.65 2.06	1.64 2.03
30	4.17 7.56	3.32 5.39	2.92 4.51	2.69 4.02	2.53 3.70	2.42 3.47	2.34 3.30	2.27 3.17	2.21 3.06	2.16 2.98	2.12 2.90	2.09 2.84	2.04 2.74	1.99 2.66	1.93 2.55	1.89 2.47	1.84 2.38	1.79 2.29	1.76 2.24	1.72 2.16	1.69 2.13	1.66 2.07	1.64 2.03	1.62 2.01

(cont.)

Reprinted by permission from *Statistical Methods*, 6th edition, by George W. Snedecor and William C. Cochran, © 1967 by the Iowa State University Press.

Table 5 continued

Level of significance 0.05 (light row) and 0.01 (dark row) points for the distribution of F

f_2 \ f_1	1	2	3	4	5	6	7	8	9	10	11	12	14	16	20	24	30	40	50	75	100	200	500	∞
32	4.15 7.50	3.30 5.34	2.90 4.46	2.67 3.97	2.51 3.66	2.40 3.42	2.32 3.25	2.25 3.12	2.19 3.01	2.14 2.94	2.10 2.86	2.07 2.80	2.02 2.70	1.97 2.62	1.91 2.51	1.86 2.42	1.82 2.34	1.76 2.25	1.74 2.20	1.69 2.12	1.67 2.08	1.64 2.02	1.61 1.98	1.59 1.96
34	4.13 7.44	3.28 5.29	2.88 4.42	2.65 3.93	2.49 3.61	2.38 3.38	2.30 3.21	2.23 3.08	2.17 2.97	2.12 2.89	2.08 2.82	2.05 2.76	2.00 2.66	1.95 2.58	1.89 2.47	1.84 2.38	1.80 2.30	1.74 2.21	1.71 2.15	1.67 2.08	1.64 2.04	1.61 1.98	1.59 1.94	1.57 1.91
36	4.11 7.39	3.26 5.25	2.86 4.38	2.63 3.89	2.48 3.58	2.36 3.35	2.28 3.18	2.21 3.04	2.15 2.94	2.10 2.86	2.06 2.78	2.03 2.72	1.98 2.62	1.93 2.54	1.87 2.43	1.82 2.35	1.78 2.26	1.72 2.17	1.69 2.12	1.65 2.04	1.62 2.00	1.59 1.94	1.56 1.90	1.55 1.87
38	4.10 7.35	3.25 5.21	2.85 4.34	2.62 3.86	2.46 3.54	2.35 3.32	2.26 3.15	2.19 3.02	2.14 2.91	2.09 2.82	2.05 2.75	2.02 2.69	1.96 2.59	1.92 2.51	1.85 2.40	1.80 2.32	1.76 2.22	1.71 2.14	1.67 2.08	1.63 2.00	1.60 1.97	1.57 1.90	1.54 1.86	1.53 1.84
40	4.08 7.31	3.23 5.18	2.84 4.31	2.61 3.83	2.45 3.51	2.34 3.29	2.25 3.12	2.18 2.99	2.12 2.88	2.07 2.80	2.04 2.73	2.00 2.66	1.95 2.56	1.90 2.49	1.84 2.37	1.79 2.29	1.74 2.20	1.69 2.11	1.66 2.05	1.61 1.97	1.59 1.94	1.55 1.88	1.53 1.84	1.51 1.81
42	4.07 7.27	3.22 5.15	2.83 4.29	2.59 3.80	2.44 3.49	2.32 3.26	2.24 3.10	2.17 2.96	2.11 2.86	2.06 2.77	2.02 2.70	1.99 2.64	1.94 2.54	1.89 2.46	1.82 2.35	1.78 2.26	1.73 2.17	1.68 2.08	1.64 2.02	1.60 1.94	1.57 1.91	1.54 1.85	1.51 1.80	1.49 1.78
44	4.06 7.24	3.21 5.12	2.82 4.26	2.58 3.78	2.43 3.46	2.31 3.24	2.23 3.07	2.16 2.94	2.10 2.84	2.05 2.75	2.01 2.68	1.98 2.62	1.92 2.52	1.88 2.44	1.81 2.32	1.76 2.24	1.72 2.15	1.66 2.06	1.63 2.00	1.58 1.92	1.56 1.88	1.52 1.82	1.50 1.78	1.48 1.75
46	4.05 7.21	3.20 5.10	2.81 4.24	2.57 3.76	2.42 3.44	2.30 3.22	2.22 3.05	2.14 2.92	2.09 2.82	2.04 2.73	2.00 2.66	1.97 2.60	1.91 2.50	1.87 2.42	1.80 2.30	1.75 2.22	1.71 2.13	1.65 2.04	1.62 1.98	1.57 1.90	1.54 1.86	1.51 1.80	1.48 1.76	1.46 1.72
48	4.04 7.19	3.19 5.08	2.80 4.22	2.56 3.74	2.41 3.42	2.30 3.20	2.21 3.04	2.14 2.90	2.08 2.80	2.03 2.71	1.99 2.64	1.96 2.58	1.90 2.48	1.86 2.40	1.79 2.28	1.74 2.20	1.70 2.11	1.64 2.02	1.61 1.96	1.56 1.88	1.53 1.84	1.50 1.78	1.47 1.73	1.45 1.70
50	4.03 7.17	3.18 5.06	2.79 4.20	2.56 3.72	2.40 3.41	2.29 3.18	2.20 3.02	2.13 2.88	2.07 2.78	2.02 2.70	1.98 2.62	1.95 2.56	1.90 2.46	1.85 2.39	1.78 2.26	1.74 2.18	1.69 2.10	1.63 2.00	1.60 1.94	1.55 1.86	1.52 1.82	1.48 1.76	1.46 1.71	1.44 1.68
55	4.02 7.12	3.17 5.01	2.78 4.16	2.54 3.68	2.38 3.37	2.27 3.15	2.18 2.98	2.11 2.85	2.05 2.75	2.00 2.66	1.97 2.59	1.93 2.53	1.88 2.43	1.83 2.35	1.76 2.23	1.72 2.15	1.67 2.06	1.61 1.96	1.58 1.90	1.52 1.82	1.50 1.78	1.46 1.71	1.43 1.66	1.41 1.64
60	4.00 7.08	3.15 4.98	2.76 4.13	2.52 3.65	2.37 3.34	2.25 3.12	2.17 2.95	2.10 2.82	2.04 2.72	1.99 2.63	1.95 2.56	1.92 2.50	1.86 2.40	1.81 2.32	1.75 2.20	1.70 2.12	1.65 2.03	1.59 1.93	1.56 1.87	1.50 1.79	1.48 1.74	1.44 1.68	1.41 1.63	1.39 1.60
65	3.99 7.04	3.14 4.95	2.75 4.10	2.51 3.62	2.36 3.31	2.24 3.09	2.15 2.93	2.08 2.79	2.02 2.70	1.98 2.61	1.94 2.54	1.90 2.47	1.85 2.37	1.80 2.30	1.73 2.18	1.68 2.09	1.63 2.00	1.57 1.90	1.54 1.84	1.49 1.76	1.46 1.71	1.42 1.64	1.39 1.60	1.37 1.56

f_2																								
70	3.98 7.01	3.13 4.92	2.74 4.08	2.50 3.60	2.35 3.29	2.32 3.07	2.14 2.91	2.07 2.77	2.01 2.67	1.97 2.59	1.93 2.51	1.89 2.45	1.84 2.35	1.79 2.28	1.72 2.15	1.67 2.07	1.62 1.98	1.56 1.88	1.53 1.82	1.47 1.74	1.45 1.69	1.40 1.62	1.37 1.56	1.35 1.53
80	3.96 6.96	3.11 4.88	2.72 4.04	2.48 3.56	2.33 3.25	2.21 3.04	2.12 2.87	2.05 2.74	1.99 2.64	1.95 2.55	1.91 2.48	1.88 2.41	1.82 2.32	1.77 2.24	1.70 2.11	1.65 2.03	1.60 1.94	1.54 1.84	1.51 1.78	1.45 1.70	1.42 1.65	1.38 1.57	1.35 1.52	1.32 1.49
100	3.94 6.90	3.09 4.82	2.70 3.98	2.46 3.51	2.30 3.20	2.19 2.99	2.10 2.82	2.03 2.69	1.97 2.59	1.92 2.51	1.88 2.43	1.85 2.36	1.79 2.26	1.75 2.19	1.68 2.06	1.63 1.98	1.57 1.89	1.51 1.79	1.48 1.73	1.42 1.64	1.39 1.59	1.34 1.51	1.30 1.46	1.28 1.43
125	3.92 6.84	3.07 4.78	2.68 3.94	2.44 3.47	2.29 3.17	2.17 2.95	2.08 2.79	2.01 2.65	1.95 2.56	1.90 2.47	1.86 2.40	1.83 2.33	1.77 2.23	1.72 2.15	1.65 2.03	1.60 1.94	1.55 1.85	1.49 1.75	1.45 1.68	1.39 1.59	1.36 1.54	1.31 1.46	1.27 1.40	1.25 1.37
150	3.91 6.81	3.06 4.75	2.67 3.91	2.43 3.44	2.27 3.13	2.16 2.92	2.07 2.76	2.00 2.62	1.94 2.53	1.89 2.44	1.85 2.37	1.82 2.30	1.76 2.20	1.71 2.12	1.64 2.00	1.59 1.91	1.54 1.83	1.47 1.72	1.44 1.66	1.37 1.56	1.34 1.51	1.29 1.43	1.25 1.37	1.22 1.33
200	3.89 6.76	3.04 4.71	2.65 3.88	2.41 3.41	2.26 3.11	2.14 2.90	2.05 2.73	1.98 2.60	1.92 2.50	1.87 2.41	1.83 2.34	1.80 2.28	1.74 2.17	1.69 2.09	1.62 1.97	1.57 1.88	1.52 1.79	1.45 1.69	1.42 1.62	1.35 1.53	1.32 1.48	1.26 1.39	1.22 1.33	1.19 1.28
400	3.86 6.70	3.02 4.66	2.62 3.83	2.39 3.36	2.23 3.06	2.12 2.85	2.03 2.69	1.96 2.55	1.90 2.46	1.85 2.37	1.81 2.29	1.78 2.23	1.72 2.12	1.67 2.04	1.60 1.92	1.54 1.84	1.49 1.74	1.42 1.64	1.38 1.57	1.32 1.47	1.28 1.42	1.22 1.32	1.16 1.24	1.13 1.19
1000	3.85 6.66	3.00 4.62	2.61 3.80	2.38 3.34	2.22 3.04	2.10 2.82	2.02 2.66	1.95 2.53	1.89 2.43	1.84 2.34	1.80 2.26	1.76 2.20	1.70 2.09	1.65 2.01	1.58 1.89	1.53 1.81	1.47 1.71	1.41 1.61	1.36 1.54	1.30 1.44	1.26 1.38	1.19 1.28	1.13 1.19	1.08 1.11
∞	3.84 6.64	2.99 4.60	2.60 3.78	2.37 3.32	2.21 3.02	2.09 2.80	2.01 2.64	1.94 2.51	1.88 2.41	1.83 2.32	1.79 2.24	1.75 2.18	1.69 2.07	1.64 1.99	1.57 1.87	1.52 1.79	1.46 1.69	1.40 1.59	1.35 1.52	1.28 1.41	1.24 1.36	1.17 1.25	1.11 1.15	1.00 1.00

Table 6 Exponential Functions

x	e^x	e^{-x}	x	e^x	e^{-x}
0.00	1.0000	1.0000	2.5	12.182	0.0821
0.05	1.0513	0.9512	2.6	13.464	0.0743
0.10	1.1052	0.9048	2.7	14.880	0.0672
0.15	1.1618	0.8607	2.8	16.445	0.0608
0.20	1.2214	0.8187	2.9	18.174	0.0550
0.25	1.2840	0.7788	3.0	20.086	0.0498
0.30	1.3499	0.7408	3.1	22.198	0.0450
0.35	1.4191	0.7047	3.2	24.533	0.0408
0.40	1.4918	0.6703	3.3	27.113	0.0369
0.45	1.5683	0.6376	3.4	29.964	0.0334
0.50	1.6487	0.6065	3.5	33.115	0.0302
0.55	1.7333	0.5769	3.6	36.598	0.0273
0.60	1.8221	0.5488	3.7	40.447	0.0247
0.65	1.9155	0.5220	3.8	44.701	0.0224
0.70	2.0138	0.4966	3.9	49.402	0.0202
0.75	2.1170	0.4724	4.0	54.598	0.0183
0.80	2.2255	0.4493	4.1	60.340	0.0166
0.85	2.3396	0.4274	4.2	66.686	0.0150
0.90	2.4596	0.4066	4.3	73.700	0.0136
0.95	2.5857	0.3867	4.4	81.451	0.0123
1.0	2.7183	0.3679	4.5	90.017	0.0111
1.1	3.0042	0.3329	4.6	99.484	0.0101
1.2	3.3201	0.3012	4.7	109.95	0.0091
1.3	3.6693	0.2725	4.8	121.51	0.0082
1.4	4.0552	0.2466	4.9	134.29	0.0074
1.5	4.4817	0.2231	5	148.41	0.0067
1.6	4.9530	0.2019	6	403.43	0.0025
1.7	5.4739	0.1827	7	1096.6	0.0009
1.8	6.0496	0.1653	8	2981.0	0.0003
1.9	6.6859	0.1496	9	8103.1	0.0001
2.0	7.3891	0.1353	10	22026	0.00005
2.1	8.1662	0.1225			
2.2	9.0250	0.1108			
2.3	9.9742	0.1003			
2.4	11.023	0.0907			

Table 7 $z = \frac{1}{2} \ln \left(\frac{1+r}{1-r}\right)$

r	z	r	z	r	z	r	z
.00	.000	.25	.255	.50	.549	.75	.973
.01	.010	.26	.266	.51	.563	.76	.996
.02	.020	.27	.277	.52	.576	.77	1.020
.03	.030	.28	.288	.53	.590	.78	1.045
.04	.040	.29	.299	.54	.604	.79	1.071
.05	.050	.30	.310	.55	.618	.80	1.099
.06	.060	.31	.321	.56	.633	.81	1.127
.07	.070	.32	.332	.57	.648	.82	1.157
.08	.080	.33	.343	.58	.662	.83	1.188
.09	.090	.34	.354	.59	.678	.84	1.221
.10	.100	.35	.365	.60	.693	.85	1.256
.11	.110	.36	.377	.61	.709	.86	1.293
.12	.121	.37	.388	.62	.725	.87	1.333
.13	.131	.38	.400	.63	.741	.88	1.376
.14	.141	.39	.412	.64	.758	.89	1.422
.15	.151	.40	.424	.65	.775	.90	1.472
.16	.161	.41	.436	.66	.793	.91	1.528
.17	.172	.42	.448	.67	.811	.92	1.589
.18	.182	.43	.460	.68	.829	.93	1.658
.19	.192	.44	.472	.69	.848	.94	1.738
.20	.203	.45	.485	.70	.867	.95	1.832
.21	.213	.46	.497	.71	.887	.96	1.946
.22	.224	.47	.510	.72	.908	.97	2.092
.23	.234	.48	.523	.73	.929	.98	2.298
.24	.245	.49	.536	.74	.950	.99	2.647

From R. A. Fisher and F. Yates, *Statistical Tables for Biological, Agricultural and Medical Research*. Edinburgh: Oliver and Boyd, Ltd., 1948. Reprinted by permission of the authors and the publisher.

Table 8 Squares, Square Roots, and Reciprocals of Numbers from 1 to 1000

N	N^2	$\sqrt{N}$	$1/N$	N	N^2	$\sqrt{N}$	$1/N$	N	N^2	$\sqrt{N}$	$1/N$
1	1	1.0000	1.000000	61	3721	7.8102	.016393	121	14641	11.0000	.00826446
2	4	1.4142	.500000	62	3844	7.8740	.016129	122	14884	11.0454	.00819672
3	9	1.7321	.333333	63	3969	7.9373	.015873	123	15129	11.0905	.00813008
4	16	2.0000	.250000	64	4096	8.0000	.015625	124	15376	11.1355	.00806452
5	25	2.2361	.200000	65	4225	8.0623	.015385	125	15625	11.1803	.00800000
6	36	2.4495	.166667	66	4356	8.1240	.015152	126	15876	11.2250	.00793651
7	49	2.6458	.142857	67	4489	8.1854	.014925	127	16129	11.2694	.00787402
8	64	2.8284	.125000	68	4624	8.2462	.014706	128	16384	11.3137	.00781250
9	81	3.0000	.111111	69	4761	8.3066	.014493	129	16641	11.3578	.00775194
10	100	3.1623	.100000	70	4900	8.3666	.014286	130	16900	11.4018	.00769231
11	121	3.3166	.090909	71	5041	8.4261	.014085	131	17161	11.4455	.00763359
12	144	3.4641	.083333	72	5184	8.4853	.013889	132	17424	11.4891	.00757576
13	169	3.6056	.076923	73	5329	8.5440	.013699	133	17689	11.5326	.00751880
14	196	3.7417	.071429	74	5476	8.6023	.013514	134	17956	11.5758	.00746269
15	225	3.8730	.066667	75	5625	8.6603	.013333	135	18225	11.6190	.00740741
16	256	4.0000	.062500	76	5776	8.7178	.013158	136	18496	11.6619	.00735294
17	289	4.1231	.058824	77	5929	8.7750	.012987	137	18769	11.7047	.00729927
18	324	4.2426	.055556	78	6084	8.8318	.012821	138	19044	11.7473	.00724638
19	361	4.3589	.052632	79	6241	8.8882	.012658	139	19321	11.7898	.00719424
20	400	4.4721	.050000	80	6400	8.9443	.012500	140	19600	11.8322	.00714286
21	441	4.5826	.047619	81	6561	9.0000	.012346	141	19881	11.8743	.00709220
22	484	4.6904	.045455	82	6724	9.0554	.012195	142	20164	11.9164	.00704225
23	529	4.7958	.043478	83	6889	9.1104	.012048	143	20449	11.9583	.00699301
24	576	4.8990	.041667	84	7056	9.1652	.011905	144	20736	12.0000	.00694444
25	625	5.0000	.040000	85	7225	9.2195	.011765	145	21025	12.0416	.00689655
26	676	5.0990	.038462	86	7396	9.2736	.011628	146	21316	12.0830	.00684932
27	729	5.1962	.037037	87	7569	9.3274	.011494	147	21609	12.1244	.00680272
28	784	5.2915	.035714	88	7744	9.3808	.011364	148	21904	12.1655	.00675676
29	841	5.3852	.034483	89	7921	9.4340	.011236	149	22201	12.2066	.00671141
30	900	5.4772	.033333	90	8100	9.4868	.011111	150	22500	12.2474	.00666667
31	961	5.5678	.032258	91	8281	9.5394	.010989	151	22801	12.2882	.00662252
32	1024	5.6569	.031250	92	8464	9.5917	.010870	152	23104	12.3288	.00657895
33	1089	5.7446	.030303	93	8649	9.6437	.010753	153	23409	12.3693	.00653595
34	1156	5.8310	.029412	94	8836	9.6954	.010638	154	23716	12.4097	.00649351
35	1225	5.9161	.028571	95	9025	9.7468	.010526	155	24025	12.4499	.00645161
36	1296	6.0000	.027778	96	9216	9.7980	.010417	156	24336	12.4900	.00641026
37	1369	6.0828	.027027	97	9409	9.8489	.010309	157	24649	12.5300	.00636943
38	1444	6.1644	.026316	98	9604	9.8995	.010204	158	24964	12.5698	.00632911
39	1521	6.2450	.025641	99	9801	9.9499	.010101	159	25281	12.6095	.00628931
40	1600	6.3246	.025000	100	10000	10.0000	.010000	160	25600	12.6491	.00625000
41	1681	6.4031	.024390	101	10201	10.0499	.00990099	161	25921	12.6886	.00621118
42	1764	6.4807	.023810	102	10404	10.0995	.00980392	162	26244	12.7279	.00617284
43	1849	6.5574	.023256	103	10609	10.1489	.00970874	163	26569	12.7671	.00613497
44	1936	6.6332	.022727	104	10816	10.1980	.00961538	164	26896	12.8062	.00609756
45	2025	6.7082	.022222	105	11025	10.2470	.00952381	165	27225	12.8452	.00606061
46	2116	6.7823	.021739	106	11236	10.2956	.00943396	166	27556	12.8841	.00602410
47	2209	6.8557	.021277	107	11449	10.3441	.00934579	167	27889	12.9228	.00598802
48	2304	6.9282	.020833	108	11664	10.3923	.00925926	168	28224	12.9615	.00595238
49	2401	7.0000	.020408	109	11881	10.4403	.00917431	169	28561	13.0000	.00591716
50	2500	7.0711	.020000	110	12100	10.4881	.00909091	170	28900	13.0384	.00588235
51	2601	7.1414	.019608	111	12321	10.5357	.00900901	171	29241	13.0767	.00584795
52	2704	7.2111	.019231	112	12544	10.5830	.00892857	172	29584	13.1149	.00581395
53	2809	7.2801	.018868	113	12769	10.6301	.00884956	173	29929	13.1529	.00578035
54	2916	7.3485	.018519	114	12996	10.6771	.00877193	174	30276	13.1909	.00574713
55	3025	7.4162	.018182	115	13225	10.7238	.00869565	175	30625	13.2288	.00571429
56	3136	7.4833	.017857	116	13456	10.7703	.00862069	176	30976	13.2665	.00568182
57	3249	7.5498	.017544	117	13689	10.8167	.00854701	177	31329	13.3041	.00564972
58	3364	7.6158	.017241	118	13924	10.8628	.00847458	178	31684	13.3417	.00561798
59	3481	7.6811	.016949	119	14161	10.9087	.00840336	179	32041	13.3791	.00558659
60	3600	7.7460	.016667	120	14400	10.9545	.00833333	180	32400	13.4164	.00555556

Reprinted by permission from J. W. Dunlap and A. K. Kurtz, *Handbook of Statistical Nomographs, Tables, and Formulas*. New York: World Book Company, 1932.

Table 8 continued

N	N^2	$\sqrt{N}$	$1/N$	N	N^2	$\sqrt{N}$	$1/N$	N	N^2	$\sqrt{N}$	$1/N$
181	32761	13.4536	.00552486	241	58081	15.5242	.00414938	301	90601	17.3494	.00332226
182	33124	13.4907	.00549451	242	58564	15.5563	.00413223	302	91204	17.3781	.00331126
183	33489	13.5277	.00546448	243	59049	15.5885	.00411523	303	91809	17.4069	.00330033
184	33856	13.5647	.00543478	244	59536	15.6205	.00409836	304	92416	17.4356	.00328947
185	34225	13.6015	.00540541	245	60025	15.6525	.00408163	305	93025	17.4642	.00327868
186	34596	13.6382	.00537634	246	60516	15.6844	.00406504	306	93636	17.4929	.00326797
187	34969	13.6748	.00534759	247	61009	15.7162	.00404858	307	94249	17.5214	.00325733
188	35344	13.7113	.00531915	248	61504	15.7480	.00403226	308	94864	17.5499	.00321675
189	35721	13.7477	.00529101	249	62001	15.7797	.00401606	309	95481	17.5784	.00323625
190	36100	13.7840	.00526316	250	62500	15.8114	.00400000	310	96100	17.6068	.00322581
191	36481	13.8203	.00523560	251	63001	15.8430	.00398406	311	96721	17.6352	.00321543
192	36864	13.8564	.00520833	252	63504	15.8745	.00396825	312	97344	17.6635	.00320513
193	37249	13.8924	.00518135	253	64009	15.9060	.00395257	313	97969	17.6918	.00319489
194	37636	13.9284	.00515464	254	64516	15.9374	.00393701	314	98596	17.7200	.00318471
195	38025	13.9642	.00512821	255	65025	15.9687	.00392157	315	99225	17.7482	.00317460
196	38416	14.0000	.00510204	256	65536	16.0000	.00390625	316	99856	17.7764	.00316456
197	38809	14.0357	.00507614	257	66049	16.0312	.00389105	317	100489	17.8045	.00315457
198	39204	14.0712	.00505051	258	66564	16.0624	.00387597	318	101124	17.8326	.00314465
199	39601	14.1067	.00502513	259	67081	16.0935	.00386100	319	101761	17.8606	.00313480
200	40000	14.1421	.00500000	260	67600	16.1245	.00384615	320	102400	17.8885	.00312500
201	40401	14.1774	.00497512	261	68121	16.1555	.00383142	321	103041	17.9165	.00311526
202	40804	14.2127	.00495050	262	68644	16.1864	.00381679	322	103684	17.9444	.00310559
203	41209	14.2478	.00492611	263	69169	16.2173	.00380228	323	104329	17.9722	.00309598
204	41616	14.2829	.00490196	264	69696	16.2481	.00378788	324	104976	18.0000	.00308642
205	42025	14.3178	.00487805	265	70225	16.2788	.00377358	325	105625	18.0278	.00307692
206	42436	14.3527	.00485437	266	70756	16.3095	.00375940	326	106276	18.0555	.00306748
207	42849	14.3875	.00483092	267	71289	16.3401	.00374532	327	106929	18.0831	.00305810
208	43264	14.4222	.00480769	268	71824	16.3707	.00373134	328	107584	18.1108	.00304878
209	43681	14.4568	.00478469	269	72361	16.4012	.00371747	329	108241	18.1384	.00303951
210	44100	14.4914	.00476190	270	72900	16.4317	.00370370	330	108900	18.1659	.00303030
211	44521	14.5258	.00473934	271	73441	16.4621	.00369004	331	109561	18.1934	.00302115
212	44944	14.5602	.00471698	272	73984	16.4924	.00367647	332	110224	18.2209	.00301205
213	45369	14.5945	.00469484	273	74529	16.5227	.00366300	333	110889	18.2483	.00300300
214	45796	14.6287	.00467290	274	75076	16.5529	.00364964	334	111556	18.2757	.00299401
215	46225	14.6629	.00465116	275	75625	16.5831	.00363636	335	112225	18.3030	.00298507
216	46656	14.6969	.00462963	276	76176	16.6132	.00362319	336	112896	18.3303	.00297619
217	47089	14.7309	.00460829	277	76729	16.6433	.00361011	337	113569	18.3576	.00296736
218	47524	14.7648	.00458716	278	77284	16.6733	.00359712	338	114244	18.3848	.00295858
219	47961	14.7986	.00456621	279	77841	16.7033	.00358423	339	114921	18.4120	.00294985
220	48400	14.8324	.00454545	280	78400	16.7332	.00357143	340	115600	18.4391	.00294118
221	48841	14.8661	.00452489	281	78961	16.7631	.00355872	341	116281	18.4662	.00293255
222	49284	14.8997	.00450450	282	79524	16.7929	.00354610	342	116964	18.4932	.00292398
223	49729	14.9332	.00448430	283	80089	16.8226	.00353357	343	117649	18.5203	.00291545
224	50176	14.9666	.00446429	284	80656	16.8523	.00352113	344	118336	18.5472	.00290698
225	50625	15.0000	.00444444	285	81225	16.8819	.00350877	345	119025	18.5742	.00289855
226	51076	15.0333	.00442478	286	81796	16.9115	.00349650	346	119716	18.6011	.00289017
227	51529	15.0665	.00440529	287	82369	16.9411	.00348432	347	120409	18.6279	.00288184
228	51984	15.0997	.00438596	288	82944	16.9706	.00347222	348	121104	18.6548	.00287356
229	52441	15.1327	.00436681	289	83521	17.0000	.00346021	349	121801	18.6815	.00286533
230	52900	15.1658	.00434783	290	84100	17.0294	.00344828	350	122500	18.7083	.00285714
231	53361	15.1987	.00432900	291	84681	17.0587	.00343643	351	123201	18.7350	.00284900
232	53824	15.2315	.00431034	292	85264	17.0880	.00342466	352	123904	18.7617	.00284091
233	54289	15.2643	.00429185	293	85849	17.1172	.00341297	353	124609	18.7883	.00283286
234	54756	15.2971	.00427350	294	86436	17.1464	.00340136	354	125316	18.8149	.00282486
235	55225	15.3297	.00425532	295	87025	17.1756	.00338983	355	126025	18.8414	.00281690
236	55696	15.3623	.00423729	296	87616	17.2047	.00337838	356	126736	18.8680	.00280899
237	56169	15.3948	.00421941	297	88209	17.2337	.00336700	357	127449	18.8944	.00280112
238	56644	15.4272	.00420168	298	88804	17.2627	.00335570	358	128164	18.9209	.00279330
239	57121	15.4596	.00418410	299	89401	17.2916	.00334448	359	128881	18.9473	.00278552
240	57600	15.4919	.00416667	300	90000	17.3205	.00333333	360	129600	18.9737	.00277778

(cont.)

Table 8 continued

N	N^2	$\sqrt{N}$	$1/N$	N	N^2	$\sqrt{N}$	$1/N$	N	N^2	$\sqrt{N}$	$1/N$
361	130321	19.0000	.00277008	421	177241	20.5183	.00237530	481	231361	21.9317	.00207900
362	131044	19.0263	.00276243	422	178084	20.5426	.00236967	482	232324	21.9545	.00207469
363	131769	19.0526	.00275482	423	178929	20.5670	.00236407	483	233289	21.9773	.00207039
364	132496	19.0788	.00274725	424	179776	20.5913	.00235849	484	234256	22.0000	.00206612
365	133225	19.1050	.00273973	425	180625	20.6155	.00235294	485	235225	22.0227	.00206186
366	133956	19.1311	.00273224	426	181476	20.6398	.00234742	486	236196	22.0454	.00205761
367	134689	19.1572	.00272480	427	182329	20.6640	.00234192	487	237169	22.0681	.00205339
368	135424	19.1833	.00271739	428	183184	20.6882	.00233645	488	238144	22.0907	.00204918
369	136161	19.2094	.00271003	429	184041	20.7123	.00233100	489	239121	22.1133	.00204499
370	136900	19.2354	.00270270	430	184900	20.7364	.00232558	490	240100	22.1359	.00204082
371	137641	19.2614	.00269542	431	185761	20.7605	.00232019	491	241081	22.1585	.00203666
372	138384	19.2873	.00268817	432	186624	20.7846	.00231481	492	242064	22.1811	.00203252
373	139129	19.3132	.00268097	433	187489	20.8087	.00230947	493	243049	22.2036	.00202840
374	139876	19.3391	.00267380	434	188356	20.8327	.00230415	494	244036	22.2261	.00202429
375	140625	19.3649	.00266667	435	189225	20.8567	.00229885	495	245025	22.2486	.00202020
376	141376	19.3907	.00265957	436	190096	20.8806	.00229358	496	246016	22.2711	.00201613
377	142129	19.4165	.00265252	437	190969	20.9045	.00228833	497	247009	22.2935	.00201207
378	142884	19.4422	.00264550	438	191844	20.9284	.00228311	498	248004	22.3159	.00200803
379	143641	19.4679	.00263852	439	192721	20.9523	.00227790	499	249001	22.3383	.00200401
380	144400	19.4936	.00263158	440	193600	20.9762	.00227273	500	250000	22.3607	.00200000
381	145161	19.5192	.00262467	441	194481	21.0000	.00226757	501	251001	22.3830	.00199601
382	145924	19.5448	.00261780	442	195364	21.0238	.00226244	502	252004	22.4054	.00199203
383	146689	19.5704	.00261097	443	196249	21.0476	.00225734	503	253009	22.4277	.00198807
384	147456	19.5959	.00260417	444	197136	21.0713	.00225225	504	254016	22.4499	.00198413
385	148225	19.6214	.00259740	445	198025	21.0950	.00224719	505	255025	22.4722	.00198020
386	148996	19.6469	.00259067	446	198916	21.1187	.00224215	506	256036	22.4944	.00197628
387	149769	19.6723	.00258398	447	199809	21.1424	.00223714	507	257049	22.5167	.00197239
388	150544	19.6977	.00257732	448	200704	21.1660	.00223214	508	258064	22.5389	.00196850
389	151321	19.7231	.00257069	449	201601	21.1896	.00222717	509	259081	22.5610	.00196464
390	152100	19.7484	.00256410	450	202500	21.2132	.00222222	510	260100	22.5832	.00196078
391	152881	19.7737	.00255754	451	203401	21.2368	.00221729	511	261121	22.6053	.00195695
392	153664	19.7990	.00255102	452	204304	21.2603	.00221239	512	262144	22.6274	.00195312
393	154449	19.8242	.00254453	453	205209	21.2838	.00220751	513	263169	22.6495	.00194932
394	155236	19.8494	.00253807	454	206116	21.3073	.00220264	514	264196	22.6716	.00194553
395	156025	19.8746	.00253165	455	207025	21.3307	.00219870	515	265225	22.6936	.00194175
396	156816	19.8997	.00252525	456	207936	21.3542	.00219298	516	266256	22.7156	.00193798
397	157609	19.9249	.00251889	457	208849	21.3776	.00218818	517	267289	22.7376	.00193424
398	158404	19.9499	.00251256	458	209764	21.4009	.00218341	518	268324	22.7596	.00193050
399	159201	19.9750	.00250627	459	210681	21.4243	.00217865	519	269361	22.7816	.00192678
400	160000	20.0000	.00250000	460	211600	21.4476	.00217391	520	270400	22.8035	.00192308
401	160801	20.0250	.00249377	461	212521	21.4709	.00216920	521	271441	22.8254	.00191939
402	161604	20.0499	.00248756	462	213444	21.4942	.00216450	522	272484	22.8473	.00191571
403	162409	20.0749	.00248139	463	214369	21.5174	.00215983	523	273529	22.8692	.00191205
404	163216	20.0998	.00247525	464	215296	21.5407	.00215517	524	274576	22.8910	.00190840
405	164025	20.1246	.00246914	465	216225	21.5639	.00215054	525	275625	22.9129	.00190476
406	164836	20.1494	.00246305	466	217156	21.5870	.00214592	526	276676	22.9347	.00190114
407	165649	20.1742	.00245700	467	218089	21.6102	.00214133	527	277729	22.9565	.00189753
408	166464	20.1990	.00245098	468	219024	21.6333	.00213675	528	278784	22.9783	.00189394
409	167281	20.2237	.00244499	469	219961	21.6564	.00213220	529	279841	23.0000	.00189036
410	168100	20.2485	.00243902	470	220900	21.6795	.00212766	530	280900	23.0217	.00188679
411	168921	20.2731	.00243309	471	221841	21.7025	.00212314	531	281961	23.0434	.00188324
412	169744	20.2978	.00242718	472	222784	21.7256	.00211864	532	283024	23.0651	.00187970
413	170569	20.3224	.00242131	473	223729	21.7486	.00211416	533	284089	23.0868	.00187617
414	171396	20.3470	.00241546	474	224676	21.7715	.00210970	534	285156	23.1084	.00187266
415	172225	20.3715	.00240964	475	225625	21.7945	.00210526	535	286225	23.1301	.00186916
416	173056	20.3961	.00240385	476	226576	21.8174	.00210084	536	287296	23.1517	.00186567
417	173889	20.4206	.00239808	477	227529	21.8403	.00209644	537	288369	23.1733	.00186220
418	174724	20.4450	.00239234	478	228484	21.8632	.00209205	538	289444	23.1948	.00185874
419	175561	20.4695	.00238663	479	229441	21.8861	.00208768	539	290521	23.2164	.00185529
420	176400	20.4939	.00238095	480	230400	21.9089	.00208333	540	291600	23.2379	.00185185

Table 8 continued

N	N²	√N	1/N	N	N²	√N	1/N	N	N²	√N	1/N
541	292681	23.2594	.00184843	601	361201	24.5153	.00166389	661	436921	25.7099	.00151286
542	293764	23.2809	.00184502	602	362404	24.5357	.00166113	662	438244	25.7294	.00151057
543	294849	23.3024	.00184162	603	363609	24.5561	.00165837	663	439569	25.7488	.00150830
544	295936	23.3238	.00183824	604	364816	24.5764	.00165563	664	440896	25.7682	.00150602
545	297025	23.3452	.00183486	605	366025	24.5967	.00165289	665	442225	25.7876	.00150376
546	298116	23.3666	.00183150	606	367236	24.6171	.00165017	666	443556	25.8070	.00150150
547	299209	23.3880	.00182815	607	368449	24.6374	.00164745	667	444889	25.8263	.00149925
548	300304	23.4094	.00182482	608	369664	24.6577	.00164474	668	446224	25.8457	.00149701
549	301401	23.4307	.00182149	609	370881	24.6779	.00164204	669	447561	25.8650	.00149477
550	302500	23.4521	.00181818	610	372100	24.6982	.00163934	670	448900	25.8844	.00149254
551	303601	23.4734	.00181488	611	373321	24.7184	.00163666	671	450241	25.9037	.00149031
552	304704	23.4947	.00181159	612	374544	24.7386	.00163399	672	451584	25.9230	.00148810
553	305809	23.5160	.00180832	613	375769	24.7588	.00163132	673	452929	25.9422	.00148588
554	306916	23.5372	.00180505	614	376996	24.7790	.00162866	674	454276	25.9615	.00148368
555	308025	23.5584	.00180180	615	378225	24.7992	.00162602	675	455625	25.9808	.00148148
556	309136	23.5797	.00179856	616	379456	24.8193	.00162338	676	456976	26.0000	.00147929
557	310249	23.6008	.00179533	617	380689	24.8395	.00162075	677	458329	26.0192	.00147710
558	311364	23.6220	.00179211	618	381924	24.8596	.00161812	678	459684	26.0384	.00147493
559	312481	23.6432	.00178891	619	383161	24.8797	.00161551	679	461041	26.0576	.00147275
560	313600	23.6643	.00178571	620	384400	24.8998	.00161290	680	462400	26.0768	.00147059
561	314721	23.6854	.00178253	621	385641	24.9199	.00161031	681	463761	26.0960	.00146843
562	315844	23.7065	.00177936	622	386884	24.9399	.00160772	682	465124	26.1151	.00146628
563	316969	23.7276	.00177620	623	388129	24.9600	.00160514	683	466489	26.1343	.00146413
564	318096	23.7487	.00177305	624	389376	24.9800	.00160256	684	467856	26.1534	.00146199
565	319225	23.7697	.00176991	625	390625	25.0000	.00160000	685	469225	26.1725	.00145985
566	320356	23.7908	.00176678	626	391876	25.0200	.00159744	686	470596	26.1916	.00145773
567	321489	23.8118	.00176367	627	393129	25.0400	.00159490	687	471969	26.2107	.00145560
568	322624	23.8328	.00176056	628	394384	25.0599	.00159236	688	473344	26.2298	.00145349
569	323761	23.8537	.00175747	629	395641	25.0799	.00158983	689	474721	26.2488	.00145138
570	324900	23.8747	.00175439	630	396900	25.0998	.00158730	690	476100	26.2679	.00144928
571	326041	23.8956	.00175131	631	398161	25.1197	.00158479	691	477481	26.2869	.00144718
572	327184	23.9165	.00164825	632	399424	25.1396	.00158228	692	478864	26.3059	.00144509
573	328329	23.9374	.00174520	633	400689	25.1595	.00157978	693	480249	26.3249	.00144300
574	329476	23.9583	.00174216	634	401956	25.1794	.00157729	694	481636	26.3439	.00144092
575	330625	23.9792	.00173913	635	403225	25.1992	.00157480	695	483025	26.3629	.00143885
576	331776	24.0000	.00173611	636	404496	25.2190	.00157233	696	484416	26.3818	.00143678
577	332929	24.0208	.00173310	637	405769	25.2389	.00156986	697	485809	26.4008	.00143472
578	334084	24.0416	.00173010	638	407044	25.2587	.00156740	698	487204	26.4197	.00143266
579	335241	24.0624	.00172712	639	408321	25.2784	.00156495	699	488601	26.4386	.00143062
580	336400	24.0832	.00172414	640	409600	25.2982	.00156250	700	490000	26.4575	.00142857
581	337561	24.1039	.00172117	641	410881	25.3180	.00156006	701	491401	26.4764	.00142653
582	338724	24.1247	.00171821	642	412164	25.3377	.00155763	702	492804	26.4953	.00142450
583	339889	24.1454	.00171527	643	413449	25.3574	.00155521	703	494209	26.5141	.00142248
584	341056	24.1661	.00171233	644	414736	25.3772	.00155280	704	495616	26.5330	.00142045
585	342225	24.1868	.00170940	645	416025	25.3969	.00155039	705	497025	26.5518	.00141844
586	343396	24.2074	.00170648	646	417316	25.4165	.00154799	706	498436	26.5707	.00141643
587	344569	24.2281	.00170358	647	418609	25.4362	.00154560	707	499849	26.5895	.00141443
588	345744	24.2487	.00170068	648	419904	25.4558	.00154321	708	501264	26.6083	.00141243
589	346921	24.2693	.00169779	649	421201	25.4755	.00154083	709	502681	26.6271	.00141044
590	348100	24.2899	.00169492	650	422500	25.4951	.00153846	710	504100	26.6458	.00140845
591	349281	24.3105	.00169205	651	423801	25.5147	.00153610	711	505521	26.6646	.00140647
592	350464	24.3311	.00168919	652	425104	25.5343	.00153374	712	506944	26.6833	.00140449
593	351649	24.3516	.00168634	653	426409	25.5539	.00153139	713	508369	26.7021	.00140252
594	352836	24.3721	.00168350	654	427716	25.5734	.00152905	714	509796	26.7208	.00140056
595	354025	24.3926	.00168067	655	429025	25.5930	.00152672	715	511225	26.7395	.00139860
596	355216	24.4131	.00167785	656	430336	25.6125	.00152439	716	512656	26.7582	.00139665
597	356409	24.4336	.00167504	657	431649	25.6320	.00152207	717	514089	26.7769	.00139470
598	357604	24.4540	.00167224	658	432964	25.6515	.00151976	718	515524	26.7955	.00139276
599	358801	24.4745	.00166945	659	434281	25.6710	.00151745	719	516961	26.8142	.00139082
600	360000	24.4949	.00166667	660	435600	25.6905	.00151515	720	518400	26.8328	.00138889

(cont.)

Table 8 continued

N	N²	√N	1/N	N	N²	√N	1/N	N	N²	√N	1/N
721	519841	26.8514	.00138696	781	609961	27.9464	.00128041	841	707281	29.0000	.00118906
722	521284	26.8701	.00138504	782	611524	27.9643	.00127877	842	708964	29.0172	.00118765
723	522729	26.8887	.00138313	783	613089	27.9821	.00127714	843	710649	29.0345	.00118624
724	524176	26.9072	.00138122	784	614656	28.0000	.00127551	844	712336	29.0517	.00118483
725	525625	26.9258	.00137931	785	616225	28.0179	.00127389	845	714025	29.0689	.00118343
726	527076	26.9444	.00137741	786	617796	28.0357	.00127226	846	715716	29.0861	.00118203
727	528529	26.9629	.00137552	787	619369	28.0535	.00127065	847	717409	29.1033	.00118064
728	529984	26.9815	.00137363	788	620944	28.0713	.00126904	848	719104	29.1204	.00117925
729	531441	27.0000	.00137174	789	622521	28.0891	.00126743	849	720801	29.1376	.00117786
730	532900	27.0185	.00136986	790	624100	28.1069	.00126582	850	722500	29.1548	.00117647
731	534361	27.0370	.00136799	791	625681	28.1247	.00126422	851	724201	29.1719	.00117509
732	535824	27.0555	.00136612	792	627264	28.1425	.00126263	852	725904	29.1890	.00117371
733	537289	27.0740	.00136426	793	628849	28.1603	.00126103	853	727609	29.2062	.00117233
734	538756	27.0924	.00136240	794	630436	28.1780	.00125945	854	729316	29.2233	.00117096
735	540225	27.1109	.00136054	795	632025	28.1957	.00125786	855	731025	29.2404	.00116959
736	541696	27.1293	.00135870	796	633616	28.2135	.00125628	856	732736	29.2575	.00116822
737	543169	27.1477	.00135685	797	635209	28.2312	.00125471	857	734449	29.2746	.00116686
738	544644	27.1662	.00135501	798	636804	28.2489	.00125313	858	736164	29.2916	.00116550
739	546121	27.1846	.00135318	799	638401	28.2666	.00125156	859	737881	29.3087	.00116414
740	547600	27.2029	.00135135	800	640000	28.2843	.00125000	860	739600	29.3258	.00116279
741	549081	27.2213	.00134953	801	641601	28.3019	.00124844	861	741321	29.3428	.00116144
742	550564	27.2397	.00134771	802	643204	28.3196	.00124688	862	743044	29.3598	.00116009
743	552049	27.2580	.00134590	803	644809	28.3373	.00124533	863	744769	29.3769	.00115875
744	553536	27.2764	.00134409	804	646416	28.3549	.00124378	864	746496	29.3939	.00115741
745	555025	27.2947	.00134228	805	648025	28.3725	.00124224	865	748225	29.4109	.00115607
746	556516	27.3130	.00134048	806	649636	28.3901	.00124069	866	749956	29.4279	.00115473
747	558009	27.3313	.00133869	807	651249	28.4077	.00123916	867	751689	29.4449	.00115340
748	559504	27.3496	.00133690	808	652864	28.4253	.00123762	868	753424	29.4618	.00115207
749	561001	27.3679	.00133511	809	654481	28.4429	.00123609	869	755161	29.4788	.00115075
750	562500	27.3861	.00133333	810	656100	28.4605	.00123457	870	756900	29.4958	.00114943
751	564001	27.4044	.00133156	811	657721	28.4781	.00123305	871	758641	29.5127	.00114811
752	565504	27.4226	.00132979	812	659344	28.4956	.00123153	872	760384	29.5296	.00114679
753	567009	27.4408	.00132802	813	660969	28.5132	.00123001	873	762129	29.5466	.00114548
754	568516	27.4591	.00132626	814	662596	28.5307	.00122850	874	763876	29.5635	.00114416
755	570025	27.4773	.00132450	815	664225	28.5482	.00122699	875	765625	29.5804	.00114286
756	571536	27.4955	.00132275	816	665856	28.5657	.00122549	876	767376	29.5973	.00114155
757	573049	27.5136	.00132100	817	667489	28.5832	.00122399	877	769129	29.6142	.00114025
758	574564	27.5318	.00131926	818	669124	28.6007	.00122249	878	770884	29.6311	.00113895
759	576081	27.5500	.00131752	819	670761	28.6182	.00122100	879	772641	29.6479	.00113766
760	577600	27.5681	.00131579	820	672400	28.6356	.00121951	880	774400	29.6648	.00113636
761	579121	27.5862	.00131406	821	674041	28.6531	.00121803	881	776161	29.6816	.00113507
762	580644	27.6043	.00131234	822	675684	28.6705	.00121655	882	777924	29.6985	.00113379
763	582169	27.6225	.00131062	823	677329	28.6880	.00121507	883	779689	29.7153	.00113250
764	583696	27.6405	.00130890	824	678976	28.7054	.00121359	884	781456	29.7321	.00113122
765	585225	27.6586	.00130719	825	680625	28.7228	.00121212	885	783225	29.7489	.00112994
766	586756	27.6767	.00130548	826	682276	28.7402	.00121065	886	784996	29.7658	.00112867
767	588289	27.6948	.00130378	827	683929	28.7576	.00120919	887	786769	29.7825	.00112740
768	589824	27.7128	.00130208	828	685584	28.7750	.00120773	888	788544	29.7993	.00112613
769	591361	27.7308	.00130039	829	687241	28.7924	.00120627	889	790321	29.8161	.00112486
770	592900	27.7489	.00129870	830	688900	28.8097	.00120482	890	792100	29.8329	.00112360
771	594441	27.7669	.00129702	831	690561	28.8271	.00120337	891	793881	29.8496	.00112233
772	595984	27.7849	.00129534	832	692224	28.8444	.00120192	892	795664	29.8664	.00112108
773	597529	27.8029	.00129366	833	693889	28.8617	.00120048	893	797449	29.8831	.00111982
774	599076	27.8209	.00129199	834	695556	28.8791	.00119904	894	799236	29.8998	.00111857
775	600625	27.8388	.00129032	835	697225	28.8964	.00119760	895	801025	29.9166	.00111732
776	602176	27.8568	.00128866	836	698896	28.9137	.00119617	896	802816	29.9333	.00111607
777	603729	27.8747	.00128700	837	700569	28.9310	.00119474	897	804609	29.9500	.00111483
778	605284	27.8927	.00128535	838	702244	28.9482	.00119332	898	806404	29.9666	.00111359
779	606841	27.9106	.00128370	839	703921	28.9655	.00119190	899	808201	29.9833	.00111235
780	608400	27.9285	.00128205	840	705600	28.9828	.00119048	900	810000	30.0000	.00111111

Table 8 continued

N	N^2	$\sqrt{N}$	$1/N$	N	N^2	$\sqrt{N}$	$1/N$	N	N^2	$\sqrt{N}$	$1/N$
901	811801	30.0167	.00110988	936	876096	30.5941	.00106838	971	942841	31.1609	.00102987
902	813604	30.0333	.00110865	937	877969	30.6105	.00106724	972	944784	31.1769	.00102881
903	815409	30.0500	.00110742	938	879844	30.6268	.00106610	973	946729	31.1929	.00102775
904	817216	30.0666	.00110619	939	881721	30.6431	.00106496	974	948676	31.2090	.00102669
905	819025	30.0832	.00110497	940	883600	30.6594	.00106383	975	950625	31.2250	.00102564
906	820836	30.0998	.00110375	941	885481	30.6757	.00106270	976	952576	31.2410	.00102459
907	822649	30.1164	.00110254	942	887364	30.6920	.00106157	977	954529	31.2570	.00102354
908	824464	30.1330	.00110132	943	889249	30.7083	.00106045	978	956484	31.2730	.00102249
909	826281	30.1496	.00110011	944	891136	30.7246	.00105932	979	958441	31.2890	.00102145
910	828100	30.1662	.00109890	945	893025	30.7409	.00105820	980	960400	31.3050	.00102041
911	829921	30.1828	.00109769	946	894916	30.7571	.00105708	981	962361	31.3209	.00101937
912	831744	30.1993	.00109649	947	896809	30.7734	.00105597	982	964324	31.3369	.00101833
913	833569	30.2159	.00109529	948	898704	30.7896	.00105485	983	966289	31.3528	.00101729
914	835396	30.2324	.00109409	949	900601	30.8058	.00105374	984	968256	31.3688	.00101626
915	837225	30.2490	.00109290	950	902500	30.8221	.00105263	985	970225	31.3847	.00101523
916	839056	30.2655	.00109170	951	904401	30.8383	.00105152	986	972196	31.4006	.00101420
917	840889	30.2820	.00109051	952	906304	30.8545	.00105042	987	974169	31.4166	.00101317
918	842724	30.2985	.00108932	953	908209	30.8707	.00104932	988	976144	31.4325	.00101215
919	844561	30.3150	.00108814	954	910116	30.8869	.00104822	989	978121	31.4484	.00101112
920	846400	30.3315	.00108696	955	912025	30.9031	.00104712	990	980100	31.4643	.00101010
921	848241	30.3480	.00108578	956	913936	30.9192	.00104603	991	982081	31.4802	.00100908
922	850084	30.3645	.00108460	957	915849	30.9354	.00104493	992	984064	31.4960	.00100806
923	851929	30.3809	.00108342	958	917764	30.9516	.00104384	993	986049	31.5119	.00100705
924	853776	30.3974	.00108225	959	919681	30.9677	.00104275	994	988036	31.5278	.00100604
925	855625	30.4138	.00108108	960	921600	30.9839	.00104167	995	990025	31.5436	.00100503
926	857476	30.4302	.00107991	961	923521	31.0000	.00104058	996	992016	31.5595	.00100402
927	859329	30.4467	.00107875	962	925444	31.0161	.00103950	997	994009	31.5753	.00103842
928	861184	30.4631	.00107759	963	927369	31.0322	.00103842	998	996004	31.5911	.00100200
929	863041	30.4795	.00107643	964	929296	31.0483	.00103734	999	998001	31.6070	.00100100
930	864900	30.4959	.00107527	965	931225	31.0644	.00103627	1000	1000000	31.6228	.00100000
931	866761	30.5123	.00107411	966	933156	31.0805	.00103520				
932	868624	30.5287	.00107296	967	935089	31.0966	.00103413				
933	870489	30.5450	.00107181	968	937024	31.1127	.00103306				
934	872356	30.5614	.00107066	969	938961	31.1288	.00103199				
935	874225	30.5778	.00106952	970	940900	31.1448	.00103093				

Answers to the Exercises

Chapter 1

Section 1.1

1. a) The set whose elements are 1 and 45
   b) The set whose elements are $a, b, 6, 7$, and 9
   c) The set whose sole element is the set whose sole element is 1
   d) The set consisting of 1 and the set whose sole element is 1
   e) The set of all animals
   f) The set of all Canadian citizens
   g) The set of Indians living in Iowa
   h) The set of even integers

2. a) True. If x is in S, then x is in T. But if x is in T, then x is in W. Hence, if x is in S, then x is in W.
   b) False. $\{1, 2\} \subset \{1, 2, 3\}$, but $\{1, 2, 3\} \not\subset \{1, 2\}$.
   c) True. If x is in S, then x is in S.
   d) False. Let $S = \{1\}$, $T = \{1, 2\}$, and $W = \{1, 3\}$.

3. a) (1, 3) (1, 4)　　b) $(a, 6)$ $(a, 7)$　　c) (q, t)
      (2, 3) (2, 4)　　　　$(b, 6)$ $(b, 7)$
      　　　　　　　　　　$(c, 6)$ $(c, 7)$

d) (1, 5) (1, 6) (1, 7) (1, 8)   e) (A, A) (A, B) (A, C) (A, D)
   (2, 5) (2, 6) (2, 7) (2, 8)   (B, A) (B, B) (B, C) (B, D)
   (3, 5) (3, 6) (3, 7) (3, 8)   (C, A) (C, B) (C, C) (C, D)
   (4, 5) (4, 6) (4, 7) (4, 8)   (D, A) (D, B) (D, C) (D, D)

4. $\{(a, b), (b, b), (c, b)\}$, range = T, domain = S, image = $\{b\}$; many correct answers: in each, the range is T and the domain is S. Be sure that each element of S appears as a first coordinate once and only once.

5. a) Fails; some people have no or more than one cousin.
   b) Fails; women cannot be fathers, and a man may have more than one, or no children.
   c) Fails; many people of same age.
   d) Fails; a person's mother, although unique, may not be a living human being.
   If we allow *any* human being, then (d) defines a function.

6. (a), (b), (e), and (f) define functions. (a) and (b) have R as their image.

Section 1.2

1. a) Not-wind (or no wind)   b) Wind and lightning
   c) Not-rain and lightning   d) Rain or lightning
   e) Not-rain or not-lightning   f) Rain and either wind or not-lightning
   g) Either rain or not-lightning, this together with wind
   h) Rain if no lightning
   i) No wind if there is both rain and lightning
   j) Rain and wind but no lightning

2. a) $A \mid (B \cap C)$ b) $A \cap \overline{C}$ c) $C \cap (\overline{B} \cap \overline{A})$
   d) $C \mid (\overline{B} \cap \overline{A})$ e) $(C \cap A) \cap \overline{B}$

3. a) All ordered triples of H (heads) and T (tails)
   b) $\{1, 2, 3, 4, 5, 6\} \times \{1, 2, 3, 4, 5, 6\}$, each ordered pair has first roll in first coordinate, second roll in second coordinate
   c) Same as in (b), except first coordinate represents what occurs on first die, second coordinate what occurs on second die
   d) $\{1, 2, \ldots, 10\}$ e) $\{1, 2, \ldots, 10\} \times \{1, 2, \ldots, 10\}$

4. a)

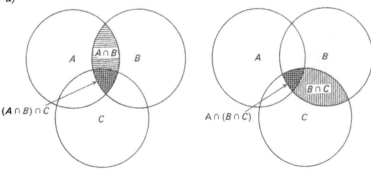

b) If x is in A, then x is not in $S - A$; hence x is in $S - (S - A) = \overline{\overline{A}}$. Therefore $A \subset \overline{\overline{A}}$. But if x is in $\overline{\overline{A}}$, then x is also in $S - A$. That is, x is not an element not in A. Therefore x is in A.

c) Both sets are represented diagramatically by the shaded portion of the figure below:

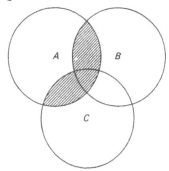

d)
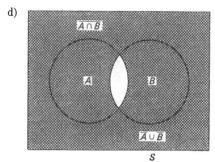

5. Many correct answers
6. a) True. That either A or B occurs is more probable than that just A occurs.
   b) True. That both A and B occur is less probable than that merely B occurs.
   c) False. This will be taken up in the next section.
   d) False. This will be taken up in the next section.
   e) True. $A \cap B$ is impossible, hence will have least probability.
   f) True

Section 1.3

1. a) $R(A) = 0.7$, $R(B) = 0.55$, $R(\overline{A}) = 0.3$, $R(\overline{B}) = 0.45$
   b) $R(\overline{A} \cup B) = 0.7$ c) $R(A \mid B) = \frac{8}{11}$
   d) i) $0.85 \geq 0.7$ iii) $0.4 \neq 0$
      ii) $0.4 \leq 0.55$ iv) $\frac{4}{7} < \frac{8}{11}$
   e) Rain with lightning is more probable than lightning with rain.
2. Many correct answers, e.g., probability of rain given certain atmospheric conditions, probability randomly chosen college student will have red hair

3. 80% of the time these atmospheric conditions were present, rain occurred; is a relative frequency
4. a) $\frac{16}{17}$;   b) $\frac{1}{17}$

Section 1.4

1. $P(A \cup B \cup C) = P(A \cap B \cap C)$ if and only if

   $P(A) + P(B) + P(C) - P(A, B) - P(B, C) - P(A, C) = 0$

   (from developing $P(A \cup B \cup C)$ and subtracting $P(A \cap B \cap C)$ from both sides). Hence

   $(P(A) - P(A, B)) + (P(B) - P(B, C)) + (P(C) - P(B, C)) = 0.$

   Since all the differences inside the parentheses are nonnegative, they must be zero. Thus, $P(A) = P(A, B) = P(A)P(B \mid A)$. Therefore $P(B \mid A) = 1$; hence B occurs if A occurs. Similarly, A occurs if B occurs, etc.

2. a) $P(A) = \frac{5}{8}$, $P(B) = \frac{9}{16}$, $P(C) = \frac{7}{16}$, not independent
   b) $P(A \cap B) = \frac{3}{8}$, $P(A \cap \overline{B}) = \frac{1}{4}$, $P(\overline{A} \cap B) = \frac{3}{16}$, $P(\overline{A} \cap \overline{B}) = \frac{3}{16}$, not independent
   c) $\frac{3}{5}$
   d) $A \cap B \cap C$ $A \cap \overline{B} \cap C$ $A \cap B \cap \overline{C}$ $A \cap \overline{B} \cap \overline{C}$
      $\frac{1}{36}$ $\frac{2}{36}$ $\frac{5}{36}$ $\frac{10}{36}$
      $\overline{A} \cap B \cap C$ $\overline{A} \cap B \cap \overline{C}$ $\overline{A} \cap \overline{B} \cap C$ $\overline{A} \cap \overline{B} \cap \overline{C}$
      $\frac{1}{36}$ $\frac{5}{36}$ $\frac{2}{36}$ $\frac{10}{36}$

3. a) $\frac{2}{5}$ b) $1/K$, $K =$ total number of bridge hands
   c) $1/M$, $M =$ total number of ways two balls can be selected from the urn (without replacement or regard to order)
   d) $24/(20 \cdot 19 \cdot 18 \cdot 17)$

4. True for $n = 2$; assume true for $n - 1$.

   $P(A_1 \cap \cdots \cap A_n)$
   $= P((A_1 \cap A_2 \cap \cdots \cap A_{n-1}) \cap A_n)$
   $= P(A_1 \cap \cdots \cap A_{n-1})P(A_n \mid (A_1 \cap \cdots \cap A_{n-1}))$
   $=$ (by induction assumption)
   $\quad P(A_1)P(A_2 \mid A_1)P(A_3 \mid A_1 \cap A_2) \cdots P(A_n \mid A_1 \cap \cdots \cap A_n)$

Chapter 2

Section 2.1

1. a) 1   b) 12   c) 60   d) 720   e) 1   f) n g) 1
2. a) 24, 720, 2
   b) $_nP_m = n(n-1) \cdots (n-m+1)$
   $= \dfrac{n(n-1) \cdots (n-m+1)(n-m)(n-m-1) \cdots 1}{(n-m)(n-m-1) \cdots 1}$
   $= \dfrac{n!}{(n-m)!}$

3. a) $_{10}P_5$ b) $_{10}P_5$ c) $_3P_3$ d) $_{20}P_5$ e) $(_8P_8)^2$

4. a) $\frac{1}{5}, \frac{1}{5}, \frac{1}{20}$ b) $\frac{1}{10}, \frac{1}{2}, \frac{1}{2}, \frac{1}{10}, \frac{1}{2}, \frac{1}{5}$

   c) $\frac{6}{15}$, $_6P_6/_{15}P_6$ (since 4, 5, 6, 7, 14, and 15 can be selected in any of $_6P_6$ orders)

Section 2.2

1. a) 5   b) 35   c) 120   d) 1   e) n f) 4950

2. a) and b) $\dfrac{C(n, k) = n!}{(n - k)!k!}$

   and

   $$C(n, n - k) = \frac{n!}{(n - (n - k))!(n - k)!} = \frac{n;}{(n - k)!k!}$$

   c) $C(n, k)$ has largest value at m if $n = 2m + 1$ and at $(n - 1)/2$ and $(n - 1)/2 + 1$ when $n = 2m$.

3. a) $C(100, 5)$ b) $C(52, 7)$ c) $C(32, 3)$
   d) $C(20, 5)$ e) $C(13, 5)C(14, 4)$ f) $C(100, 10)C(25, 3)C(10, 5)$

4. a) $\frac{3}{10}, \frac{3}{10}, 1/C(10, 3), \frac{7}{10}, P(4) + P(5) - P(4 \cap 5) = \frac{6}{10} - \dfrac{C(2, 2)C(8, 1)}{C(10, 3)}$

   b) $\frac{7}{15}, \dfrac{C(8, 2)}{C(15, 2)}, \dfrac{C(8, 1)C(7, 1)}{C(15, 2)}, 0, \dfrac{C(8, 3)C(7, 2)}{C(15, 5)}$,

   $\dfrac{C(8, 3)C(7, 2) + C(8, 2)C(7, 3) + C(8, 1)C(7, 4) + C(8, 0)C(7, 5)}{C(15, 5)}$,

   $1 - P$ (at least 2)

   c) $\frac{6}{18}, \dfrac{C(5, 1)C(6, 1)}{C(18, 2)}, \dfrac{C(7, 2)C(5, 3)C(6, 2)}{C(18, 7)}, 0$

Section 2.3

1. a) 81   b) 64   c) 1024   d) n e) 1   f) n^n

2. $C(n, m) \leq {_nP_m} \leq A(n, m)$

3. a) 2^6, all strings aren't equiprobable since it is more probable that one would draw red than blue.
   b) $A(6, 5)$ c) $A(30, 2)$
   d) $C(30, 15)$, since once jobs have been assigned to one person, the remainder must go to the other person.

4. a) $(\frac{1}{2})^5$, $C(5, 2)(\frac{1}{2})^5$, $(C(5, 2) + C(5, 3) + C(5, 4) + C(5, 5))(\frac{1}{2})^5$
   b) $(\frac{1}{6})^6$, $C(6, 4)(\frac{1}{6})^4(\frac{5}{6})^2$, $(\frac{5}{6})^6$, $C(6, 3)C(3, 3)(\frac{1}{6})^6(\frac{5}{6})^0 = C(6, 3)(\frac{1}{6})^6$
   c) $\frac{1}{3}, (\frac{1}{3})^{10}, (\frac{1}{3})^2, (\frac{2}{3})^{10}$

Section 2.4

1. a) $\dfrac{(\frac{1}{2})(C(4, 4)/C(8, 4))}{(\frac{1}{2})(C(4, 4)/C(8, 4)) + (\frac{1}{2})(C(7, 4)/C(19, 4))}$,

   $\dfrac{(\frac{1}{2})((C(6, 2)C(7, 2))/C(19, 4))}{(\frac{1}{2})(C(6, 2)C(7, 2))/C(19, 4) + (\frac{1}{2})(C(4, 2)C(4, 2))/C(8, 4)}$, 1

b) $\dfrac{(\frac{1}{3})(\frac{2}{7})^5}{(\frac{1}{3})(\frac{2}{7})^5 + (\frac{2}{3})(\frac{1}{6})^5}$, $\dfrac{(\frac{1}{3})(\frac{5}{7})^5}{(\frac{1}{3})(\frac{5}{7})^5 + (\frac{2}{3})(\frac{5}{6})^5}$

c) $P(A) = (0.7)(0.1) + (0.98)(0.1)$, $P(B_1 \mid A) = \dfrac{(0.7)(0.1)}{(0.7)(0.1) + (0.98)(0.1)}$

d) $(0.3)^3 + (0.2)^3 + (0.5)^3$, $\dfrac{(\frac{1}{3})(0.2)^3}{(\frac{1}{3})(0.2)^3 + (\frac{1}{3})(0.3)^3 + (\frac{1}{3})(0.5)^3}$

2. Divide numerator and denominator by $P(B_1)$.
3. a) Yes, A needs to occur with B_1, B_2, or B_3.
   b) and c) No information

Chapter 3

Section 3.1

1. Discrete: (a), (c)   Continuous: (b)
   Neither discrete nor continuous: (d), (e) and (f), since none are intervals, R, or half-lines
2. "Finite" means that the set is in one–one correspondence with $\{1, 2, \ldots, n\}$ for some integer n.
3. a) Define $A + B$ to be $\{a + b \mid a \text{ in } A, b \text{ in } B\}$. Then $A + B$ is the maximum admissible range of $X + Y$. If A and B are countable, then $A + B$ is also countable. If X and Y are continuous, $A + B$ is still maximum possible admissible range of $X + Y$, but since admissible range of $X + Y$ may not be all of $A + B$, $X + Y$ may not be continuous.
   b) $AB = \{ab \mid a \text{ in } A, b \text{ in } B\}$. AB is countable if A and B are. XY may not be continuous even if X and Y are.
   c) $A + B = \{3\frac{1}{2}, 5, \frac{13}{4}, 4\frac{1}{2}, 6, \frac{21}{4}, 5\frac{1}{2}, 7, \frac{25}{4}\}$;
      $AB = \{\frac{3}{2}, 6, \frac{15}{4}, 2, 8, 5, \frac{5}{2}, 10, \frac{25}{4}\}$
   d) $A + B = (0, 3)$; $AB = (-1, 2)$
   e) $g(X) = X^2 + 3X$;
      admissible range is $\{x^2 + 3x \mid x \text{ is in } A$, the admissible range of $X\}$
4. a) Number of not-improved   b) % of improved
   c) difference of temperature on tenth day and average temperature
5. Many correct answers

Section 3.2

1. a) 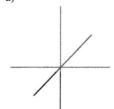 b) 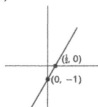 c)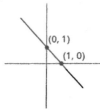

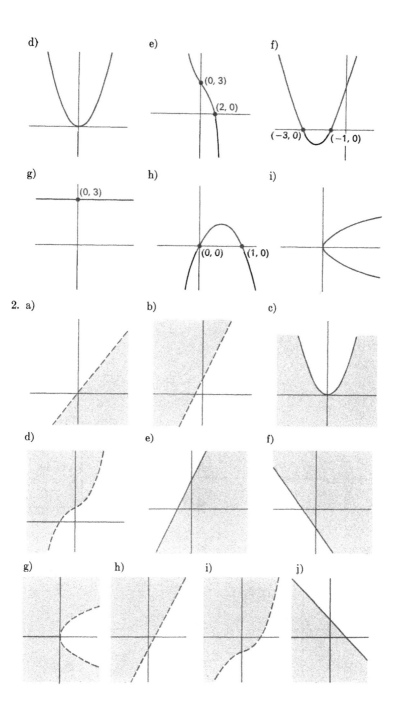

3. a)

b)

c)

d) No solution

4. a)

b) (0, 3), (2, 0)

c) (3, 8)

d)

e)

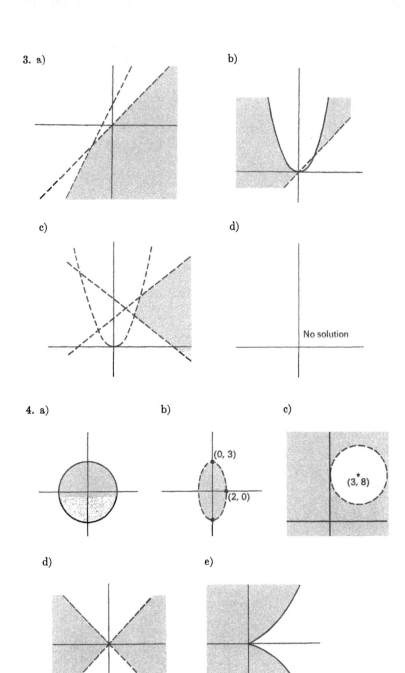

Section 3.3

1.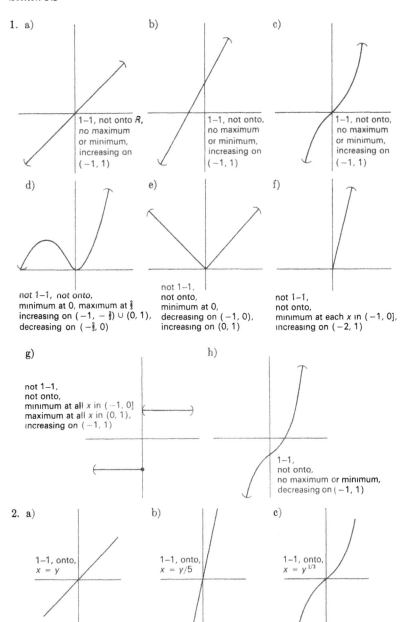

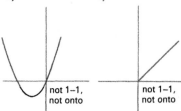

3. $(g - h)(x) = g(x) - h(x)$
   a) "Add" "directed" segment $f(x)$ to "directed" segment $g(x)$ (see figure below).

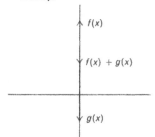

b)

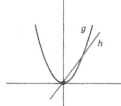 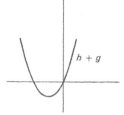

c) If $x \leq x'$, then $g(x) \leq g(x')$ and $h(x) \leq h(x')$; therefore
$(g + h)(x) = g(x) + h(x) \leq g(x') + h(x') = (g + h)(x')$.

d) Yes. Difference between $h(x)$ and $g(x)$ is increasing. Since h and g are increasing, h must be increasing faster.

e) $h + g$ is decreasing, but not necessarily more slowly than g.

Section 3.4

1. a) Use $p = q$; $|x - 0| < p$ implies $|f(x) - f(0)| = |x| < p$.
   b) Use $q = p/4$; $|x - 5| < p/4$ implies $4|x - 5| < p$, which in turn implies that $|4x - 20| = |f(x) - f(4)| < p$.

c) Use $q = p/6$.
d) Use $q = p^{1/2}$. Then $|x - 0| < p^{1/2}$ implies $|x| < p^{1/2}$ implies $x^2 < p$ implies $|x^2 - 0| = |f(x) - f(0)| < p$.
e) Use $p = q$.

2. a) For each a, use $q = p$. b) For each a, use $q = p/2$.
   c) Use $q = $ minimum of $\sqrt{a^2 + p} - a$ and $a - \sqrt{a^2 - p}$.
   d) For each a, use $q = $ minimum of $|a - (\sqrt{a} - p)^2|$ and $|(\sqrt{a} + p)^2 - a|$.
   e) For each a, use $q = $ min of $|a - \sqrt[3]{a^3 - p}|$ and $|\sqrt[3]{a^3 + p} - a|$.

3. a) Discontinuous at $x = 0$. For $p = 1$, we cannot find q such that $|x| < q$ implies $|1/x| < p$.
   b) Discontinuous at 2; use $p = 1$.
   c) Discontinuous at 0 and 1; use $p = 1$ for either point.

4. Assume that f is continuous and U is an open subset of R. Denote $\{x \mid f(x)$ is in $U\}$ by $f^{-1}(U)$. If x is in $f^{-1}(U)$, then $f(x)$ is in U. Since U is open, there is $p > 0$ such that $(f(x) - p, f(x) + p) \subset U$. Since f is continuous, there is $q > 0$ such that $|x - y| < q$ implies $f(y)$ is in $(f(x) - p, f(x) + p)$; hence $|x - y| < q$ implies that y is in $f^{-1}(U)$. But $\{y \mid |x - y| < q\}$ is $(x - q, x + q)$; therefore there is an open interval about x in $f^{-1}(U)$. Consequently, $f^{-1}(U)$ is open. Suppose now that $f^{-1}(U)$ is open whenever U is open. Suppose $p > 0$. Then for any x, $(f(x) - p, f(x) + p)$ is an open set. Hence $f^{-1}(f(x) - p, f(x) + p)$ is open and contains x. Then $(x - q, x + q)$ is a subset of

$$f^{-1}((f(x) - p, f(x) + p))$$

for some $q > 0$. Use this q to show the continuity of f. That is, $|x - y| < q$ implies $|f(x) - f(y)| < p$.

5. a) $f + g$ is continuous; for any a and $p > 0$, there are $q_1, q_2 > 0$ such that $|x - a| < q_1$ implies $|f(x) - f(a)| < p/2$, and $|x - a| < q_2$ implies $|g(x) - g(a)| < p/2$. Use $q = $ minimum (q_1, q_2) to show the continuity of $f + g$.
   b) True. If "break" occurred at 0, then f would have to be discontinuous at 0 from either the right or left side. We exclude both possibilities by assuming that f is continuous from both right and left.

Section 3.5

1. a) $1^2 + 2^2 + 3^2 + 4^2$
   b) $f(x_0) + f(x_1) + f(x_2) + \cdots + f(x_7)$
   c) $(-\frac{1}{3}) + (-\frac{1}{2}) + (-1) + 1 + \frac{1}{2} + \frac{1}{3} + \frac{1}{4} + \frac{1}{5}$
   d) $f(x^0) + f(x^7) + f(x^8) + f(x^{15}) + f(x^{25})$
   e) $1 + \frac{1}{3} + (\frac{1}{3})^2 + (\frac{1}{3})^3 + (\frac{1}{3})^4 + \cdots$
   f) $\frac{1}{2} + (\frac{1}{2})^3 + (\frac{1}{2})^5 + (\frac{1}{2})^7 + (\frac{1}{2})^9 + (\frac{1}{2})^{11}$
   g) $(\frac{1}{1})^2 + (\frac{1}{3})^2 + (\frac{1}{5})^2 + \cdots + (\frac{1}{15})^2$

2. a) $\sum_{i=1}^{6} i$ b) $\sum_{j \neq 8}^{9} x^j$ c) $\sum_{n=1}^{\infty} (\frac{1}{6})^n$ d) $\sum_{n \neq 0}^{\infty} f(x^{-n})$

3. a) $\sum_{j=0}^{3} (j+2)$ b) $\sum_{j=1}^{12} f(x_{j-6})$ c) $\sum_{j=1}^{\infty} x^j$

4. a) $(g(1) + \cdots + g(m)) + (f(1) + \cdots + f(m))$
$$= (g(1) + f(1)) + \cdots (g(m) + f(m)) = \sum_{k=1}^{m} (g(k) + f(k))$$

b) $q(g(1) + \cdots + g(m)) = qg(1) + \cdots + qg(m) = \sum_{j=1}^{m} qg(j)$

Section 3.6

1. In the proof of Proposition 4, it was shown that $P(x < y) = F(y)$, whence $P(y' \leq x \leq y) + P(x < y') = P(x \leq y)$.
Therefore
$$P(y' \leq x \leq y) = P(x \leq y) - P(x < y') = F(y) - F(y').$$
The other parts have similar proofs.

a) $f(0) = f(1) = \frac{1}{2}$; $F(y) = \begin{cases} 0, & y < 0 \\ \frac{1}{2}, & 0 \leq y < 1 \\ 1, & 1 \leq y \end{cases}$

b) $f(x) = \frac{1}{5}$, $x = 1, 2, 3, 4, 5$; $F(y) = \begin{cases} 0, & y < 1 \\ \frac{1}{5}, & 1 \leq y < 2 \\ \frac{2}{5}, & 2 \leq y < 3 \\ \frac{3}{5}, & 3 \leq y < 4 \\ \frac{4}{5}, & 4 \leq y < 5 \\ 1, & 5 \leq y \end{cases}$

c) $f(0) = f(2) = \frac{1}{4}$, $f(1) = \frac{1}{2}$; $F(y) = \begin{cases} 0, & y < 0 \\ \frac{1}{4}, & 0 \leq y < 1 \\ \frac{3}{4}, & 1 \leq y < 2 \\ 1, & 2 \leq y \end{cases}$

d) $f(x) = (C(4, x)C(3 - x))/C(9, 3)$, $x = 0, 1, 2, 3$; $F(y) = \sum_{\substack{x \leq y \\ x=0,1,2, \text{or} 3}} f(x)$

e) $f(x) = \frac{1}{7}$, $x = 1, 2, 4, 5, 6$, and $f(3) = \frac{2}{7}$; $F(y) = \sum_{\substack{x \leq y \\ x=1,2,3,4,5,\text{or} 6}} f(x)$

3. True since on any trial something will occur.
4. Possible distribution functions: (a), (b), (f)

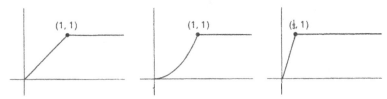

5. a) 0;   b) $\frac{1}{4}$;   c) $\frac{3}{4}$;   d) 0;
   e) $\frac{3}{4}$;   f) $\frac{1}{2}$;   g) $\frac{3}{8}$;   h) $\frac{1}{4}$

Chapter 4

Section 4.1

1. For any $p > 0$, set $q = 1$ (or anything else greater than 0). Then for any a, $0 < |x - a| < q$ gives $|f(x) - k| = |k - k| = 0 < p$.

2. For any $p > 0$, there is $q > 0$ such that $0 < |x - a| < q$ implies $|f(x) - L| < p$. Also $0 < |x - a| < q$ implies
$$|rf(x) - rL| = |r| \, |f(x) - L| < |r|p.$$
Choose q' so that $|f(x) - L| < p/|r|$. Then $0 < |x - a| < q'$ implies $|rf(x) - rL| < p$.

3. a) 6   b) 1   c) 2   d) $\frac{7}{3}$ e) 7
   f) 0   g) 1   h) $1/2\sqrt{a}$ i) $3a^2$ j) $-1/a^2$

4. a) $\lim_{y \to 1} \left(\frac{y+5}{y} \right)$ b) $\lim_{y \to 0} \frac{(y-1)^2 + 1}{y^2}$
   c) $\lim_{y \to -5} \frac{(x+y+5)^2 - (y+5)}{y+5}$

5. $\frac{f(0+h) - f(0)}{h} = -1$ if h is negative and $+1$ if h is positive.

6. a) $\lim_{x \to a^+} f(x) = 1$ if given any $p > 0$, there is $q > 0$ such that if $x > a$ and $|x - a| < q$, then $|f(x) - L| < p$.
   b) $\lim_{x \to 0^-} f(x) = 2$ and $\lim_{x \to 0^+} f(x) = 0$ for Exercise 19 of Chapter 3. In Exercise 1 of Chapter 4, $\lim_{x \to 0^+} f(x) = \lim_{x \to 0^-} f(x) = 4$.

Section 4.2

1. a) 8   b) 2   c) 62
   For $f(x) = x^3$: a) 3   b) 0   c) 300

2. $\frac{f(x+h) - f(x)}{h} = \frac{k-k}{h} = \frac{0}{h} = 0$

3. a) $y = 3x + 1$
   b) $y = mx + b$, line tangent to linear function is the linear function
   c) $y + 7 = -2(x - 1)$ d) $y = 2(x - 1)$

4. a) $f'(x) = -2x^{-3}$ for all $x \neq 0$
   b) $f'(x) = (\frac{1}{2})x^{-1/2}$ for $x > 0$
   c) $f'(x) = 3x^2$ for all x
   d) $f'(x) = 4x - 3 - x^{-2}$ for $x \neq 0$
   e) $f'(x) = (\frac{1}{2})(x - 1)^{-1/2}$ for $x > 1$

5. a) $F'(0) = F'(4) = 0$, since tangents at $x = 0$ and $x = 4$ are parallel to x-axis.
   b) Since $F'(x)$ measures rate of increase and F is an increasing function, $F'(x)$ should be nonnegative.

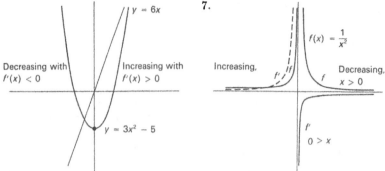

Section 4.3

1. a) $f'(x) = 2x$ b) $f'(x) = (\frac{4}{5})x^{-1/5}$
   c) $g'(x) = 2x - 4x^3$ d) $m'(x) = 15x^4 + 8x + 10$
   e) $p'(x) = (3x^2 + 7x + 3)(36x + 4) + (9x^4 + 4x + 1)(6x + 7)$
   f) $q'(x) = -4(x - 7)^{-2}$
   g) $r'(x) = (\frac{1}{2})x^{-1/2} + (\frac{1}{3})x^{-2/3} - 17x^{-18} - 0.67x^{-1.67}$
   h) $t'(x) = -2ax^{-3} + b(4x^3 + 3)$
   i) $f'(x) = \dfrac{x^{1/2}(2x - 8) - (\frac{1}{2})x^{-1/2}(x - 3)(x - 5)}{x}$, makes sense for $x > 0$
   j) $\dfrac{(3x^2 + 4x + 1)(2x - \frac{1}{2}x^{-3/2}) - (x^2 + x^{-1/2})(6x + 4)}{(3x^2 + 4x + 1)^2}$
   k) $h(x) = 1$
2. $y - 4 = 24(x - 1)$
3. $y - 18 = (-\frac{5}{16})(x - 4)$
4. $(f - g)(x) = f(x) - g(x)$
   $$\dfrac{(f - g)(a + h) - (f - g)(a)}{h} = \dfrac{f(a + h) - f(a)}{h} - \dfrac{g(a + h) - g(a)}{h}$$
   leads to $(f - g)'(a) = f'(a) - g'(a)$.
   Two functions with the same derivative for all x differ by a constant.
5. a) $F(x) = 3x$ b) $F(x) = x^2$ c) $F(x) = x^3/3$
   d) $F(x) = x^2 + 7x$ e) $G(x) = -x^{-1}$
6. a) $f'(x) = 3(g(x))^2 g'(x)$.
   b) True for $n = 1$; assume true for $n - 1$. $f(x) = (g(x))^{n-1}g(x)$; hence
   $f'(x) = (g(x))^{n-1}g'(x) + g(x)((n - 1)(g(x))^{n-2})g'(x)$
   $= n(g(x))^{n-1}g'(x)$.
   c) $f'(x) = -(g(x))^{-2}g'(x)$.
   d) $f(x) = (g(x))^{-(n-1)}g(x)^{-1}$. Assume that if $h(x) = (g(x))^{-(n-1)}$, then $h'(x) = -(n - 1)(g(x))^{-n}g'(x)$.

Now $f(x) = h(x)g(x)$; hence
$f'(x) = h(x)(-1)(g(x))^{-2}g'(x) + (g(x))^{-1}(-(n-1))(g(x))^{-n}g'(x)$
$= -n(g(x))^{-n-1}g'(x)$

e) i) $f'(x) = 17(x^3 + 4)^{16}(3x^2)$
   ii) $g'(x) = -4(x^2 + 3x + 1)^{-5}(2x + 3)$
   iii) $h'(x) = (x^2 + 3)^{18}(-15)(x^3 + 1)^{-16}(3x^2)$
        $+ (x^3 + 1)^{-15}(18)(x^2 + 3)^{17}(2x)$
   iv) $m'(x) = (x^{1/2} + x^{1/3})^{23}(1678)(x + 1)^{1677}$
        $+ (x + 1)^{1678}(23)(x^{1/2} + x^{1/3})^{22}(x^{-1/2}/2 + x^{-2/3}/3)$

Section 4.4

1. a) $f'(x) = 67(x + 25)^{66}$
   b) $g'(x) = 96(x^3 + 5x + 78)^{95}(3x^2 + 5)$
   c) $h'(x) = (x^3 + 2x^2 + 3x + 1)^{15}(10)(x^4 + x^{1/2})^9(4x^3 + x^{-1/2}/2)$
      $+ (x^4 + x^{1/2})(15)(x^3 + 2x^2 + 3x + 1)^{14}(3x^2 + 4x + 3);$
   d) $k'(x) = (\frac{1}{3})(x^{1/3} + 1)^{-2/3}(x^{-2/3}/3)$
   e) $f'(x) = (\frac{1}{2})((x^{1/2} + 1)^{1/2} + 1)^{-1/2}(\frac{1}{2})(x^{1/2} + 1)^{-1/2}(\frac{1}{2})(x^{-1/2})$
   f) $f'(x) = e^{x+4}$
   g) $f'(x) = e^{x^3+(1-x^2)^2}(3x^2 + 2(1 - x^2)(-2x))$
   h) $h'(x) = e^{f(x)}f'(x)$, where $f(x)$ is as in (e).

2. $f'(x) = g'(m(n(x)))(m(n(x)))' = g'(m(n(x)))m'(n(x))n'(x)$

3. $f''(x) = g'(m(x))m''(x) + (m'(x))^2 g''(m(x))$
   $f'''(x) = g'(m(x))m'''(x) + m''(x)g''(m(x))m'(x)$
   $+ (m'(x))^2 g'''(m(x))m'(x) + g''(m(x))2m'(x)m''(x)$
   a) $f'(x) = 3.4(x + 3)^{2.4}, f''(x) = (3.4)(2.4)(x + 3)^{1.4}$
   b) $f'(x) = 6.7(x^3 + 4x + 5)^{5.7}(3x^2 + 4),$
      $f''(x) = 6.7(x^3 + 4x + 5)^{5.7}(6x) + (3x^2 + 4)^2(6.7)(5.7)(x^3 + 4x + 5)^{4.7}$
   c) $g'(x) = e^{(x^2+3x)}(2x + 3),$
      $g''(x) = e^{(x^2+3x)}(2) + (2x + 3)^2 e^{(x^2+3x)}$
   d) $h'(x) = e^{e^x}e^x, \ h''(x) = e^{e^x}e^x + (e^x)^2 e^{e^x}$

4. a) $f'(x) = -\dfrac{2f(x)}{x}$ b) $f'(x) = \dfrac{-f(x) - 1}{x}$
   c) $f'(x) = 0$ d) $f'(x) = \dfrac{-f(x) - 2(f(x))^2 x - 3x^2}{x + 2x^2 f(x)}$
   e) $f'(x) = \dfrac{-e^x f(x) + e^{-x}(f(x))^{-1}}{e^x - e^{-x}(f(x))^{-2}}$

5. $y' = -b^2 x/a^2 y$

6. $y' = \dfrac{-2Ax - By - D}{Bx + 2Cy + E}, \dfrac{y - 4}{x - 3} = \dfrac{-6A - 4B - D}{3B + 8C + E}$

Chapter 5

Section 5.1

1. Repeat the proof of Proposition 1, substituting $f(a + h) \geq f(a)$ for $f(a + h) \leq f(a)$ and interchanging *nonnegative* and *nonpositive*.

2. a) None
   b) Critical point, absolute and relative minima
   c) Critical point, absolute and relative minima
   d) Absolute and relative minima
   e) Critical point, relative minimum
   f) Critical point, relative maximum g) None
3. a) 9 b) 7, 3, $\frac{233}{69}$ c) None d) None e) None f) 3
4. a) True b) True c) False, e.g., look at $h(x) = (x^2 - 2)/x^2$
   d) False, look at $f(x) = x$, $g(x) = -x$
   e) False, look at $f(x) = g(x) = -x^2$
   f) False, look at $f(x) = x^2$, $g(x) = x^2 + 2$ g) True

Section 5.2

1. The proof of Proposition 4 can be obtained from the proof of Proposition 3 by replacing < with > and *positive* with *negative* wherever appropriate.

2. a)   b)   c)

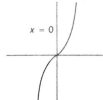

d)   e)   f)

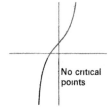
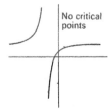

g)   h)

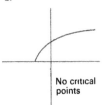
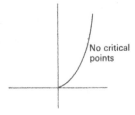

3. $B^2 < 3AC$ implies no critical points.
4. $A > 0$ implies that f is decreasing for $x < -B/2A$ and increasing for $x > -B/2A$; if $A < 0$, then f is increasing for $x < -B/2A$ and decreasing for $x > -B/2A$.
5. a) False, $f(x) = x^3$ b) True
   c) False, $f(x) = -x$, $g(x) = -x$, interval $(-1, 1)$
   d) True   e) True
6. Consider $f \circ g$, g decreasing and f increasing. If $x < y$, then $g(x) \geq g(y)$. Then $f \circ g(y) = f(g(y)) \geq f(g(x)) = f \circ g(x)$ since f is increasing. Therefore $f \circ g$ is decreasing. If f and g are both increasing, then $x < y$ means $g(x) \leq g(y)$; hence $f(g(x)) \leq f(g(y))$.

Section 5.3

1. $f(-1) = f(1) = 1$; hence $\dfrac{f(-1) - f(1)}{(-1) - (-1)} = 0$.

   But there is no x' in $[-1, 1]$ such that $f'(x') = 0$.

2. For any n, $f(x) = x^n$ is continuous. If a is any positive real number, then there is a positive real number b such that $f(b) > a$. If n is even, then $f(-b) = f(b) > a$. Using Proposition 9, there is a number x' in $[0, b]$ such that $f(x') = x'^n = a$. If n is even, then $-x'$ also works.

3. a) False, $f(x) = x^3$
   b) True, essentially the same proof as for Exercise 2
   c) False, let $f(x) = x$, $g(x) = -x$, $a = -1$, $b = 1$
   d) True by Proposition 9
   e) False, let $f(x) = e^x$
   f) True, if false, then by the Mean Value Theorem we would have x'' in (x, x') with $g'(x'') = 0$

4. a) $\frac{1}{2}$ b) $(33 + \sqrt{273})/12$ c) Any point   d) $\frac{1}{4}(\sqrt{2} - 1)^2$

5.

	Largest	Smallest
a)	4	0
b)	0	$f(x')$
c)	$f(b)$	$f(a)$
d)	$\sqrt{2}$	1

Section 5.4

1. a) $x = 0$, point of inflection   b) $x = 5$, point of inflection
   c) $(1 \pm \sqrt{6})/3$; $-$ gives minimum, $+$ gives maximum, point of inflection at $x = 1$
   d) $x = 0$, maximum, $x = \pm\sqrt{4/3}$ points of inflection
   e) None   f) $x = 2$, maximum, $x = 3$, minimum

2. $f'(x) = n(x-a)^{n-1}g(x) + (x-a)^n g'(x)$
   $= (x-a)^{n-1}(ng(x) + (x-a)g'(x))$;
   hence $f'(a) = 0$ if $n \geq 2$. Not necessarily. Let $f(x) = x^3(x+1)$; then $a = 0$ and f has a minimum at a.
3. a) $f''(x) = 6x$, convex $x < 0$, concave $x > 0$
   b) $f''(x) = 2$, concave all x
   c) $f''(x) = x^{-3}$, concave $x > 0$, convex $x < 0$
   d) Concave $(-\infty, 1-2\sqrt{2})$, convex $(1-2\sqrt{2}, 1+2\sqrt{2})$ and $(1+2\sqrt{2}, \infty)$
4. a) True, as may be seen informally from the diagram on the right:
   b) True; follows from $(f+g)' = f' + g'$
   c) False, $f(x) = x^2$, $g(x) = (x-1)^2$, fg is convex on part of $(0, 1)$
   d) True since point of inflection is a maximum of f'

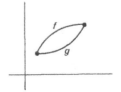

Section 5.5

1. Area $= h/3 = F(a+h) - F(a)$; area $= \frac{1}{2}(a + (a+h))h$
   $= F(a+h) - F(a)$

2. a) $f(x) = 1$ for each x in $[0, 1]$; each x in $[0, 1]$ a mode, symmetric about $x = \frac{1}{2}$

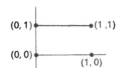

b) $f(x) = 3x^2$, x in $[0\ 1]$; mode 1, skewed to left

c) $f(x) = 1$, x in $[1, 2]$; symmetric about $x = \frac{3}{2}$; all x modes

d) $f(x) = -x^{-3}$, x in $(-\infty, -1]$; mode -1, skewed to left

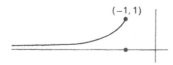

e) $f(x) = 18x^2 + 2x$, x in $[0, \frac{1}{2}]$, mode $\frac{1}{2}$, skewed to left

3. $p = \sqrt[3]{3/2}$ 4. $F(x) = x^2/3 + 2x/3$

Chapter 6

Section 6.1

1. a) 1, 2, 3, 4, 5   b) 5, 5, 5, 5, 5   c) $0, \frac{1}{4}, \frac{1}{8}, \frac{15}{16}, \frac{35}{36}$
   d) $1, 2^{1/3}, 3^{1/3}, 4^{1/3}, 5^{1/3}$ e) 4, 21, 50, 91, 144   f) 2, 3, 4, 5, 6
   g) 2, 2, 2, 2, 2   h) 2, 3, 6, 10, 17

2. a) $s_n \to 3$. For any $p > 0$, let $M = 1$. Then for $n > M$,
   $$|s_n - 3| = |3 - 3| = 0 < p.$$
   b) Not convergent, is eventually greater than preassigned L
   c) $s_n \to 0$. For $p > 0$, choose $M > 1/p^2$.
   d) $s_n \to 1$. For $p > 0$, choose $M > 1/p$.
   e) Not convergent, s_n eventually smaller than any preassigned L
   f) Not convergent. If convergent, then the limit would have to be 3 or -3. But for any M, there are n, $n' > M$ such that $|s_n - 3| = 6$ and $|s_{n'} - 3| = 6$.

3. a) Since $s_n \to L$ and $t_n \to L'$, we have that for any $p > 0$, there are M_1, M_2 such that $n > M_1$ gives $|s_n - L| < p/2$ and $n > M_2$ gives
   $$|s_n - L'| < p/2.$$

Choose $M = \max(M_1, M_2)$. Then for $n > M$,
$$|(s_n + t_n) - (L + L')| \leq |s_n - L| + |t_n - L'| < p/2 + p/2 = p.$$

d) For $p > 0$, there is M such that $n > M$ implies $|s_n - L| < p/|r|$, provided $r \neq 0$. (If $r = 0$, the proposition is trivial.) Then for $n > M$,
$$|rs_n - rL| = |r| |s_n - L| < |r|(p/|r|) = p.$$

4. For any $p > 0$, let $M = 1$. Compare with 2(a).

5. Suppose each $s_n > 0$, but $s_n \to L < 0$. Choose $p = |L|$. Then for any M, $|s_n - L| \geq p$, contradicting $s_n \to L$.

6. a) 1, 3, 4, 10, 15 b) 1, 2, 3, 4, 5 c) $-1, 0, -1, 0, -1$
 d) $-1, -\frac{3}{4}, -\frac{3}{4} + (-\frac{1}{3})^3, (-\frac{3}{4}) + (-\frac{1}{3})^3 + (-\frac{1}{4})^4,$
 $-\frac{3}{4} + (-\frac{1}{3})^3 + (-\frac{1}{4})^4 + (\frac{1}{5})^5$
 e) $0, \frac{1}{4}, \frac{1}{4} + (\frac{2}{3})^2, \frac{1}{4} + (\frac{2}{3})^2 + (\frac{3}{4})^2, \frac{1}{4} + (\frac{2}{3})^2 + (\frac{3}{4})^2 + (\frac{4}{5})^2$

7. Suppose $\sum_{n=1}^{\infty} s_n = L$. Then given any $p > 0$, there exists M such that $n > M$ implies $|t_n - L| < p$. Now
$$|s_n| = |t_{n+1} - t_n| \leq |t_{n+1} - L| + |L - t_n| < 2p.$$
Since p is arbitrarily small, $|s_n|$ must be getting arbitrarily small; that is, $s_n \to 0$.

8. a) Outlined in text
 b) $S_n = a(1 - r^{n+1})/(1 - r)$; as $n \to \infty$, S_n must converge. The only way this can happen is if $r^{n+1} \to 0$, and this will occur if and only $|r| < 1$.
 c) i) $\frac{1}{2}$, ii) 4, iii) 1

Section 6.2

1. a) Diverges, $s_n \to 0$ b) Diverges, compare with $\sum_{n=1}^{\infty} 1/n$
   c) Diverges, $s_n \to 0$ d) Diverges, comparison with $\sum 1/n$
   e) Converges, Proposition 8   f) Converges, ratio test, limit greater than 1
   g) Diverges, $s_n \to 0$ h) Converges, Proposition 6
   i) Diverges, compare with $\sum 1/n$, or use Proposition 8
   j) Converges, Proposition 8   k) Diverges, Proposition 8
   l) Diverges, $s_n \to 0$

2. $S \to L$ and $T \to L'$ mean S_n (the nth partial sum) $\to L$ and $T_n \to L'$. Hence, by Proposition 2, $S_n + T_n \to L + L'$. But $S_n + T_n$ is the nth partial sum of $S + T$. Hence $S + T \to L + L'$. If $S_n \to L$, then $rS_n \to rL$ [(d) of Proposition 2]; hence $rS \to rL$. $S - T = \sum_{n=1}^{\infty}(s_n - t_n)$. Not true; for example, let $S = \sum_{n=1}^{\infty}(1/n)$ and $T = -S$. True, for if $S + T \to L$ and $S \to L'$, then T must converge to $L - L'$.

3. $\sum_{n=1}^{\infty} 1/(kn) = (1/k)\sum_{n=1}^{\infty}(1/n)$. If $\sum_{n=1}^{\infty} 1/(kn) \to L$, then $\sum_{n=1}^{\infty}(1/n) \to kL$.

4. $\frac{637}{999}$ 5. $\frac{93}{99}$

Section 6.3

1. a) $1/x$
   b) $e^{3x^2+1}(6x)$
   c) $(3x^2 + e^x)/(x^3 + e^x)$
   d) $(\ln 5)5^x$
   e) $(\ln 5)5^{3x^4+5}(12x^3)$
   f) $\left(\dfrac{1}{\ln 5}\right)\left(\dfrac{1+2x}{x+x^2}\right)$
   g) $(\ln 3)3^x + (\ln 4)4^x$
   h) $e^{e^x + x}$
   i) $1/(x \ln x)$
   j) $x^x(1 + \ln x)$
   k) $e^{\log_3 x}(1/(x \ln 3))$
   l) $1/\ln 3$

2. The Maclaurin series expansion for $P(x)$ is $P(x)$. The best "polynomial approximation" for $P(x)$ is $P(x)$.

3. By the Chain Rule, $f'(x) = L'(g(x))g'(x) = (1/g(x))g'(x)$.

4. a) $1 - x + x^2/2! - x^3/3! + x^4/5! \cdots$
   b) $-\tfrac{1}{2} + x^2/2^2 - x^3/2^3 + x^4/2^4 - \cdots$
   c) $x + x^2 + x^3/2! + x^4/3! + \cdots$
   d) $1 + 4x + x^2/2! + x^3/3! + \cdots$
   e) $1 + x/\ln 4 + (1/\ln 4)^2(x^2/2!) + (1/\ln 4)^3(x^3/3!) + \cdots$

5. a), c), d), and e) converge for any x b) Converges for x in $(-2, 2)$

6. a) $xy = a^{\log_a x} a^{\log_a y} = a^{\log_a x + \log_a y}$ b) $x^r = (a^{\log_a x})^r = a^{r \log_a x}$
   c) $x = a^{\log_a x} = (b^{\log_b a})^{\log_a x} = b^{\log_b a \log_a x} = b^{\log_b x}$; therefore $\log_b x = \log_b a \log_a x$.

Section 6.4

1. a) $\sum_{i=0}^{\infty}(1-x)^i$, interval of convergence is $(0, 2)$
   b) $e^{-1}\sum_{n=0}^{\infty}(-1)^n(x-1)^n/n!$, converges for all x
   c) Using $x^2 = (x-1)^2 + 2(x-1) + 1$, we have
   $$\ln x + x^2 = 1 + 3(x-1) + (x-1)^2/2 + (x-1)^3/3 - (x-1)^4/4 + \cdots;$$
   interval of convergence is $(0, 2]$
   d) $\sum_{n=0}^{\infty} x^{n+1}/n!$, converges for all x
   e) $x^3 + 3x + 7$, converges for all x

2. $e = 2.7183$, $e^{-1} = 0.43429$; $n = 32$ will work

3. $f^{(n)}(0) = 0$ for any n

4. We develop $\log_{10} x$ about 10 and get
$$\log_{10} x = 1 + (1/\ln 10) \sum_{n=1}^{\infty} (x-10)^n/(n \cdot 10^n).$$
Since this series is alternating for $x = 3$, the error is no greater than its $(n+1)$ — term. We want
$$\left|\frac{(1/\ln 10)(7-10)^{n+1}}{n \cdot 10^n}\right| < 0.00001 \quad \text{or} \quad \frac{(1/\ln 10) \cdot 3^{n+1}}{n \cdot 10^n} < 0.00001.$$
Using logarithms, we find that $n = 10$ will work.

5. $\sum_{n=1}^{\infty}(-1)^{n+1} x^{2n-1}/(2n-1)!$ converges for all x; errors less than $1/n!$ for terminating after nth term.

6. The series follows from $L^{(n)}(x) = (-1)^{n-1}x^{-n}(n-1)!$ for $n \geq 1$; hence $L^{(n)}(1) = (-1)^{n-1}(n-1)!$.

Chapter 7

Section 7.1

1. $(1/n)^3 \sum_{k=1}^{n} k^2 - (1/n)^3 \sum_{k=1}^{n-1} k^2 = (1/n)^3 \left(\sum_{k=1}^{n} k^2 - \sum_{k=1}^{n-1} k^2 \right)$
$= (1/n)^3 n^2 = 1/n$; $\lim_{n \to \infty} (1/n) = 0$

2. $\sum_{k=1}^{n} f(r_k)(3/n) = \sum_{k=1}^{n} (6k/n)(3/n) = (18/n^2) \sum_{k=1}^{n} k$
$= (18/n^2)(n/2)(n+1) = 9((n+1)/n)$

$\lim_{n \to \infty} 9((n+1)/n) = 9$, which is also the area of the triangle in question

3. $\sum_{k=1}^{n} f(r_k)(2/n) = \sum_{k=1}^{n} 10(8+2k/n)(2/n) = \sum_{k=1}^{n} (160/n + 40k/n^2)$
$= 160 + (40/n^2) \sum_{k=1}^{n} k = 160 + (40/n^2)(n/2)(n+1)$
$= 160 + 20((n+1)/n)$.

$\lim (160 + 20((n+1)/n)) = 180 =$ area of the trapezoid in question

4. $\sum_{i=1}^{n} f(r_i) \left(\frac{b-a}{n} \right) = \sum_{i=1}^{n} k \left(\frac{b-a}{n} \right) = nk \left(\frac{b-a}{n} \right)$
$= k(b-a) =$ area of rectangle

5. a) $\int_a^b kf(x)\, dx = \lim_{n \to \infty} \sum_{i=1}^{n} kf(r_i) \left(\frac{b-a}{n} \right) = k \lim_{n \to \infty} \sum_{i=1}^{n} f(r_i) \left(\frac{b-a}{n} \right)$
$= k \int_a^b f(x)\, dx$

b) $\int_a^b (f+g)(x)\, dx = \lim_{n \to \infty} \sum_{i=1}^{n} (f+g)(r_i) \left(\frac{b-a}{n} \right)$
$= \lim_{n \to \infty} \sum_{i=1}^{n} (f(r_i)) \left(\frac{b-a}{n} \right) + g(r_i) \left(\frac{b-a}{n} \right)$
$= \lim_{n \to \infty} \sum_{i=1}^{n} f(r_i) \left(\frac{b-a}{n} \right) + \sum_{i=1}^{n} g(r_i) \left(\frac{b-a}{n} \right)$
$= \lim_{n \to \infty} \sum_{i=1}^{n} f(r_i) \left(\frac{b-a}{n} \right) + \lim_{n \to \infty} \sum_{i=1}^{n} g(r_i) \left(\frac{b-a}{n} \right)$
$= \int_a^b f(x)\, dx + \int_a^b g(x)\, dx$

c) $\int_a^b f(x)\,dx = \lim_{n\to\infty} \sum_{i=1}^n f(r_i)\left(\frac{b-a}{n}\right) \leq \lim_{n\to\infty} \sum_{i=1}^n M\left(\frac{b-a}{n}\right)$
$= \lim_{n\to\infty} M(b-a) = M(b-a)$

6. All statements are true.

Section 7.2

1. a) $\frac{3}{2}$ b) $\frac{75}{2}$ c) $-\frac{141}{2}$
   d) 0 (area under x-axis = area above x-axis)
   e) $e - 1$ f) $\ln 16 - \ln 7$ g) $(3^4/4 + 39) - (\frac{58}{3})$
   h) $\frac{8}{15} + 4$ i) $(\frac{2}{3})(6^{3/2} - 5^{3/2})$ j) $((\frac{3}{4})2^{2/3} + 2(2^{1/2})) - (\frac{3}{4} + 2)$
   k) $(5.1^{1.9}/1.9 + 5.1^{0.3}/0.3 + 4(5.1^{1.65}/1.65))$
   $\qquad - (4^{1.9}/1.9 + 4^{0.3}/0.3 + 4(4^{1.65}/1.65))$
   l) $e - 1$ m) $5^{18}/18 - 4^{18}/18$ n) $5^{18}/54 - 4^{18}/54$ o) 2

2. Set $F(x) = x^{n+1}/(n+1)$, $n \neq -1$. Then $F'(x) = x^n$. Therefore
$$\int_a^b x^n\,dx = F(b) - F(a).$$
If $n = -1$, then
$$\int_a^b x^{-1}\,dx = \ln|b| - \ln|a|.$$

3. a) $k = 2$, $F(x) = x^2$ b) No k can be found
   c) $k = 4/(4^4 - 3^4)$, $F(x) = kx^4/4 - k3^4/4$, k as given
   d) $k = \frac{3}{2}(2^{3/2} - 1)$, $F(x) = (2k/3)(x^{3/2} - 1)$, k as given
   e) $k = -3$, $F(x) = -x^3 + x^2 + x$
   f) $k = 1/(e-1)$, $F(x) = k(e^x - 1)$, k as given   g) No k can be found

Section 7.3

1. a) $x^5/5 + k$ b) $x^2/2 + 2x^{3/2}/3 + k$
   c) $\ln x - 3x^{-1} + x^{-2}/2 + k$ d) $e^x - e^{-x} + k$
   e) $ax + bx^2/2 + cx^3/3 + k$ f) $ae^x - b\ln x + x^{79}/79 + k$
   g) $-(e^{-3x}/3) + (e^{2x}/2) + k$ h) $(e^{-1} + e^8)x + k$

2. a) $5^4/4 + 10^4/4$ b) 100   c) $\frac{1}{6}$
   d) Set $F(x) = x^4/2 - (7x^3/3) - 7x^2 - 5x$; then
   $A = |F(-10) - F(-1)| + |F(-1) - F(-\frac{1}{2})|$
   $\quad + |F(-\frac{1}{2}) - F(5)| + |F(5) - F(10)|$

3. a) 1   b) 4   c) $\frac{25}{2} + (\frac{1000}{3} - 200 - \frac{125}{3} + 100)$ d) 3

Section 7.4

1. a) $\int_0^1 (\frac{3}{2})e^y\,dy$ b) $\int_{-5}^{-9} y\,dy$ c) $(\frac{1}{3})\int_{4^3+7}^{16^3+7} y^{1/2}\,dy$

    d) $\int_{4}^{18} (1/y)\, dy$ e) $\int_{1}^{e} e^{y}\, dy$

2. a) $(\frac{3}{2})(e-1)$ b) -15 c) $e-2$
   d) $(9\ln 3 - 3) + \frac{1}{9}$ e) $2\ln 2$ f) $(\frac{1}{2})(\ln 4 - \ln 3)$
   g) $\frac{399}{35}$ h) 0.5 i) $-2e^{-1} + 1$

3. a) $\sin x + k$ b) $-3\cos x + k$
   c) $2\sin^2 x + k$, or $-2\cos^2 x + k'$ d) $-\ln \cos x + k$
   e) $-(\frac{1}{6})(\cos x)^6 + k$ f) $\sin x - x\cos x + k$
   g) $\cos x + x\sin x + k$ h) $2x\sin x + 2\cos x - x^2\cos x + k$
   i) $(e^x/2)(\sin x - \cos x) + k$ j) $-(\frac{5}{2})\cos x^2 + k$
   k) $x/2 - (\sin 2x)/4 + k$

Section 7.5

1. a) Diverges   b) 2   c) -1 d) $-3^{-4}/4$
   e) Diverges   f) Diverges   g) 1
2. Converges for $w > 1$
3. Converges for $w > 1$, diverges for $w \leq 1$
4. a) Diverges   b) Converges to $L \leq 3\pi/2$
   c) Converges to $L \leq 1$ d) Converges to $L \leq 2$
5. a) $1/\ln 2$ b) $(\frac{1}{3})(\tan^{-1}(\frac{2}{3}) - \tan^{-1}(\frac{1}{3}))$ c) $\sin^{-1}(\frac{1}{4})$
   d) $(1/\sqrt{3})(\ln(\sqrt{3} + \sqrt{2.99}) - \ln(0.1\sqrt{3} + \sqrt{0.02}))$
   e) $(\frac{1}{3})((y/2)\sqrt{y^2 + 15} + (\frac{15}{2})\ln(y + \sqrt{y^2 + 15}))|_{29}^{149}$
   f) $\frac{1}{2}$ g) 1   h) Diverges

Section 7.6

1. a) All estimates give the actual value 0
   b) All estimates give the actual value 6
   c) TR gives 535, SR, series and actual value 520
   d) Actual value and all estimates are 0
   e) Actual value about 0.69, TR $\doteq$ 0.73, SR $\doteq$ 0.92; series $\doteq$ 0.685
   f) TR $\doteq$ 0.62, SR $\doteq$ 0.75, series $\doteq$ 0.71, actual value $\doteq$ 0.75

Chapter 8

Section 8.1

1.

	Mean	Mode(s)	Median	Best measure	Graph
a)	2	1, 3	2	Mean or median	(points at 1, 2, 3)

	Mean	Mode(s)	Median	Best measure	Graph
b)	4	All x	4	Mean or median	
c)	3	4	3	Mean or median	
d)	11.8	1	1	Mean	
e)	2	2	2	All same	
f)	$\frac{1}{2}$	All x	$\frac{1}{2}$	Mean or median	
g)	$\frac{3}{4}$	1	$(\frac{2}{3})^{1/3}$	Mean or median	
h)	-2	-1	$-(2^{1/2})$	Mean or median	
i)	$e-1$	1	$e^{1/2}$	Mean or median	

2. a) For continuous case:
$$\int_a^b f(x)\,dx = 2\int_K^b f(x)\,dx = 1.$$

The first equality comes from the fact that the area to the right of K is the mirror image of, and hence is equal to, the area on the left of K. Reasoning is similar for discrete case, but sums are used instead of integrals.

b) We have $f(-x) = f(x)$. Setting $g(x) = xf(x)$, we have $g(x) = -g(-x)$. For discrete case

$$\sum_{x \leq 0} g(x) = -\sum_{x \geq 0} g(x).$$

Hence

$$m_x = \sum_{x \leq 0} g(x) + \sum_{x \geq 0} g(x) = 0.$$

For continuous case:

$$\int_A g(x)\,dx = \int_a^0 g(x)\,dx + \int_0^b g(x)\,dx$$
$$= \int_a^0 g(x)\,dx - \int_a^0 g(x)\,dx = 0 = m_x.$$

3. Modes are $1, -1$; mean = median = 0

4. Mean = median = $(b-a)/2$; all admissible values are modes, hence mean of modes is also $(b-a)/2$

5. a) Obtain $\sum_{n=1}^{\infty} 2^n(\tfrac{1}{2}^n) = 1+1+1+1+\cdots$

   b) Obtain an expression involving $\int_1^\infty dy/y = \ln|_1^\infty$, which does not converge.

Section 8.2

1. For all x, $m_{m_x-x} = 0$ and $\sigma_x = (\sigma_x^2)^{1/2}$

	m_x	q_x	p_x	m_{p_x-x}	m_{q_x-x}	$m_{|m_x-x|}$	σ_x^2
a)	2	2	2	0	0	0.6	0.6
b)	2	2	2	0	0	0.02	0.02
c)	2.3	2.5	3	0.7	0.2	0.7	0.472
d)	0	0	0	0	0	0.1	0.16
e)	$\tfrac{4}{5}$	$\sqrt[n]{\tfrac{1}{2}}$	1	$\tfrac{1}{5}$	$\sqrt[n]{\tfrac{1}{2}} - \tfrac{4}{5}$	—	$\tfrac{2}{75}$
f)	$\dfrac{n}{n+1}$	$\sqrt[n]{\tfrac{1}{2}}$	1	$\dfrac{1}{n+1}$	$q_x - \dfrac{n}{n+1}$	—	$\dfrac{n}{(n+1)^2(n+2)}$
g)	1	$\ln 2$	None	—	$\ln 2 - 1$	—	1
h)	$e-1$	$e^{1/2}$	1	$2-e$	$e^{1/2} - (e-1)$	—	$2e - \dfrac{e^2}{2} - \dfrac{3}{2}$

2. $\int_A (m_x - x)^2 f(x)\, dx = \int_A (m_x^2 - 2m_x x + x^2) f(x)\, dx$
$$= m_x^2 \int_A f(x)\, dx - 2m_x \int_A x f(x)\, dx + \int_A x^2 f(x)\, dx$$
$$= m_x^2 - 2m_x^2 + m_{x^2} = m_{x^2} - (m_x)^2$$

3. $m_x = (n+1)/2$, each admissible value is a mode, $\sigma_x^2 = (n^2 - 1)/12$

4. a) 0.4   b) 0.98   c) 0.3   d) 0.9

5. a) Not necessarily; see (b) of Exercise 1 where $t_x = 0$ b) True
   c) Depends on interpretation of "more closely concentrated," but it would be reasonable to say this is true

Section 8.3

1. a) $M(t; x) = (1/t)(e^t - 1)$, which is not defined at 0
   b) $M(t; x) = 1/(1+t)$, $m_x = -1$, $\sigma_x = 1$, $S_k = -1$,
   $P(|m_x - x| > 5) \leq \frac{1}{16}$
   c) $M(t; x) = \frac{1}{2} + 3e^t/(3 - e^t)$, $m_x = \frac{9}{4}$, $\sigma_x^2 = \frac{23}{16}$, $P(|m_x - x| > 5) < \frac{1}{16}$

2. $M(t; w) = \sum e^{t(ax+b)} g(ax + b)$, where g is the density function of W. But $g(ax + b) = f(x)$; hence $M(t; w) = e^{tb}\sum e^{(at)x} f(x) = e^{tb} M(at; x)$

3. $M(t; x) = \int_A e^{tx} f(x)\, dx$ (continuous case) or $\sum_A e^{tx} f(x)$ (discrete case). If $t = 0$, we are left with $\int_A f(x)\, dx = 1$ or $\sum_A f(x) = 1$

4. We give the proof for the continuous case. For the discrete case, replace $\int_A dx$ with $\sum_A$.
$$\int_A \left(\frac{x - m_x}{\sigma_x}\right) f(x)\, dx = \left(\frac{1}{\sigma_x}\right)\left(\int_A x f(x)\, dx\right) - \left(\frac{m_x}{\sigma_x}\right)\int_A f(x)\, dx$$
$$= \left(\frac{m_x}{\sigma_x}\right) - \left(\frac{m_x}{\sigma_x}\right)(1) = 0.$$
And
$$\int_A \left(\frac{x - m_x}{\sigma_x}\right)^2 f(x)\, dx = \left(\frac{1}{\sigma_x^2}\right)\int_A (x - m_x)^2 f(x)\, dx = \frac{\sigma_x^2}{\sigma_x^2} = 1.$$
If $w = (x - m_x)/\sigma_x$, then
$$M(t; w) = e^{-m_x t/\sigma_x}\bigl(M(t/\sigma_x; x)\bigr).$$
We find that $M'(0; w) = 0$ and $M''(0; w) = 1$.

5. $(4b^2 - 9)|y|^2 + 18b|y| - 5b^2 = 0$

Chapter 9

Section 9.1

1. a) $f(x) = \frac{1}{100}$, $x = 1, \ldots, 100$; $m_x = $ median $= 50.5$, $\sigma_x^2 = \frac{9999}{12}$
   b) $f(x) = \dfrac{C(40, x)C(160, 3 - x)}{C(200, 3)}$

   c) $f(0) = \frac{1}{8}, f(1) = \frac{7}{8}$

   d) $f(x) = \dfrac{C(13, x)C(39, 13 - x)}{C(52, 13)}$

   e) $f(x) = \frac{1}{13}, x = 2, 3, \ldots, 14$; $m_x = $ median $= 8, \sigma_x^2 = 14$

   f) $f(0) = \frac{9}{13}, f(1) = f(2) = f(3) = f(4) = \frac{1}{13}$

   g) $f(x) = \dfrac{C(4, x)C(48, 5 - x)}{C(52, 5)}$

2. a) $f(x) = \dfrac{C(14, x)C(21, 10 - x)}{C(35, 10)}$, $g(y) = \dfrac{C(16, y)\dot{C}(19, 10 - y)}{C(35, 10)}$

   b) $h(z) = \dfrac{C(30, z)C(5, 10 - z)}{C(35, 10)}$

Section 9.2

1.

	$f(x)$	$M(t; x)$	m_x	mode	σ_x^2
a)	$C(10, x)(\frac{1}{2})^{10}$	$(e^t/2 + \frac{1}{2})^{10}$	5	5	$\frac{5}{2}$
b)	$C(10, x)(\frac{2}{3})^x(\frac{1}{3})^{10-x}$	$(2e^t/3 + \frac{1}{3})^{10}$	$\frac{20}{3}$	3	$\frac{20}{9}$
c)	$C(10, x)(0.9999)^x(0.0001)^{10-x}$	$(0.9999e^t + 0.0001)^{10}$	9.999	10	0.0009999
d)	$C(10, x)(0.05)^x(0.95)^{10-x}$	$(e^t/20 + \frac{19}{20})^{10}$	0.5	0	$\frac{190}{400}$
e)	Answer same as for (a)				

2. a) True, since $m_x = np$

   b) True, $m_x = np$

   c) True if n is at least 2, for then $np - np^2 = \sigma_x^2$ is increasing as p increases from 0 to 1

   d) True, $\sigma_x^2 = np(1 - p)$

   e) False, $f(0) \geq g(0)$

3. a) $\frac{5}{9}$ b) 1

4. $\dfrac{f(x)}{f(x - 1)}$ reduces to $\dfrac{n - x + 1}{x}\left(\dfrac{p}{1 - p}\right)$. We find that this equals 1 when $x = pn + p$. It follows that the ratio is less than 1 for x less than $np + p$ and greater than 1 for x greater than $pn + p = (n + 1)p$.

Section 9.3

1. a) $f(x) = (\frac{1}{3})^{x-1}(\frac{2}{3})$, mode 1, $m_x = \frac{3}{2}$

   b) $f(x) = (\frac{1}{3})^{x-1}(\frac{2}{3})$, mode 1, $m_x = \frac{3}{2}$

   c) $f(x) = \dfrac{C(99, x - 1)}{C(100, x - 1)} \cdot \dfrac{1}{100 - x + 1}$

   d) $f(x) = \dfrac{C(48, x)}{C(49, x)}\left(\dfrac{1}{49 - x}\right)$, admissible range is $\{1, 2, 3, \ldots, 48\}$

   e) $f(x) = (\frac{11}{12})^x(\frac{1}{12})$, mode 0, $m_x = 11$

2. For $p = 1$,
$$f(x) = \frac{C(n-1, x-1)}{C(n, x-1)}\left(\frac{1}{n-x+1}\right) = \frac{n-x+1}{n} \cdot \frac{1}{n-x+1} = \frac{1}{n};$$
mean is less than $(n+1)/2$, skewed to right.

3. If x is the trial on which the mth occurrence of A takes place, then in the first $x-1$ trials, A occurs exactly $m-1$ times. The number of ways $m-1$ occurrences of A can happen is $C(x-1, m-1)$. The density function of X then is
$$f(x) = C(x-1, m-1)p^{m-1}(1-p)^{x-1-(m-1)}p$$
$$= C(x-1, m-1)p^m(1-p)^{x-m}.$$
The admissible range is $m, m+1, \ldots$, and $m_x = m/p$.
a) $f(x) = C(x-1, 2)(\frac{1}{2})^x$
b) $f(x) = C(x-1, 1)(0.99)^2(0.01)^{x-2}$
c) $C(x-1, 1)(\frac{1}{6})^2(\frac{5}{6})^{x-2}$

4. Since p lies between 0 and 1, inclusive, so does $1-p$. Therefore $(1-p)^{x-1}p$ is at most p for x greater than 1. Mode is not meaningful as norm.

Section 9.4

1. 0.0021
2. $f(x+1)/f(x)$ reduces to $\mu t/(x+1)$. This is less than 1 when $x > \mu t - 1$, is when $x = \mu t - 1$, and is greater than 1 when $x < \mu t - 1$. If μt is an integer, then mode is at $\mu t - 1$; otherwise, mode is at $[\mu t - 1]$.
3. $S_k = \dfrac{\mu t - [\mu t - 1]}{\mu t > 0}$, skewed to right
4. $\displaystyle\sum_{x=0}^{\infty} \frac{(\mu t)^x}{x!} e^{-\mu t} = e^{-\mu t} \sum_{x=0}^{\infty} \frac{(\mu t)^x}{x!} = e^{-\mu t}e^{\mu t} = 1$
5. a) 0.00674, $(5^{50}/50!)e^{-5}$, true probability of no misprints on 50 consecutive pages is approximately 0.00423
   b) 0.54881, true probability about 0.526; 0.09, 0
6. (a), possibly (b), neither (c) nor (d)

Chapter 10

Section 10.1

1. a) 0.4987   b) 0.0678   c) 0.6826   d) 1   e) 0.0919
2. a) 0.2586   b) 0.1151   c) Negligible
   d) 0.3830   e) 0.94   f) 0.5
3. $1.96\sigma_x$ for 95%; $2.57\sigma_x$ for 99%
4. $m_x \doteq 15.1$ 5. 0.001; $b \doteq 11.5$ 6. Approximately 50%
7. $\bar{x} = 71$; $s_{\bar{x}}^2 = \frac{28}{6}$; population is the set of all heights of college freshman boys; $f(x) = (1/(0.5\sqrt{2\pi}))e^{-(x-70)^2/2}$; approximately 0.5

Section 10.2

1. 0.3413
2. $b \doteq 2.4$
3. Symmetric since $f(t:n) = f(-t;n)$. Median, mode, and mean are all 0.
   a) 2.015   b) 1.895   c) 1.740   d) 2.201   e) 2.131
4. 0.7, $b \doteq 2.08$
5. Normal distribution gives 0.3413; t-distribution gives approximately 0.35.
6. a) $b \doteq 3.01$ b) $b \doteq 2.29$ c) $b \doteq 1.9$
   b) gets smaller as n increases

Section 10.3

1. We compute the value of t in (22) to be about 0.05; there are 23 degrees of freedom. The probability of obtaining a value of t at least as large as the one observed is very large.
2. a) Observed value of t [from (22)] is 1.24. Probability of obtaining a value of t outside $(-1, 24, 1.24)$ is greater than 0.20.
   b) Observed value of t here is 1.61; probability of t outside $(-1.61, 1.61)$ is slightly less than 0.20.
   c) Observed value of standard normal variate from Proposition 7 is 1.3. Probability of value outside $(-1.3, 1.3)$ is about 0.185.
3. Many correct answers
4. Using Proposition 8 with $d = 0$, we get an observed value of $t = 2.22$. Probability of getting a value of t at least as large as 2.22 is less than 0.05. (We only consider $t \geq 2.22$ since $t \leq -2.22$ would imply less production with music.) The value of t obtained therefore seems somewhat inconsistent with the assumption that $d = 0$; hence we can feel some justification in supposing that production is improved by the music.

Section 10.4

1. $P(\chi^2 \geq 16) < 0.001; P(\chi^2 \leq 10) < 0.01$
2. $P(\chi^2 \geq 30) < 0.001$
3. (Uncorrected) $P(\chi^2 \geq 3.6) > 0.05$;
   (corrected) $P(\chi^2 > \text{corrected value}) > 0.05$
4. Under the assumption that the number of repairs for each manufacturer is the same, we obtain a value of χ^2 of about 3.2. $P(\chi^2 > 3.2)$, with 4 degrees of freedom, is greater than 0.5. Hence we cannot accept the manufacturer's claim that his cars require fewer repairs.

Section 10.5

1. $e^{-5/4} - e^{-6/4}; 0.2$
2. $f(t) = (\frac{4}{3})e^{-4t/3}$

a) 0   b) 0.999   c) $(60^{60}/60!)e^{-80}$ d) $\sum_{i=0}^{20} (60^i/i!)e^{-80}$

3. Using the data given in the table below (computed assuming that X is exponentially distributed), we obtain a value of $\chi^2 = 4.24$.

Interval	$0 \leq x \leq 1$	$1 < x \leq 2$	$2 < x \leq 3$	$3 < x \leq 4$	$4 < x \leq 5$	$5 < x$
Observed	5	7	2	0	1	0
Expected	7	3.6	2.1	1.1	.5	.5

The value of χ^2 is not sufficiently large to discredit the assumption that X is exponentially distributed (at least at the 0.05 level of significance).

4. $f_1 = f_2 = 6$; $F = 3.58$

Chapter 11

Section 11.1

1. There may be many correct tests of those hypotheses which can be tested statistically. We give only one such test for each part.
   a) Roll the die 10 times and record the number of 1's that occur. Null hypothesis: The die is not biased. X is the number of 1's;

   $f(x) = C(10, x)(\frac{1}{6})^x(\frac{5}{6})^{10-x}$.

   b) On ten randomly chosen days, ask the woman to predict whether or not it will rain. If the probability of rain on any given day is p and the null hypothesis is, "The woman's chances are no better than chance would allow," and X is the number of correct predictions of rain, then

   $f(x) = C(10, x)p^x(1 - p)^{10-x}$.

   c) Matter of opinion, not statistically testable
   d) In a situation such as this with a small population, it would probably be fairly simple to ask each '65 Harvard English major about his preference. We might, however, decide to sample 10 of the English majors with a null hypothesis: Cooper and Dickens are favored equally. If X is the number of Cooper fans, then $f(x) = C(10, x)(\frac{1}{2})^{10}$. One could also use the χ^2-distribution.
   e) Have the man draw 10 cards which he feels are aces (replacing each draw and carefully shuffling the deck before the next draw is made). Null hypothesis: Man has no better probability of drawing an ace than mere chance would allow, X is the number of aces drawn;

   $f(x) = C(10, x)(\frac{1}{13})^x(\frac{12}{13})^{10-x}$.

   f) Roll the die 120 times, obtaining a table of observed values for number of times each face appears. Use assumption (null hypothesis) that die is fair to obtain expected values. Use χ^2 with 5 degrees of freedom.

g) Tabulate the heights of 500 randomly selected male citizens of the United States. Count the number of heights obtained in unit intervals on both sides of 68; that is, form a table such as the one below:

height	65–66	66–67	67–68	68–69	69–70	70–71
number						

Standardize the data obtained and calculate the expected number in each standardized interval, assuming $(X - 68)/5$ is a standard normal variate. Use χ^2 to test this assumption.

h) Not statistically testable

2. $C(10, 9)(0.6)^9(0.4) + (0.6)^{10}$, $C(10, 9)(0.9)^9(0.1) + (0.9)^{10}$, p must be about 0.95

Section 11.2

1. a) $[1.771, \infty)$, $[2.650, \infty)$
   b) $[0, 1.48] \cup [5.62, \infty)$
   c) $[0, 0.676] \cup [7.01, \infty)$
2. a) Assuming equal variances and applying Proposition 7 (for the t-distribution with 15 degrees of freedom), we can accept hypothesis that m_x equals m_y.
   b) Obtain $F \doteq 1.62$ with $f_1 = f_2 = 5$. This value is well outside critical region for rejecting $\sigma_x^2 = \sigma_y^2$.
3. Can easily reject null hypothesis at 0.05 level of significance that cereals are equally popular.

Section 11.3

1. There are several correct answers for this problem; we suggest only one for each part.
   a) People are assigned numbers from 1 to 1000. Select people having numbers from first 100 numbers in table of random numbers.
   b) Assign each person a number from 1 to 1000. Put the nth person in Group A if and only if the nth number in the table is even.
   c) Number professors from 1 to 100. Select 10 whose numbers appear first in the table.
2. No
3. Many correct answers

Section 11.4

1. $\{0, 1, 9, 10\}$
2. The answer to this question is itself subject to statistical testing.

3. The maximum sum n-tuple is (0, 4, 2, 2, 6, 0, 2, 2, 6, 2). To get 18, two of the 2's must be made negative, or the 4 must be made negative. There are $C(5, 2) = 10$ ways to make two 2's negative and only one way to make the 4 negative; hence there are 11 ways to get 18. $f(-18) = \frac{11}{256}$; 16; $f(16) = \frac{17}{256}$. Make each ordered 10-tuple correspond to that (unique) 10-tuple in which all the signs are different; this leads to $f(x) = f(-x)$.

4. See Example 4 to set up a contingency table.

5. Many correct answers

6. Many correct answers

7. Better than 0.9

8. Test I is least sensitive; Test III is most sensitive.

Chapter 12

Section 12.1

1. a) $f_1(1) = 0.1, f_1(2) = 0.2, f_1(3) = 0.3, f_1(4) = 0.4$;

   $f_2(10) = 0.6, f_2(11) = 0.4$

   b) $F(2, 11) = 0.3, F(10, 4) = 0.6, F(11, 4) = 1$
   c) $g(1 \mid x_2 = 10) = 0.1, g(2 \mid x_2 = 10) = 0.2$,

   $g(3 \mid x_2 = 10) = 0.3, g(4 \mid x_2 = 10) = 0.4$

   d) X_1 and X_2 are independent

2. a) $\{(x_1, \ldots, x_n) \mid x_i \leq a_i, i = 1, \ldots, n\}$ is a set of mutually exclusive events; hence the probability that at least one of these events will occur is the sum of their respective probabilities which is the sum given in (a) of Proposition 1.

   b) It is certain that something will happen on any trial.

   c) $x_i = \{(y_1, \ldots, x_i, \ldots, y_n) \mid y_j \text{ is in } A_j, i \neq j)\} = B$. Hence

   $f(x_i)$ (the probability of x_i) = $\sum_B f(y_1, \ldots, x_i, \ldots, y_n)$.

   d) Since $f(x_1, x_2, \ldots, x_n) = P(x_1, x_2, \ldots, x_n)$ and $f_i(x_i) = P(x_i)$, in order to have independence, we must have

   $f(x_1, \ldots, x_n) = f_1(x_1) f_2(x_2) \cdots f_n(x_n)$.

3. a) $f(11) = 0.06, f(12) = 0.16, f(13) = 0.26, f(14) = 0.36, f(15) = 0.16$;
   b) $f(0) = 0.48, f(1) = 0.2, f(2) = 0.11, f(3) = 0.08, f(4) = 0.03, f(6) = 0.1$

4. For independence, entry for (i, j) would have to be $f_1(i)f_2(j)$, $i = 0, 1, 2, 3$ and $j = 0, 1, 2$.

   $g(0 \mid x_2 = 2) = \frac{2}{9}, g(1 \mid x_2 = 2) = \frac{1}{18}, g(2 \mid x_2 = 2) = \frac{1}{6}, g(3 \mid x_2 = 2) = \frac{5}{9}$

5. $f(x, y) = C(10, x)C(10, y)(\frac{1}{6})^{x+y}(\frac{5}{16})^{20-x-y}$

6. $C(10, x)C(10 - x, y)(\frac{1}{6})^{x+y}(\frac{5}{16})^{10-x-y} = f(x, y)$, no independence since maximum possible value of y depends on the value of x

Section 12.2

1.

	$\dfrac{\partial f}{\partial x_1}$	$\dfrac{\partial f}{\partial x_2}$	$\dfrac{\partial^2 f}{\partial x_1 \partial x_2}$	$\dfrac{\partial^2 f}{\partial x_2^2}\Big|_{(1,1,\dots,1)}$
a)	x_2	x_1	1	0
b)	$2x_1$	1	0	0
c)	$x_2 e^{x_1 x_2}$ $- x_2 e^{-x_1 x_2}$	$x_1(e^{x_1 x_2} - e^{-x_1 x_2})$	$(x_1 x_2 + 1)(e^{x_1 x_2})$ $+ (x_1 x_2 - 1)(e^{-x_1 x_2})$	$e - e^{-1}$
d)	$x_2 x_3$	$x_1 x_3$	x_3	0
e)	$1 + x_3(x_1+x_2)^{-2}$	$x_3(x_1+x_2)^{-2}$	$-2x_3(x_1+x_2)^{-3}$	$-\tfrac{1}{4}$

2. a) $(0,0)$, minimum b) None
   c) $(0,0)$, neither maximum nor minimum
   d) $(0,0)$, neither maximum nor minimum
   e) $(-\tfrac{9}{8}, \tfrac{3}{4})$ minimum, $(0,0)$, no information

3. 0

4. $V(x, y, z) = xyz$ and $(*)\, 54 = 2(xy + yz + zx)$. Because of $(*)$ we can consider z to be a function of x and y. Hence

$$\frac{\partial V}{\partial x} = yz + xy\left(\frac{\partial z}{\partial x}\right) \quad \text{and} \quad \frac{\partial V}{\partial y} = xz + xy\left(\frac{\partial z}{\partial y}\right).$$

From $(*)$ we have therefore

$$0 = 2\left(y + z + x\left(\frac{\partial z}{\partial x}\right) + y\left(\frac{\partial z}{\partial x}\right)\right)$$

and

$$0 = 2\left(x + z + x\left(\frac{\partial z}{\partial y}\right) + y\left(\frac{\partial z}{\partial y}\right)\right).$$

Therefore

$$\frac{\partial z}{\partial x} = -\frac{y+z}{x+y} \quad \text{and} \quad \frac{\partial z}{\partial y} = -\frac{x+z}{x+y},$$

from which we obtain

$$\frac{\partial V}{\partial x} = yz - \left(\frac{xy(y+z)}{(x+y)}\right) = 0$$

(since we are looking for critical points) and

$$\frac{\partial V}{\partial y} = xy - \left(\frac{xy(x+z)}{(x+y)}\right) = 0.$$

It follows then that $x = y = z$.

5. $x = y = z = 10$

Section 12.3

1. a) $\frac{1}{4}$ b) $\frac{3}{2}(50e^8 - 37e^7)$
   c) $(10^4/4) - (10^2/2) + (10^3/3)$ d) $\frac{1}{4}$
   e) $(\frac{1}{3})12^6 + (\frac{4}{5})12^5 + (\frac{3}{2})12^4 + (\frac{4}{3})12^3 + (\frac{1}{2})12^2$

2. a) $\int_0^1 \int_0^1 xy \, dx \, dy$ b) $\int_0^1 \int_{y^{1/3}}^{y^{1/4}} y \, dx \, dy$ c) $\int_0^\infty \int_{-\ln y/2}^{-\ln y} e^{-y} dx \, dy$

   d) $\int_0^1 \int_0^{(1-y^2)^{1/2}} dx \, dy$ e) $\int_0^1 \int_y^{y^{1/3}} xe^y \, dx \, dy + \int_{-1}^0 \int_{y^{1/3}}^y xe^y \, dx \, dy$

   f) $\int_{-8}^0 \int_{y^{1/3}}^{-2} xy^2 \, dx \, dy + \int_0^4 \int_{y^{1/2}}^{-2} xy^2 \, dx \, dy + \int_0^1 \int_{y^{1/3}}^{y^{1/2}} xy^2 \, dx \, dy$

   $\qquad\qquad + \int_1^{64} \int_{y^{1/2}}^{y^{1/3}} xy^2 \, dx \, dy$

3. a) 45   b) $\frac{1}{2}$

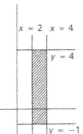

c) 1   d) π

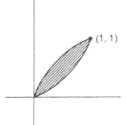
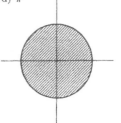

e)

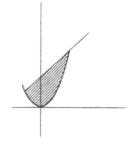

$F(-1 + 5) - F(-1 - 5)$,
where $F(x) = 2x^3/3 + 2x^2 - x^6/6$

4. $f(y) = 1/(1+y)^2$, $f(x) = e^{-x}$, $\int_0^\infty \int_0^\infty xe^{-x(1+y)}\,dx\,dy = \int_0^\infty f(y)\,dy = 1$

Chapter 13

Section 13.1

1. From $\sum_{i=1} (y_i - mx_i - b)^2 = f(m, b)$, we obtain
   $\partial f/\partial m = \sum 2(y_i - mx_i - b)(-x_i) = 0,$ and
   $\partial f/\partial b = \sum (-2)(y_i - mx_i - b) = 0,$
   whence we obtain
   $b \sum x_i + m(\sum x_i^2) = \sum x_i y_i,$ and $\sum y_i = m \sum x_i + nb.$
   We then have the straightforward solution of two linear equations in two unknowns.

2. a)

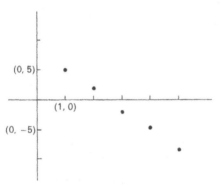

b) $Y_e = -3.7X + 8.7$ c) $-13.5, 19.8$

3. a)

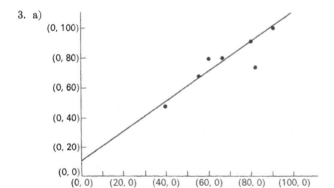

b) $Y_e = 0.94X + 12$ c) No

Section 13.2

1. $\hat{s}_e = (2.3/5)^{1/2}$, $s = (2.3/3)^{1/2}$, SSR = 136.9, SSE = 12.3, TV = 139.2
2. $\hat{s}_e = (297.6/7)^{1/2}$, $s = (297.6/5)^{1/2}$, SSR = 1373.9, SSE = 297.6
   TV = 1671.5

Section 13.3

1 and 2. We prove that if $Y = aY' + k$ and $X = a'X' + k'$, $a, a' \neq 0$, then $r^2 = r'^2$. Since $Y = aY' + k$, we have $Y_e = aY'_e + k$, $\bar{y} = a\bar{y}' + k$, and $y_i = ay'_i + k$. From (13), we obtain

$$r^2 = \frac{\sum((ay'_{e_i} + k) - (a\bar{y}_i + k))^2}{\sum((ay'_i + k) - (a\bar{y} + k))^2} = \frac{a^2\sum(y'_{e_i} - \bar{y})^2}{a^2\sum(y'_i - \bar{y})^2} = \frac{\sum(y'_{e_i} - \bar{y}')^2}{\sum(y'_i - \bar{y})^2} = r'^2$$

3. $r^2 = +0.99$ 4. $r^2 = 0.82$

Section 13.4

1. Significantly different
2. Significantly different
3. $-0.05 \leq \rho \leq 0.70$
4. $0.71 \leq \rho \leq 0.93$
5. Hypothesis is tenable at 0.05 level of significance.
6. $|r|$ would have to be virtually 1
7. 25
8. 0.16 for sample of 18, 0.054 for sample of 66

Indexes

General Index

absolute maximum, 95
absolute minimum, 95
absolute value, 53
accumulation, 33
admissible value, 39
alternating series, 128, 139
antiderivative, 156
area under a graph, 142 ff.
average, 176

Bayes' Theorem, 35
Bernoulli trial, 194
binomial distribution, 194
binomial expansion, 196
bivariate population, 271

Camp-Meidell inequality, 186
Cartesian product, 3
Central Limit Theorem, 210
Chain Rule, 89
chi-square distribution, 226
coefficient of correlation, 280
coefficient of determination, 280
combination, 28
Comparison Test, 127
complement, 8
composition of functions, 88
conditional density function, 254
conditioned events, 9
confidence interval, 284
conic section, 93
contingency table, 228
continuous function, 53

continuous variate, 40
convergence, absolute, 128
   conditional, 128
   of a sequence, 122
   of a series, 124
coordinate, 2
correlation, 275
   linear, 276
   nonlinear, 276
critical point, 97, 259
critical region, 237

definite integral, 147
degrees of freedom, 218, 226, 233
density function, discrete case, 64
   continuous case, 115, 152
dependent events, 19
derivative, first, 81, 256
   second and third, 92, 109
differentiable, 82
differential, 159
discrete variate, 40
distribution, 62
   binomial, 195
   chi-square, 226
   exponential, 231
   F, 232
   geometric, 202
   hypergeometric, 193
   negative binomial, 203
   of number of trials until success, 199 ff.

Poisson, 205, 231
rectangular, 191
standard normal, 210
Student's t, 218
distribution function, 64
  discrete approximated by continuous, 66
domain, 5

e, 134
element of a set, 1
empty set, 8
equivalent events, 9
error, Type 1 and 2, 239
events, 7
  certain, 14
  conditional, 9
  impossible, 14
  independent, 19
  mutually exclusive, 15
  simple, 7
expected value, 176
explained variation, 279
exponential distribution, 232
exponential function, 90

F-distribution, 232
factorial, 133
first derivative, 81
First Mean Value Theorem, 107
function, 2 ff
  conditional density, 254
  concave, 113
  convex, 113
  decreasing, 50
  density, 64
  differentiable, 82
  exponential, 90
  Gamma, 168
  implicit, 90
  increasing, 50
  integrable, 147
  joint density, 252, 261, 269
  joint distribution, 252, 268
  logarithmic, 135
  marginal density, 252
  moment-generating, 187
  multivariate, 250 ff.
  one-one, 50
  onto, 50
  raised to a power, 83
Fundamental Theorem of the Calculus, 151

Gamma function, 168
geometric distribution, 202
geometric progression, 125
geometric series, 58, 126
graph, 42

half-line, 40
higher derivative, 109
hypergeometric distribution, 193
hypothesis, alternative, 236
  null, 236
  statistical, 236

image, 5
implicit function, 90
independent events, 19
independent variates, 252
index of summation, 58
infinity, 58
integral, definite, 147
  improper, 165
  indefinite, 156
Integral Test for Convergence, 166
integration, by change of variable, 159
  by parts, 161
  by series, 171
intersection, 8
interval, 39–40
interval of convergence, 138

joint density function, 252, 261
joint distribution function, 252

level of significance, 237
limit of a function, 71
limits of integration, 155
line of regression, 274
logarithm, 134–35

Maclaurin series, 133
Magnitude Test, 246
marginal density function, 252
maximum, 50, 95
mean, 176
   sample, 211
mean square deviation, 182
measures of central tendency, 174
median, 177
minimum, 50, 95
mode, 117, 174
moment-generating function, 187
multivariate function, 250 ff.
mutually exclusive events, 7, 15

natural logarithm, 134
norm, 174
null hypothesis, 236

ordered m-tuple, 24
ordered pair, 2

partial derivative, 256 ff.
partial sum, 124
Pearson's measure of skewness, 186
permutation, 25
point, 1
point of inflection, 112
Poisson distribution, 205
Poisson process, 204
polynomial function, 75
population, 211
   bivariate, 271
population mean, 211
population parameter, 211
population variance, 211
power series, 132 ff.
probability, 6, 11
   assumptions about, 14
probability distribution, 62
product, Cartesian, 3
   of functions, 74
   of sequence, 123

quotient, of functions, 74
   of sequences, 123

random, 23
random variable, 39
range, 55
Ratio Test, 128
rectangular distribution, 191
relative error, 140
relative frequency, 11
relative maximum, 95
relative minimum, 95
Rolle's Theorem, 106

saddle point, 259
sample mean, 211
sample space, 7
sample variance, 211
sampling, 22 ff.
scatter diagram, 272
selection, with replacement and regard to order, 31
   without replacement but with regard to order, 22 ff.
   without replacement or regard to order, 27 ff.
sequence, 120
   limit of a, 122
series, 124
   alternating, 128, 139
   convergence of, 124
   divergence of, 124
   Maclaurin, 133
   power, 132
   Taylor's, 138
set, 1
Sign Test, 245
simple event, 7
Simpson's Rule, 171
skewness, 118, 186
slope, of function, 81
   of line, 77
standard deviation, 182
standard error of estimate, 276
   modified, 277
standard score, 188
statistic, 216
subset, 6

sum, of functions, 74
  of sequences, 123
  of series, 124
  of squares for error, 277
  of squres for regression, 277
summation notation, 57 ff.
symmetric distribution, 117

t-distribution, 216 ff.
tangent line, 79, 81
Tchebycheff's Lemma, 185
test, one- and two-tailed, 237
test variate, 240
total variation, 277
Trapezoidal Rule, 169

trial, 7
Type 1 and 2 error, 239

uncorrelated, 275
unexplained variation, 279
unimodal distribution, 119
union, 8
unordered pair, 3

variance, 182
  sample, 211
variates, 39
  independent, 252
volume, 263

Yates' Correction for Continuity, 230

Index of Symbols

Symbol	Page
(s, t)	2
$S \times T$	3
$t = f(s)$	3
$\{\ \}, \{\ |\ \}$	4, 5
$S \subset T$	6
$A \cap B$	8
$A \cup B$	8
$\overline{A}$	8
$A = B$	9
$A \mid B$	9
$P(A)$	11
$R(A)$	11
$S_1 \times S_2 \times \cdots \times S_t$	24
$(a_1, a_2, \ldots, a_m)$	24
$_nP_m$	25
$n!$	26
$C(n, m)$	28
$\binom{n}{m}$	28
$A(n, m)$	33
$X, X(s)$	39
$x_1, x_2, \ldots, x_n$	39
(a, b)	38
$[a, b]$	39
$(a, \infty), (-\infty, a)$	40

Symbol	Page
R	48
$|b|$	53
∞	40, 58
E_B	61
$\lim_{x \to a} f(x) = L$	71
$f + g$	74
fg	74
f/g	74
rf	74
$f'(a)$	81
f^r	83
$f \circ g$	88
e^x	90, 133
N	120
$\{s_n\}, n$ in N	120
$s_n \to L$	122
$s + t$	123
st	123
s/t	123
rs	123
$\sum_{n=1}^{\infty} s_n$	124
$f^{(n)}$	133
$\ln x$	134
$\log_a x$	135

Symbol	Page
$\int_a^b f(x)\, dx$	147, 165
$\int f(x)\, dx$	156
dy	159
$\Gamma(x)$	168
m_x	176
σ_x^2, σ_x	182
s_k	186
$M(t; x)$	187
$\widetilde{X}, \tilde{x}$	211
$s_{\tilde{x}}^2$	211
$\exp x$	212
s	217, 277
χ^2	226
F	232
$\partial f/\partial x_i |_{(a_1, \ldots, a_m)}$	257
$\partial^2 f/\partial x_i\, \partial x_j$	258
(X, Y)	271
Y_e	274
$\hat{s}_e$	276
SSR	277
SSE	277
TV	277
r	280
ρ	282
$n(X, Y)$	282

A Brief Table of Integrals

1. $\int u\, dv = uv - \int v\, du$

2. $\int a^u\, du = \dfrac{a^u}{\ln a} + C, \quad a \neq 1, \quad a > 0$

3. $\int \cos u\, du = \sin u + C$

4. $\int \sin u\, du = -\cos u + C$

5. $\int (ax+b)^n\, dx = \dfrac{(ax+b)^{n+1}}{a(n+1)} + C, \quad n \neq -1$

6. $\int (ax+b)^{-1}\, dx = \dfrac{1}{a} \ln |ax+b| + C$

7. $\int x(ax+b)^n\, dx = \dfrac{(ax+b)^{n+1}}{a^2}\left[\dfrac{ax+b}{n+2} - \dfrac{b}{n+1}\right] + C, \quad n \neq -1, -2$

8. $\int x(ax+b)^{-1}\, dx = \dfrac{x}{a} - \dfrac{b}{a^2} \ln |ax+b| + C$

9. $\int x(ax+b)^{-2}\, dx = \dfrac{1}{a^2}\left[\ln |ax+b| + \dfrac{b}{ax+b}\right] + C$

10. $\int \dfrac{dx}{x(ax+b)} = \dfrac{1}{b} \ln \left|\dfrac{x}{ax+b}\right| + C$

11. $\int (\sqrt{ax+b})^n\, dx = \dfrac{2}{a} \dfrac{(\sqrt{ax+b})^{n+2}}{n+2} + C, \quad n \neq -2$

12. $\int \dfrac{\sqrt{ax+b}}{x}\, dx = 2\sqrt{ax+b} + b \int \dfrac{dx}{x\sqrt{ax+b}}$

13. (a) $\int \dfrac{dx}{x\sqrt{ax+b}} = \dfrac{2}{\sqrt{-b}} \tan^{-1} \sqrt{\dfrac{ax+b}{-b}} + C, \quad \text{if } b < 0$

    (b) $\int \dfrac{dx}{x\sqrt{ax+b}} = \dfrac{1}{\sqrt{b}} \ln \left|\dfrac{\sqrt{ax+b} - \sqrt{b}}{\sqrt{ax+b} + \sqrt{b}}\right| + C, \quad \text{if } b > 0$

14. $\int \dfrac{\sqrt{ax+b}}{x^2}\, dx = -\dfrac{\sqrt{ax+b}}{x} + \dfrac{a}{2} \int \dfrac{dx}{x\sqrt{ax+b}} + C$

15. $\int \dfrac{dx}{x^2\sqrt{ax+b}} = -\dfrac{\sqrt{ax+b}}{bx} - \dfrac{a}{2b} \int \dfrac{dx}{x\sqrt{ax+b}} + C$

16. $\int \dfrac{dx}{a^2 + x^2} = \dfrac{1}{a} \tan^{-1} \dfrac{x}{a} + C$

17. $\int \dfrac{dx}{(a^2+x^2)^2} = \dfrac{x}{2a^2(a^2+x^2)} + \dfrac{1}{2a^3} \tan^{-1} \dfrac{x}{a} + C$

18. $\int \dfrac{dx}{a^2 - x^2} = \dfrac{1}{2a} \ln \left|\dfrac{x+a}{x-a}\right| + C$

19. $\displaystyle\int \frac{dx}{(a^2-x^2)^2} = \frac{x}{2a^2(a^2-x^2)} + \frac{1}{2a^2}\int \frac{dx}{a^2-x^2}$

20. $\displaystyle\int \frac{dx}{\sqrt{a^2+x^2}} = \ln|x+\sqrt{a^2+x^2}| + C$

21. $\displaystyle\int \sqrt{a^2+x^2}\, dx = \frac{x}{2}\sqrt{a^2+x^2} + \frac{a^2}{2}(\ln|x+\sqrt{a^2+x^2}|) + C$

22. $\displaystyle\int x^2\sqrt{a^2+x^2}\, dx = \frac{x(a^2+2x^2)\sqrt{a^2+x^2}}{8} - \frac{a^4}{8}(\ln|x+\sqrt{a^2+x^2}|) + C$

23. $\displaystyle\int \frac{\sqrt{a^2+x^2}}{x^2}\, dx = (\ln|x+\sqrt{a^2+x^2}|) - \frac{\sqrt{a^2+x^2}}{x} + C$

24. $\displaystyle\int \frac{x^2}{\sqrt{a^2+x^2}}\, dx = -\frac{a^2}{2}(\ln|x+\sqrt{a^2+x^2}|) + \frac{x\sqrt{a^2+x^2}}{2} + C$

25. $\displaystyle\int \frac{dx}{x\sqrt{a^2+x^2}} = -\frac{1}{a}\ln\left|\frac{a+\sqrt{a^2+x^2}}{x}\right| + C$

26. $\displaystyle\int \frac{dx}{x^2\sqrt{a^2+x^2}} = -\frac{\sqrt{a^2+x^2}}{a^2 x} + C$

27. $\displaystyle\int \frac{dx}{\sqrt{a^2-x^2}} = \sin^{-1}\frac{x}{a} + C$

28. $\displaystyle\int \sqrt{a^2-x^2}\, dx = \frac{x}{2}\sqrt{a^2-x^2} + \frac{a^2}{2}\sin^{-1}\frac{x}{a} + C$

29. $\displaystyle\int x^2\sqrt{a^2-x^2}\, dx = \frac{a^4}{8}\sin^{-1}\frac{x}{a} - \frac{1}{8}x\sqrt{a^2-x^2}\,(a^2-2x^2) + C$

30. $\displaystyle\int \frac{\sqrt{a^2-x^2}}{x}\, dx = \sqrt{a^2-x^2} - a\ln\left|\frac{a+\sqrt{a^2-x^2}}{x}\right| + C$

31. $\displaystyle\int \frac{\sqrt{a^2-x^2}}{x^2}\, dx = -\sin^{-1}\frac{x}{a} - \frac{\sqrt{a^2-x^2}}{x} + C$

32. $\displaystyle\int \frac{x^2}{\sqrt{a^2-x^2}}\, dx = \frac{a^2}{2}\sin^{-1}\frac{x}{a} - \frac{1}{2}x\sqrt{a^2-x^2} + C$

33. $\displaystyle\int \frac{dx}{x\sqrt{a^2-x^2}} = -\frac{1}{a}\ln\left|\frac{a+\sqrt{a^2-x^2}}{x}\right| + C$

34. $\displaystyle\int \frac{dx}{x^2\sqrt{a^2-x^2}} = -\frac{\sqrt{a^2-x^2}}{a^2 x} + C$

35. $\displaystyle\int \frac{dx}{\sqrt{x^2-a^2}} = \ln|x+\sqrt{x^2-a^2}| + C$

36. $\displaystyle\int \sqrt{x^2-a^2}\, dx = \frac{x}{2}\sqrt{x^2-a^2} - \frac{a^2}{2}(\ln|x+\sqrt{x^2-a^2}|) + C$

37. $\displaystyle\int (\sqrt{x^2-a^2})^n\, dx = \frac{x(\sqrt{x^2-a^2})^n}{n+1} - \frac{na^2}{n+1}\int (\sqrt{x^2-a^2})^{n-2}\, dx,\quad n\neq -1$

38. $\int \frac{dx}{(\sqrt{x^2-a^2})^n} = \frac{x(\sqrt{x^2-a^2})^{2-n}}{(2-n)a^2} - \frac{n-3}{(n-2)a^2} \int \frac{dx}{(\sqrt{x^2-a^2})^{n-2}}$, $n \neq 2$

39. $\int x(\sqrt{x^2-a^2})^n \, dx = \frac{(\sqrt{x^2-a^2})^{n+2}}{n+2} + C$, $n \neq -2$

40. $\int x^2 \sqrt{x^2-a^2} \, dx = \frac{x}{8}(2x^2-a^2)\sqrt{x^2-a^2} - \frac{a^4}{8}(\ln|x+\sqrt{x^2-a^2}|) + C$

41. $\int \frac{\sqrt{x^2-a^2}}{x} \, dx = \sqrt{x^2-a^2} - a\sec^{-1}\left|\frac{x}{a}\right| + C$

42. $\int \frac{\sqrt{x^2-a^2}}{x^2} \, dx = (\ln|x+\sqrt{x^2-a^2}|) - \frac{\sqrt{x^2-a^2}}{x} + C$

43. $\int \frac{x^2}{\sqrt{x^2-a^2}} \, dx = \frac{a^2}{2}(\ln|x+\sqrt{x^2-a^2}|) + \frac{x}{2}\sqrt{x^2-a^2} + C$

44. $\int \frac{dx}{x\sqrt{x^2-a^2}} = \frac{1}{a}\sec^{-1}\left|\frac{x}{a}\right| + C = \frac{1}{a}\cos^{-1}\left|\frac{a}{x}\right| + C$

45. $\int \frac{dx}{x^2\sqrt{x^2-a^2}} = \frac{\sqrt{x^2-a^2}}{a^2 x} + C$

46. $\int \frac{dx}{\sqrt{2ax-x^2}} = \sin^{-1}\left(\frac{x-a}{a}\right) + C$

47. $\int \sqrt{2ax-x^2} \, dx = \frac{x-a}{2}\sqrt{2ax-x^2} + \frac{a^2}{2}\sin^{-1}\left(\frac{x-a}{a}\right) + C$

48. $\int (\sqrt{2ax-x^2})^n \, dx = \frac{(x-a)(\sqrt{2ax-x^2})^n}{n+1} + \frac{na^2}{n+1}\int (\sqrt{2ax-x^2})^{n-2} \, dx$

49. $\int \frac{dx}{(\sqrt{2ax-x^2})^n} = \frac{(x-a)(\sqrt{2ax-x^2})^{2-n}}{(n-2)a^2} + \frac{(n-3)}{(n-2)a^2}\int \frac{dx}{(\sqrt{2ax-x^2})^{n-2}}$

50. $\int x\sqrt{2ax-x^2} \, dx = \frac{(x+a)(2x-3a)\sqrt{2ax-x^2}}{6} + \frac{a^3}{2}\sin^{-1}\frac{x-a}{a} + C$

51. $\int \frac{\sqrt{2ax-x^2}}{x} \, dx = \sqrt{2ax-x^2} + a\sin^{-1}\frac{x-a}{a} + C$

52. $\int \frac{\sqrt{2ax-x^2}}{x^2} \, dx = -2\sqrt{\frac{2a-x}{x}} - \sin^{-1}\left(\frac{x-a}{a}\right) + C$

53. $\int \frac{x \, dx}{\sqrt{2ax-x^2}} = a\sin^{-1}\frac{x-a}{a} - \sqrt{2ax-x^2} + C$

54. $\int \frac{dx}{x\sqrt{2ax-x^2}} = -\frac{1}{a}\sqrt{\frac{2a-x}{x}} + C$

55. $\int \sin ax \, dx = -\frac{1}{a}\cos ax + C$ 56. $\int \cos ax \, dx = \frac{1}{a}\sin ax + C$

57. $\int \sin^2 ax \, dx = \frac{x}{2} - \frac{\sin 2ax}{4a} + C$ 58. $\int \cos^2 ax \, dx = \frac{x}{2} + \frac{\sin 2ax}{4a} + C$

59. $\int x \sin ax \, dx = \frac{1}{a^2} \sin ax - \frac{x}{a} \cos ax + C$

60. $\int x \cos ax \, dx = \frac{1}{a^2} \cos ax + \frac{x}{a} \sin ax + C$

61. $\int x^n \sin ax \, dx = -\frac{x^n}{a} \cos ax + \frac{n}{a} \int x^{n-1} \cos ax \, dx$

62. $\int x^n \cos ax \, dx = \frac{x^n}{a} \sin ax - \frac{n}{a} \int x^{n-1} \sin ax \, dx$

63. $\int e^{ax} \, dx = \frac{1}{a} e^{ax} + C$

64. $\int b^{ax} \, dx = \frac{1}{a} \frac{b^{ax}}{\ln b} + C, \quad b > 0, \; b \neq 1$

65. $\int x e^{ax} \, dx = \frac{e^{ax}}{a^2} (ax - 1) + C$

66. $\int x^n e^{ax} \, dx = \frac{1}{a} x^n e^{ax} - \frac{n}{a} \int x^{n-1} e^{ax} \, dx$

67. $\int x^n b^{ax} \, dx = \frac{x^n b^{ax}}{a \ln b} - \frac{n}{a \ln b} \int x^{n-1} b^{ax} \, dx, \quad b > 0, \; b \neq 1$

68. $\int e^{ax} \sin bx \, dx = \frac{e^{ax}}{a^2 + b^2} (a \sin bx - b \cos bx) + C$

69. $\int e^{ax} \cos bx \, dx = \frac{e^{ax}}{a^2 + b^2} (a \cos bx + b \sin bx) + C$

70. $\int \ln ax \, dx = x \ln ax - x + C$

71. $\int x^n \ln ax \, dx = \frac{x^{n+1}}{n+1} \ln ax - \frac{x^{n+1}}{(n+1)^2} + C, \quad n \neq -1$

72. $\int x^{-1} \ln ax \, dx = \frac{1}{2} (\ln ax)^2 + C$

73. $\int \frac{dx}{x \ln ax} = \ln |\ln ax| + C$

74. $\int_0^\infty x^{n-1} e^{-x} \, dx = \Gamma(n) = (n-1)!, \quad n > 0$

75. $\int_0^\infty e^{-ax^2} \, dx = \frac{1}{2} \sqrt{\frac{\pi}{a}}, \quad a > 0$

76. $\int_0^{\pi/2} \sin^n x \, dx = \int_0^{\pi/2} \cos^n x \, dx = \frac{1 \cdot 3 \cdot 5 \cdots (n-1)}{2 \cdot 4 \cdot 6 \cdots n} \cdot \frac{\pi}{2},$ if n is an even integer ≥ 2,

$= \frac{2 \cdot 4 \cdot 6 \cdots (n-1)}{3 \cdot 5 \cdot 7 \cdots n},$ if n is an odd integer ≥ 3

A CATALOG OF SELECTED
DOVER BOOKS
IN SCIENCE AND MATHEMATICS

Astronomy

BURNHAM'S CELESTIAL HANDBOOK, Robert Burnham, Jr. Thorough guide to the stars beyond our solar system. Exhaustive treatment. Alphabetical by constellation: Andromeda to Cetus in Vol. 1; Chamaeleon to Orion in Vol. 2; and Pavo to Vulpecula in Vol. 3. Hundreds of illustrations. Index in Vol. 3. 2,000pp. 6⅛ x 9¼.
Vol. I: 0-486-23567-X
Vol. II: 0-486-23568-8
Vol. III: 0-486-23673-0

EXPLORING THE MOON THROUGH BINOCULARS AND SMALL TELESCOPES, Ernest H. Cherrington, Jr. Informative, profusely illustrated guide to locating and identifying craters, rills, seas, mountains, other lunar features. Newly revised and updated with special section of new photos. Over 100 photos and diagrams. 240pp. 8¼ x 11. 0-486-24491-1

THE EXTRATERRESTRIAL LIFE DEBATE, 1750–1900, Michael J. Crowe. First detailed, scholarly study in English of the many ideas that developed from 1750 to 1900 regarding the existence of intelligent extraterrestrial life. Examines ideas of Kant, Herschel, Voltaire, Percival Lowell, many other scientists and thinkers. 16 illustrations. 704pp. 5⅜ x 8½. 0-486-40675-X

THEORIES OF THE WORLD FROM ANTIQUITY TO THE COPERNICAN REVOLUTION, Michael J. Crowe. Newly revised edition of an accessible, enlightening book recreates the change from an earth-centered to a sun-centered conception of the solar system. 242pp. 5⅜ x 8½. 0-486-41444-2

A HISTORY OF ASTRONOMY, A. Pannekoek. Well-balanced, carefully reasoned study covers such topics as Ptolemaic theory, work of Copernicus, Kepler, Newton, Eddington's work on stars, much more. Illustrated. References. 521pp. 5⅜ x 8½.
0-486-65994-1

A COMPLETE MANUAL OF AMATEUR ASTRONOMY: TOOLS AND TECHNIQUES FOR ASTRONOMICAL OBSERVATIONS, P. Clay Sherrod with Thomas L. Koed. Concise, highly readable book discusses: selecting, setting up and maintaining a telescope; amateur studies of the sun; lunar topography and occultations; observations of Mars, Jupiter, Saturn, the minor planets and the stars; an introduction to photoelectric photometry; more. 1981 ed. 124 figures. 25 halftones. 37 tables. 335pp. 6½ x 9¼. 0-486-40675-X

AMATEUR ASTRONOMER'S HANDBOOK, J. B. Sidgwick. Timeless, comprehensive coverage of telescopes, mirrors, lenses, mountings, telescope drives, micrometers, spectroscopes, more. 189 illustrations. 576pp. 5⅝ x 8¼. (Available in U.S. only.)
0-486-24034-7

STARS AND RELATIVITY, Ya. B. Zel'dovich and I. D. Novikov. Vol. 1 of *Relativistic Astrophysics* by famed Russian scientists. General relativity, properties of matter under astrophysical conditions, stars, and stellar systems. Deep physical insights, clear presentation. 1971 edition. References. 544pp. 5⅝ x 8¼. 0-486-69424-0

CATALOG OF DOVER BOOKS

Chemistry

THE SCEPTICAL CHYMIST: THE CLASSIC 1661 TEXT, Robert Boyle. Boyle defines the term "element," asserting that all natural phenomena can be explained by the motion and organization of primary particles. 1911 ed. viii+232pp. 5⅜ x 8½.
0-486-42825-7

RADIOACTIVE SUBSTANCES, Marie Curie. Here is the celebrated scientist's doctoral thesis, the prelude to her receipt of the 1903 Nobel Prize. Curie discusses establishing atomic character of radioactivity found in compounds of uranium and thorium; extraction from pitchblende of polonium and radium; isolation of pure radium chloride; determination of atomic weight of radium; plus electric, photographic, luminous, heat, color effects of radioactivity. ii+94pp. 5⅜ x 8½. 0-486-42550-9

CHEMICAL MAGIC, Leonard A. Ford. Second Edition, Revised by E. Winston Grundmeier. Over 100 unusual stunts demonstrating cold fire, dust explosions, much more. Text explains scientific principles and stresses safety precautions. 128pp. 5⅜ x 8½. 0-486-67628-5

THE DEVELOPMENT OF MODERN CHEMISTRY, Aaron J. Ihde. Authoritative history of chemistry from ancient Greek theory to 20th-century innovation. Covers major chemists and their discoveries. 209 illustrations. 14 tables. Bibliographies. Indices. Appendices. 851pp. 5⅜ x 8½. 0-486-64235-6

CATALYSIS IN CHEMISTRY AND ENZYMOLOGY, William P. Jencks. Exceptionally clear coverage of mechanisms for catalysis, forces in aqueous solution, carbonyl- and acyl-group reactions, practical kinetics, more. 864pp. 5⅜ x 8½.
0-486-65460-5

ELEMENTS OF CHEMISTRY, Antoine Lavoisier. Monumental classic by founder of modern chemistry in remarkable reprint of rare 1790 Kerr translation. A must for every student of chemistry or the history of science. 539pp. 5⅜ x 8½. 0-486-64624-6

THE HISTORICAL BACKGROUND OF CHEMISTRY, Henry M. Leicester. Evolution of ideas, not individual biography. Concentrates on formulation of a coherent set of chemical laws. 260pp. 5⅜ x 8½. 0-486-61053-5

A SHORT HISTORY OF CHEMISTRY, J. R. Partington. Classic exposition explores origins of chemistry, alchemy, early medical chemistry, nature of atmosphere, theory of valency, laws and structure of atomic theory, much more. 428pp. 5⅜ x 8½. (Available in U.S. only.) 0-486-65977-1

GENERAL CHEMISTRY, Linus Pauling. Revised 3rd edition of classic first-year text by Nobel laureate. Atomic and molecular structure, quantum mechanics, statistical mechanics, thermodynamics correlated with descriptive chemistry. Problems. 992pp. 5⅜ x 8½. 0-486-65622-5

FROM ALCHEMY TO CHEMISTRY, John Read. Broad, humanistic treatment focuses on great figures of chemistry and ideas that revolutionized the science. 50 illustrations. 240pp. 5⅜ x 8½. 0-486-28690-8

Engineering

DE RE METALLICA, Georgius Agricola. The famous Hoover translation of greatest treatise on technological chemistry, engineering, geology, mining of early modern times (1556). All 289 original woodcuts. 638pp. 6¾ x 11. 0-486-60006-8

FUNDAMENTALS OF ASTRODYNAMICS, Roger Bate et al. Modern approach developed by U.S. Air Force Academy. Designed as a first course. Problems, exercises. Numerous illustrations. 455pp. 5⅜ x 8½. 0-486-60061-0

DYNAMICS OF FLUIDS IN POROUS MEDIA, Jacob Bear. For advanced students of ground water hydrology, soil mechanics and physics, drainage and irrigation engineering and more. 335 illustrations. Exercises, with answers. 784pp. 6⅛ x 9¼. 0-486-65675-6

THEORY OF VISCOELASTICITY (Second Edition), Richard M. Christensen. Complete consistent description of the linear theory of the viscoelastic behavior of materials. Problem-solving techniques discussed. 1982 edition. 29 figures. xiv+364pp. 6⅛ x 9¼. 0-486-42880-X

MECHANICS, J. P. Den Hartog. A classic introductory text or refresher. Hundreds of applications and design problems illuminate fundamentals of trusses, loaded beams and cables, etc. 334 answered problems. 462pp. 5⅜ x 8½. 0-486-60754-2

MECHANICAL VIBRATIONS, J. P. Den Hartog. Classic textbook offers lucid explanations and illustrative models, applying theories of vibrations to a variety of practical industrial engineering problems. Numerous figures. 233 problems, solutions. Appendix. Index. Preface. 436pp. 5⅜ x 8½. 0-486-64785-4

STRENGTH OF MATERIALS, J. P. Den Hartog. Full, clear treatment of basic material (tension, torsion, bending, etc.) plus advanced material on engineering methods, applications. 350 answered problems. 323pp. 5⅜ x 8½. 0-486-60755-0

A HISTORY OF MECHANICS, René Dugas. Monumental study of mechanical principles from antiquity to quantum mechanics. Contributions of ancient Greeks, Galileo, Leonardo, Kepler, Lagrange, many others. 671pp. 5⅜ x 8½. 0-486-65632-2

STABILITY THEORY AND ITS APPLICATIONS TO STRUCTURAL MECHANICS, Clive L. Dym. Self-contained text focuses on Koiter postbuckling analyses, with mathematical notions of stability of motion. Basing minimum energy principles for static stability upon dynamic concepts of stability of motion, it develops asymptotic buckling and postbuckling analyses from potential energy considerations, with applications to columns, plates, and arches. 1974 ed. 208pp. 5⅜ x 8½.
0-486-42541-X

METAL FATIGUE, N. E. Frost, K. J. Marsh, and L. P. Pook. Definitive, clearly written, and well-illustrated volume addresses all aspects of the subject, from the historical development of understanding metal fatigue to vital concepts of the cyclic stress that causes a crack to grow. Includes 7 appendixes. 544pp. 5⅜ x 8½. 0-486-40927-9

ROCKETS, Robert Goddard. Two of the most significant publications in the history of rocketry and jet propulsion. "A Method of Reaching Extreme Altitudes" (1919) and "Liquid Propellant Rocket Development" (1936). 128pp. 5⅜ x 8½. 0-486-42537-1

STATISTICAL MECHANICS: PRINCIPLES AND APPLICATIONS, Terrell L. Hill. Standard text covers fundamentals of statistical mechanics, applications to fluctuation theory, imperfect gases, distribution functions, more. 448pp. 5⅜ x 8½. 0-486-65390-0

ENGINEERING AND TECHNOLOGY 1650–1750: ILLUSTRATIONS AND TEXTS FROM ORIGINAL SOURCES, Martin Jensen. Highly readable text with more than 200 contemporary drawings and detailed engravings of engineering projects dealing with surveying, leveling, materials, hand tools, lifting equipment, transport and erection, piling, bailing, water supply, hydraulic engineering, and more. Among the specific projects outlined-transporting a 50-ton stone to the Louvre, erecting an obelisk, building timber locks, and dredging canals. 207pp. 8⅜ x 11¼. 0-486-42232-1

THE VARIATIONAL PRINCIPLES OF MECHANICS, Cornelius Lanczos. Graduate level coverage of calculus of variations, equations of motion, relativistic mechanics, more. First inexpensive paperbound edition of classic treatise. Index. Bibliography. 418pp. 5⅜ x 8½. 0-486-65067-7

PROTECTION OF ELECTRONIC CIRCUITS FROM OVERVOLTAGES, Ronald B. Standler. Five-part treatment presents practical rules and strategies for circuits designed to protect electronic systems from damage by transient overvoltages. 1989 ed. xxiv+434pp. 6⅛ x 9¼. 0-486-42552-5

ROTARY WING AERODYNAMICS, W. Z. Stepniewski. Clear, concise text covers aerodynamic phenomena of the rotor and offers guidelines for helicopter performance evaluation. Originally prepared for NASA. 537 figures. 640pp. 6⅛ x 9¼. 0-486-64647-5

INTRODUCTION TO SPACE DYNAMICS, William Tyrrell Thomson. Comprehensive, classic introduction to space-flight engineering for advanced undergraduate and graduate students. Includes vector algebra, kinematics, transformation of coordinates. Bibliography. Index. 352pp. 5⅜ x 8½. 0-486-65113-4

HISTORY OF STRENGTH OF MATERIALS, Stephen P. Timoshenko. Excellent historical survey of the strength of materials with many references to the theories of elasticity and structure. 245 figures. 452pp. 5⅜ x 8½. 0-486-61187-6

ANALYTICAL FRACTURE MECHANICS, David J. Unger. Self-contained text supplements standard fracture mechanics texts by focusing on analytical methods for determining crack-tip stress and strain fields. 336pp. 6⅛ x 9¼. 0-486-41737-9

STATISTICAL MECHANICS OF ELASTICITY, J. H. Weiner. Advanced, self-contained treatment illustrates general principles and elastic behavior of solids. Part 1, based on classical mechanics, studies thermoelastic behavior of crystalline and polymeric solids. Part 2, based on quantum mechanics, focuses on interatomic force laws, behavior of solids, and thermally activated processes. For students of physics and chemistry and for polymer physicists. 1983 ed. 96 figures. 496pp. 5⅜ x 8½. 0-486-42260-7

CATALOG OF DOVER BOOKS

Mathematics

FUNCTIONAL ANALYSIS (Second Corrected Edition), George Bachman and Lawrence Narici. Excellent treatment of subject geared toward students with background in linear algebra, advanced calculus, physics and engineering. Text covers introduction to inner-product spaces, normed, metric spaces, and topological spaces; complete orthonormal sets, the Hahn-Banach Theorem and its consequences, and many other related subjects. 1966 ed. 544pp. 6⅛ x 9¼. 0-486-40251-7

ASYMPTOTIC EXPANSIONS OF INTEGRALS, Norman Bleistein & Richard A. Handelsman. Best introduction to important field with applications in a variety of scientific disciplines. New preface. Problems. Diagrams. Tables. Bibliography. Index. 448pp. 5⅜ x 8½. 0-486-65082-0

VECTOR AND TENSOR ANALYSIS WITH APPLICATIONS, A. I. Borisenko and I. E. Tarapov. Concise introduction. Worked-out problems, solutions, exercises. 257pp. 5⅛ x 8¼. 0-486-63833-2

AN INTRODUCTION TO ORDINARY DIFFERENTIAL EQUATIONS, Earl A. Coddington. A thorough and systematic first course in elementary differential equations for undergraduates in mathematics and science, with many exercises and problems (with answers). Index. 304pp. 5⅜ x 8½. 0-486-65942-9

FOURIER SERIES AND ORTHOGONAL FUNCTIONS, Harry F. Davis. An incisive text combining theory and practical example to introduce Fourier series, orthogonal functions and applications of the Fourier method to boundary-value problems. 570 exercises. Answers and notes. 416pp. 5⅜ x 8½. 0-486-65973-9

COMPUTABILITY AND UNSOLVABILITY, Martin Davis. Classic graduate-level introduction to theory of computability, usually referred to as theory of recurrent functions. New preface and appendix. 288pp. 5⅜ x 8½. 0-486-61471-9

ASYMPTOTIC METHODS IN ANALYSIS, N. G. de Bruijn. An inexpensive, comprehensive guide to asymptotic methods–the pioneering work that teaches by explaining worked examples in detail. Index. 224pp. 5⅜ x 8½ 0-486-64221-6

APPLIED COMPLEX VARIABLES, John W. Dettman. Step-by-step coverage of fundamentals of analytic function theory–plus lucid exposition of five important applications: Potential Theory; Ordinary Differential Equations; Fourier Transforms; Laplace Transforms; Asymptotic Expansions. 66 figures. Exercises at chapter ends. 512pp. 5⅜ x 8½. 0-486-64670-X

INTRODUCTION TO LINEAR ALGEBRA AND DIFFERENTIAL EQUATIONS, John W. Dettman. Excellent text covers complex numbers, determinants, orthonormal bases, Laplace transforms, much more. Exercises with solutions. Undergraduate level. 416pp. 5⅜ x 8½. 0-486-65191-6

RIEMANN'S ZETA FUNCTION, H. M. Edwards. Superb, high-level study of landmark 1859 publication entitled "On the Number of Primes Less Than a Given Magnitude" traces developments in mathematical theory that it inspired. xiv+315pp. 5⅜ x 8½. 0-486-41740-9

CATALOG OF DOVER BOOKS

CALCULUS OF VARIATIONS WITH APPLICATIONS, George M. Ewing. Applications-oriented introduction to variational theory develops insight and promotes understanding of specialized books, research papers. Suitable for advanced undergraduate/graduate students as primary, supplementary text. 352pp. 5⅜ x 8½.
0-486-64856-7

COMPLEX VARIABLES, Francis J. Flanigan. Unusual approach, delaying complex algebra till harmonic functions have been analyzed from real variable viewpoint. Includes problems with answers. 364pp. 5⅜ x 8½. 0-486-61388-7

AN INTRODUCTION TO THE CALCULUS OF VARIATIONS, Charles Fox. Graduate-level text covers variations of an integral, isoperimetrical problems, least action, special relativity, approximations, more. References. 279pp. 5⅜ x 8½.
0-486-65499-0

COUNTEREXAMPLES IN ANALYSIS, Bernard R. Gelbaum and John M. H. Olmsted. These counterexamples deal mostly with the part of analysis known as "real variables." The first half covers the real number system, and the second half encompasses higher dimensions. 1962 edition. xxiv+198pp. 5⅜ x 8½. 0-486-42875-3

CATASTROPHE THEORY FOR SCIENTISTS AND ENGINEERS, Robert Gilmore Advanced-level treatment describes mathematics of theory grounded in the work of Poincaré, R. Thom, other mathematicians. Also important applications to problems in mathematics, physics, chemistry and engineering. 1981 edition. References. 28 tables. 397 black-and-white illustrations. xvii + 666pp. 6⅛ x 9¼.
0-486-67539-4

INTRODUCTION TO DIFFERENCE EQUATIONS, Samuel Goldberg. Exceptionally clear exposition of important discipline with applications to sociology, psychology, economics. Many illustrative examples; over 250 problems. 260pp. 5⅜ x 8½.
0-486-65084-7

NUMERICAL METHODS FOR SCIENTISTS AND ENGINEERS, Richard Hamming. Classic text stresses frequency approach in coverage of algorithms, polynomial approximation, Fourier approximation, exponential approximation, other topics. Revised and enlarged 2nd edition. 721pp. 5⅜ x 8½. 0-486-65241-6

INTRODUCTION TO NUMERICAL ANALYSIS (2nd Edition), F. B. Hildebrand. Classic, fundamental treatment covers computation, approximation, interpolation, numerical differentiation and integration, other topics. 150 new problems. 669pp. 5⅜ x 8½. 0-486-65363-3

THREE PEARLS OF NUMBER THEORY, A. Y. Khinchin. Three compelling puzzles require proof of a basic law governing the world of numbers. Challenges concern van der Waerden's theorem, the Landau-Schnirelmann hypothesis and Mann's theorem, and a solution to Waring's problem. Solutions included. 64pp. 5¾ x 8¼.
0-486-40026-3

THE PHILOSOPHY OF MATHEMATICS: AN INTRODUCTORY ESSAY, Stephan Körner. Surveys the views of Plato, Aristotle, Leibniz & Kant concerning propositions and theories of applied and pure mathematics. Introduction. Two appendices. Index. 198pp. 5⅜ x 8½. 0-486-25048-2

CATALOG OF DOVER BOOKS

INTRODUCTORY REAL ANALYSIS, A.N. Kolmogorov, S. V. Fomin. Translated by Richard A. Silverman. Self-contained, evenly paced introduction to real and functional analysis. Some 350 problems. 403pp. 5⅜ x 8½. 0-486-61226-0

APPLIED ANALYSIS, Cornelius Lanczos. Classic work on analysis and design of finite processes for approximating solution of analytical problems. Algebraic equations, matrices, harmonic analysis, quadrature methods, much more. 559pp. 5⅜ x 8½. 0-486-65656-X

AN INTRODUCTION TO ALGEBRAIC STRUCTURES, Joseph Landin. Superb self-contained text covers "abstract algebra": sets and numbers, theory of groups, theory of rings, much more. Numerous well-chosen examples, exercises. 247pp. 5⅜ x 8½. 0-486-65940-2

QUALITATIVE THEORY OF DIFFERENTIAL EQUATIONS, V. V. Nemytskii and V.V. Stepanov. Classic graduate-level text by two prominent Soviet mathematicians covers classical differential equations as well as topological dynamics and ergodic theory. Bibliographies. 523pp. 5⅜ x 8½. 0-486-65954-2

THEORY OF MATRICES, Sam Perlis. Outstanding text covering rank, nonsingularity and inverses in connection with the development of canonical matrices under the relation of equivalence, and without the intervention of determinants. Includes exercises. 237pp. 5⅜ x 8½. 0-486-66810-X

INTRODUCTION TO ANALYSIS, Maxwell Rosenlicht. Unusually clear, accessible coverage of set theory, real number system, metric spaces, continuous functions, Riemann integration, multiple integrals, more. Wide range of problems. Undergraduate level. Bibliography. 254pp. 5⅜ x 8½. 0-486-65038-3

MODERN NONLINEAR EQUATIONS, Thomas L. Saaty. Emphasizes practical solution of problems; covers seven types of equations. ". . . a welcome contribution to the existing literature...."–*Math Reviews.* 490pp. 5⅜ x 8½. 0-486-64232-1

MATRICES AND LINEAR ALGEBRA, Hans Schneider and George Phillip Barker. Basic textbook covers theory of matrices and its applications to systems of linear equations and related topics such as determinants, eigenvalues and differential equations. Numerous exercises. 432pp. 5⅜ x 8½. 0-486-66014-1

LINEAR ALGEBRA, Georgi E. Shilov. Determinants, linear spaces, matrix algebras, similar topics. For advanced undergraduates, graduates. Silverman translation. 387pp. 5⅜ x 8½. 0-486-63518-X

ELEMENTS OF REAL ANALYSIS, David A. Sprecher. Classic text covers fundamental concepts, real number system, point sets, functions of a real variable, Fourier series, much more. Over 500 exercises. 352pp. 5⅜ x 8½. 0-486-65385-4

SET THEORY AND LOGIC, Robert R. Stoll. Lucid introduction to unified theory of mathematical concepts. Set theory and logic seen as tools for conceptual understanding of real number system. 496pp. 5⅜ x 8½. 0-486-63829-4

CATALOG OF DOVER BOOKS

TENSOR CALCULUS, J.L. Synge and A. Schild. Widely used introductory text covers spaces and tensors, basic operations in Riemannian space, non-Riemannian spaces, etc. 324pp. 5⅜ x 8¼. 0-486-63612-7

ORDINARY DIFFERENTIAL EQUATIONS, Morris Tenenbaum and Harry Pollard. Exhaustive survey of ordinary differential equations for undergraduates in mathematics, engineering, science. Thorough analysis of theorems. Diagrams. Bibliography. Index. 818pp. 5⅜ x 8½. 0-486-64940-7

INTEGRAL EQUATIONS, F. G. Tricomi. Authoritative, well-written treatment of extremely useful mathematical tool with wide applications. Volterra Equations, Fredholm Equations, much more. Advanced undergraduate to graduate level. Exercises. Bibliography 238pp. 5⅜ x 8½. 0-486-64828-1

FOURIER SERIES, Georgi P. Tolstov. Translated by Richard A. Silverman. A valuable addition to the literature on the subject, moving clearly from subject to subject and theorem to theorem. 107 problems, answers. 336pp. 5⅜ x 8½. 0-486-63317-9

INTRODUCTION TO MATHEMATICAL THINKING, Friedrich Waismann. Examinations of arithmetic, geometry, and theory of integers; rational and natural numbers; complete induction; limit and point of accumulation; remarkable curves; complex and hypercomplex numbers, more. 1959 ed. 27 figures. xii+260pp. 5⅜ x 8½.
0-486-63317-9

POPULAR LECTURES ON MATHEMATICAL LOGIC, Hao Wang. Noted logician's lucid treatment of historical developments, set theory, model theory, recursion theory and constructivism, proof theory, more. 3 appendixes. Bibliography. 1981 edition. ix + 283pp. 5⅜ x 8½. 0-486-67632-3

CALCULUS OF VARIATIONS, Robert Weinstock. Basic introduction covering isoperimetric problems, theory of elasticity, quantum mechanics, electrostatics, etc. Exercises throughout. 326pp. 5⅜ x 8½. 0-486-63069-2

THE CONTINUUM: A CRITICAL EXAMINATION OF THE FOUNDATION OF ANALYSIS, Hermann Weyl. Classic of 20th-century foundational research deals with the conceptual problem posed by the continuum. 156pp. 5⅜ x 8½.
0-486-67982-9

CHALLENGING MATHEMATICAL PROBLEMS WITH ELEMENTARY SOLUTIONS, A. M. Yaglom and I. M. Yaglom. Over 170 challenging problems on probability theory, combinatorial analysis, points and lines, topology, convex polygons, many other topics. Solutions. Total of 445pp. 5⅜ x 8½. Two-vol. set.
Vol. I: 0-486-65536-9 Vol. II: 0-486-65537-7

INTRODUCTION TO PARTIAL DIFFERENTIAL EQUATIONS WITH APPLICATIONS, E. C. Zachmanoglou and Dale W. Thoe. Essentials of partial differential equations applied to common problems in engineering and the physical sciences. Problems and answers. 416pp. 5⅜ x 8½. 0-486-65251-3

THE THEORY OF GROUPS, Hans J. Zassenhaus. Well-written graduate-level text acquaints reader with group-theoretic methods and demonstrates their usefulness in mathematics. Axioms, the calculus of complexes, homomorphic mapping, p-group theory, more. 276pp. 5⅜ x 8½. 0-486-40922-8

CATALOG OF DOVER BOOKS

Math–Decision Theory, Statistics, Probability

ELEMENTARY DECISION THEORY, Herman Chernoff and Lincoln E. Moses. Clear introduction to statistics and statistical theory covers data processing, probability and random variables, testing hypotheses, much more. Exercises. 364pp. 5⅜ x 8½. 0-486-65218-1

STATISTICS MANUAL, Edwin L. Crow et al. Comprehensive, practical collection of classical and modern methods prepared by U.S. Naval Ordnance Test Station. Stress on use. Basics of statistics assumed. 288pp. 5⅜ x 8½. 0-486-60599-X

SOME THEORY OF SAMPLING, William Edwards Deming. Analysis of the problems, theory and design of sampling techniques for social scientists, industrial managers and others who find statistics important at work. 61 tables. 90 figures. xvii +602pp. 5⅜ x 8½. 0-486-64684-X

LINEAR PROGRAMMING AND ECONOMIC ANALYSIS, Robert Dorfman, Paul A. Samuelson and Robert M. Solow. First comprehensive treatment of linear programming in standard economic analysis. Game theory, modern welfare economics, Leontief input-output, more. 525pp. 5⅜ x 8½. 0-486-65491-5

PROBABILITY: AN INTRODUCTION, Samuel Goldberg. Excellent basic text covers set theory, probability theory for finite sample spaces, binomial theorem, much more. 360 problems. Bibliographies. 322pp. 5⅜ x 8½. 0-486-65252-1

GAMES AND DECISIONS: INTRODUCTION AND CRITICAL SURVEY, R. Duncan Luce and Howard Raiffa. Superb nontechnical introduction to game theory, primarily applied to social sciences. Utility theory, zero-sum games, n-person games, decision-making, much more. Bibliography. 509pp. 5⅜ x 8½. 0-486-65943-7

INTRODUCTION TO THE THEORY OF GAMES, J. C. C. McKinsey. This comprehensive overview of the mathematical theory of games illustrates applications to situations involving conflicts of interest, including economic, social, political, and military contexts. Appropriate for advanced undergraduate and graduate courses; advanced calculus a prerequisite. 1952 ed. x+372pp. 5⅜ x 8½. 0-486-42811-7

FIFTY CHALLENGING PROBLEMS IN PROBABILITY WITH SOLUTIONS, Frederick Mosteller. Remarkable puzzlers, graded in difficulty, illustrate elementary and advanced aspects of probability. Detailed solutions. 88pp. 5⅜ x 8½. 65355-2

PROBABILITY THEORY: A CONCISE COURSE, Y. A. Rozanov. Highly readable, self-contained introduction covers combination of events, dependent events, Bernoulli trials, etc. 148pp. 5⅜ x 8¼. 0-486-63544-9

STATISTICAL METHOD FROM THE VIEWPOINT OF QUALITY CONTROL, Walter A. Shewhart. Important text explains regulation of variables, uses of statistical control to achieve quality control in industry, agriculture, other areas. 192pp. 5⅜ x 8½. 0-486-65232-7

CATALOG OF DOVER BOOKS

Math–Geometry and Topology

ELEMENTARY CONCEPTS OF TOPOLOGY, Paul Alexandroff. Elegant, intuitive approach to topology from set-theoretic topology to Betti groups; how concepts of topology are useful in math and physics. 25 figures. 57pp. 5⅜ x 8½.　0-486-60747-X

COMBINATORIAL TOPOLOGY, P. S. Alexandrov. Clearly written, well-organized, three-part text begins by dealing with certain classic problems without using the formal techniques of homology theory and advances to the central concept, the Betti groups. Numerous detailed examples. 654pp. 5⅜ x 8½.　0-486-40179-0

EXPERIMENTS IN TOPOLOGY, Stephen Barr. Classic, lively explanation of one of the byways of mathematics. Klein bottles, Moebius strips, projective planes, map coloring, problem of the Koenigsberg bridges, much more, described with clarity and wit. 43 figures. 210pp. 5⅜ x 8½.　0-486-25933-1

THE GEOMETRY OF RENÉ DESCARTES, René Descartes. The great work founded analytical geometry. Original French text, Descartes's own diagrams, together with definitive Smith-Latham translation. 244pp. 5⅜ x 8½.　0-486-60068-8

EUCLIDEAN GEOMETRY AND TRANSFORMATIONS, Clayton W. Dodge. This introduction to Euclidean geometry emphasizes transformations, particularly isometries and similarities. Suitable for undergraduate courses, it includes numerous examples, many with detailed answers. 1972 ed. viii+296pp. 6⅛ x 9¼.　0-486-43476-1

PRACTICAL CONIC SECTIONS: THE GEOMETRIC PROPERTIES OF ELLIPSES, PARABOLAS AND HYPERBOLAS, J. W. Downs. This text shows how to create ellipses, parabolas, and hyperbolas. It also presents historical background on their ancient origins and describes the reflective properties and roles of curves in design applications. 1993 ed. 98 figures. xii+100pp. 6½ x 9¼.　0-486-42876-1

THE THIRTEEN BOOKS OF EUCLID'S ELEMENTS, translated with introduction and commentary by Sir Thomas L. Heath. Definitive edition. Textual and linguistic notes, mathematical analysis. 2,500 years of critical commentary. Unabridged. 1,414pp. 5⅜ x 8½. Three-vol. set.
　　　Vol. I: 0-486-60088-2　Vol. II: 0-486-60089-0　Vol. III: 0-486-60090-4

SPACE AND GEOMETRY: IN THE LIGHT OF PHYSIOLOGICAL, PSYCHOLOGICAL AND PHYSICAL INQUIRY, Ernst Mach. Three essays by an eminent philosopher and scientist explore the nature, origin, and development of our concepts of space, with a distinctness and precision suitable for undergraduate students and other readers. 1906 ed. vi+148pp. 5⅜ x 8½.　0-486-43909-7

GEOMETRY OF COMPLEX NUMBERS, Hans Schwerdtfeger. Illuminating, widely praised book on analytic geometry of circles, the Moebius transformation, and two-dimensional non-Euclidean geometries. 200pp. 5⅜ x 8½.　0-486-63830-8

DIFFERENTIAL GEOMETRY, Heinrich W. Guggenheimer. Local differential geometry as an application of advanced calculus and linear algebra. Curvature, transformation groups, surfaces, more. Exercises. 62 figures. 378pp. 5⅜ x 8½.　0-486-63433-7

CATALOG OF DOVER BOOKS

History of Math

THE WORKS OF ARCHIMEDES, Archimedes (T. L. Heath, ed.). Topics include the famous problems of the ratio of the areas of a cylinder and an inscribed sphere; the measurement of a circle; the properties of conoids, spheroids, and spirals; and the quadrature of the parabola. Informative introduction. clxxxvi+326pp. 5⅜ x 8½.
0-486-42084-1

A SHORT ACCOUNT OF THE HISTORY OF MATHEMATICS, W. W. Rouse Ball. One of clearest, most authoritative surveys from the Egyptians and Phoenicians through 19th-century figures such as Grassman, Galois, Riemann. Fourth edition. 522pp. 5⅜ x 8½. 0-486-20630-0

THE HISTORY OF THE CALCULUS AND ITS CONCEPTUAL DEVELOPMENT, Carl B. Boyer. Origins in antiquity, medieval contributions, work of Newton, Leibniz, rigorous formulation. Treatment is verbal. 346pp. 5⅜ x 8½. 0-486-60509-4

THE HISTORICAL ROOTS OF ELEMENTARY MATHEMATICS, Lucas N. H. Bunt, Phillip S. Jones, and Jack D. Bedient. Fundamental underpinnings of modern arithmetic, algebra, geometry and number systems derived from ancient civilizations. 320pp. 5⅜ x 8½. 0-486-25563-8

A HISTORY OF MATHEMATICAL NOTATIONS, Florian Cajori. This classic study notes the first appearance of a mathematical symbol and its origin, the competition it encountered, its spread among writers in different countries, its rise to popularity, its eventual decline or ultimate survival. Original 1929 two-volume edition presented here in one volume. xxviii+820pp. 5⅜ x 8½. 0-486-67766-4

GAMES, GODS & GAMBLING· A HISTORY OF PROBABILITY AND STATISTICAL IDEAS, F. N. David. Episodes from the lives of Galileo, Fermat, Pascal, and others illustrate this fascinating account of the roots of mathematics. Features thought-provoking references to classics, archaeology, biography, poetry. 1962 edition. 304pp. 5⅜ x 8½. (Available in U.S. only.) 0-486-40023-9

OF MEN AND NUMBERS: THE STORY OF THE GREAT MATHEMATICIANS, Jane Muir. Fascinating accounts of the lives and accomplishments of history's greatest mathematical minds–Pythagoras, Descartes, Euler, Pascal, Cantor, many more. Anecdotal, illuminating. 30 diagrams. Bibliography. 256pp 5⅜ x 8½. 0-486-28973-7

HISTORY OF MATHEMATICS, David E. Smith. Nontechnical survey from ancient Greece and Orient to late 19th century; evolution of arithmetic, geometry, trigonometry, calculating devices, algebra, the calculus. 362 illustrations. 1,355pp. 5⅜ x 8½. Two-vol. set. Vol. I: 0-486-20429-4 Vol. II: 0-486-20430-8

A CONCISE HISTORY OF MATHEMATICS, Dirk J. Struik. The best brief history of mathematics. Stresses origins and covers every major figure from ancient Near East to 19th century. 41 illustrations. 195pp. 5⅜ x 8½. 0-486-60255-9

CATALOG OF DOVER BOOKS

Physics

OPTICAL RESONANCE AND TWO-LEVEL ATOMS, L. Allen and J. H. Eberly. Clear, comprehensive introduction to basic principles behind all quantum optical resonance phenomena. 53 illustrations. Preface. Index. 256pp. 5⅜ x 8½. 0-486-65533-4

QUANTUM THEORY, David Bohm. This advanced undergraduate-level text presents the quantum theory in terms of qualitative and imaginative concepts, followed by specific applications worked out in mathematical detail. Preface. Index. 655pp. 5⅜ x 8½. 0-486-65969-0

ATOMIC PHYSICS (8th EDITION), Max Born. Nobel laureate's lucid treatment of kinetic theory of gases, elementary particles, nuclear atom, wave-corpuscles, atomic structure and spectral lines, much more. Over 40 appendices, bibliography. 495pp. 5⅜ x 8½. 0-486-65984-4

A SOPHISTICATE'S PRIMER OF RELATIVITY, P. W. Bridgman. Geared toward readers already acquainted with special relativity, this book transcends the view of theory as a working tool to answer natural questions: What is a frame of reference? What is a "law of nature"? What is the role of the "observer"? Extensive treatment, written in terms accessible to those without a scientific background. 1983 ed. xlviii+172pp. 5⅜ x 8½. 0-486-42549-5

AN INTRODUCTION TO HAMILTONIAN OPTICS, H. A. Buchdahl. Detailed account of the Hamiltonian treatment of aberration theory in geometrical optics. Many classes of optical systems defined in terms of the symmetries they possess. Problems with detailed solutions. 1970 edition. xv + 360pp. 5⅜ x 8½. 0-486-67597-1

PRIMER OF QUANTUM MECHANICS, Marvin Chester. Introductory text examines the classical quantum bead on a track: its state and representations; operator eigenvalues; harmonic oscillator and bound bead in a symmetric force field; and bead in a spherical shell. Other topics include spin, matrices, and the structure of quantum mechanics; the simplest atom; indistinguishable particles; and stationary-state perturbation theory. 1992 ed. xiv+314pp. 6⅛ x 9¼. 0-486-42878-8

LECTURES ON QUANTUM MECHANICS, Paul A. M. Dirac. Four concise, brilliant lectures on mathematical methods in quantum mechanics from Nobel Prize-winning quantum pioneer build on idea of visualizing quantum theory through the use of classical mechanics. 96pp. 5⅜ x 8½. 0-486-41713-1

THIRTY YEARS THAT SHOOK PHYSICS: THE STORY OF QUANTUM THEORY, George Gamow. Lucid, accessible introduction to influential theory of energy and matter. Careful explanations of Dirac's anti-particles, Bohr's model of the atom, much more. 12 plates. Numerous drawings. 240pp. 5⅜ x 8½. 0-486-24895-X

ELECTRONIC STRUCTURE AND THE PROPERTIES OF SOLIDS: THE PHYSICS OF THE CHEMICAL BOND, Walter A. Harrison. Innovative text offers basic understanding of the electronic structure of covalent and ionic solids, simple metals, transition metals and their compounds. Problems. 1980 edition. 582pp. 6⅛ x 9¼. 0-486-66021-4

CATALOG OF DOVER BOOKS

HYDRODYNAMIC AND HYDROMAGNETIC STABILITY, S. Chandrasekhar. Lucid examination of the Rayleigh-Benard problem; clear coverage of the theory of instabilities causing convection. 704pp. 5⅜ x 8¼. 0-486-64071-X

INVESTIGATIONS ON THE THEORY OF THE BROWNIAN MOVEMENT, Albert Einstein. Five papers (1905–8) investigating dynamics of Brownian motion and evolving elementary theory. Notes by R. Fürth. 122pp. 5⅜ x 8½. 0-486-60304-0

THE PHYSICS OF WAVES, William C. Elmore and Mark A. Heald. Unique overview of classical wave theory. Acoustics, optics, electromagnetic radiation, more. Ideal as classroom text or for self-study. Problems. 477pp. 5⅜ x 8½. 0-486-64926-1

GRAVITY, George Gamow. Distinguished physicist and teacher takes reader-friendly look at three scientists whose work unlocked many of the mysteries behind the laws of physics: Galileo, Newton, and Einstein. Most of the book focuses on Newton's ideas, with a concluding chapter on post-Einsteinian speculations concerning the relationship between gravity and other physical phenomena. 160pp. 5⅜ x 8½. 0-486-42563-0

PHYSICAL PRINCIPLES OF THE QUANTUM THEORY, Werner Heisenberg. Nobel Laureate discusses quantum theory, uncertainty, wave mechanics, work of Dirac, Schroedinger, Compton, Wilson, Einstein, etc. 184pp. 5⅜ x 8½. 0-486-60113-7

ATOMIC SPECTRA AND ATOMIC STRUCTURE, Gerhard Herzberg. One of best introductions; especially for specialist in other fields. Treatment is physical rather than mathematical. 80 illustrations. 257pp. 5⅜ x 8½. 0-486-60115-3

AN INTRODUCTION TO STATISTICAL THERMODYNAMICS, Terrell L. Hill. Excellent basic text offers wide-ranging coverage of quantum statistical mechanics, systems of interacting molecules, quantum statistics, more. 523pp. 5⅜ x 8½.
0-486-65242-4

THEORETICAL PHYSICS, Georg Joos, with Ira M. Freeman. Classic overview covers essential math, mechanics, electromagnetic theory, thermodynamics, quantum mechanics, nuclear physics, other topics. First paperback edition. xxiii + 885pp. 5⅜ x 8½. 0-486-65227-0

PROBLEMS AND SOLUTIONS IN QUANTUM CHEMISTRY AND PHYSICS, Charles S. Johnson, Jr. and Lee G. Pedersen. Unusually varied problems, detailed solutions in coverage of quantum mechanics, wave mechanics, angular momentum, molecular spectroscopy, more. 280 problems plus 139 supplementary exercises. 430pp. 6½ x 9¼. 0-486-65236-X

THEORETICAL SOLID STATE PHYSICS, Vol. 1: Perfect Lattices in Equilibrium; Vol. II: Non-Equilibrium and Disorder, William Jones and Norman H. March. Monumental reference work covers fundamental theory of equilibrium properties of perfect crystalline solids, non-equilibrium properties, defects and disordered systems. Appendices. Problems. Preface. Diagrams. Index. Bibliography. Total of 1,301pp. 5⅜ x 8½. Two volumes. Vol. I: 0-486-65015-4 Vol. II: 0-486-65016-2

WHAT IS RELATIVITY? L. D. Landau and G. B. Rumer. Written by a Nobel Prize physicist and his distinguished colleague, this compelling book explains the special theory of relativity to readers with no scientific background, using such familiar objects as trains, rulers, and clocks. 1960 ed. vi+72pp. 5⅜ x 8½. 0-486-42806-0

CATALOG OF DOVER BOOKS

A TREATISE ON ELECTRICITY AND MAGNETISM, James Clerk Maxwell. Important foundation work of modern physics. Brings to final form Maxwell's theory of electromagnetism and rigorously derives his general equations of field theory. 1,084pp. 5⅜ x 8½. Two-vol. set. Vol. I: 0-486-60636-8 Vol. II: 0-486-60637-6

QUANTUM MECHANICS: PRINCIPLES AND FORMALISM, Roy McWeeny. Graduate student-oriented volume develops subject as fundamental discipline, opening with review of origins of Schrödinger's equations and vector spaces. Focusing on main principles of quantum mechanics and their immediate consequences, it concludes with final generalizations covering alternative "languages" or representations. 1972 ed. 15 figures. xi+155pp. 5⅜ x 8½. 0-486-42829-X

INTRODUCTION TO QUANTUM MECHANICS With Applications to Chemistry, Linus Pauling & E. Bright Wilson, Jr. Classic undergraduate text by Nobel Prize winner applies quantum mechanics to chemical and physical problems. Numerous tables and figures enhance the text. Chapter bibliographies. Appendices. Index. 468pp. 5⅜ x 8½. 0-486-64871-0

METHODS OF THERMODYNAMICS, Howard Reiss. Outstanding text focuses on physical technique of thermodynamics, typical problem areas of understanding, and significance and use of thermodynamic potential. 1965 edition. 238pp. 5⅜ x 8½. 0-486-69445-3

THE ELECTROMAGNETIC FIELD, Albert Shadowitz. Comprehensive undergraduate text covers basics of electric and magnetic fields, builds up to electromagnetic theory. Also related topics, including relativity. Over 900 problems. 768pp. 5⅜ x 8¼. 0-486-65660-8

GREAT EXPERIMENTS IN PHYSICS: FIRSTHAND ACCOUNTS FROM GALILEO TO EINSTEIN, Morris H. Shamos (ed.). 25 crucial discoveries: Newton's laws of motion, Chadwick's study of the neutron, Hertz on electromagnetic waves, more. Original accounts clearly annotated. 370pp. 5⅜ x 8½. 0-486-25346-5

EINSTEIN'S LEGACY, Julian Schwinger. A Nobel Laureate relates fascinating story of Einstein and development of relativity theory in well-illustrated, nontechnical volume. Subjects include meaning of time, paradoxes of space travel, gravity and its effect on light, non-Euclidean geometry and curving of space-time, impact of radio astronomy and space-age discoveries, and more. 189 b/w illustrations. xiv+250pp. 8⅜ x 9¼. 0-486-41974-6

STATISTICAL PHYSICS, Gregory H. Wannier. Classic text combines thermodynamics, statistical mechanics and kinetic theory in one unified presentation of thermal physics. Problems with solutions. Bibliography. 532pp. 5⅜ x 8½. 0-486-65401-X

Paperbound unless otherwise indicated. Available at your book dealer, online at **www.doverpublications.com**, or by writing to Dept. GI, Dover Publications, Inc., 31 East 2nd Street, Mineola, NY 11501. For current price information or for free catalogues (please indicate field of interest), write to Dover Publications or log on to **www.doverpublications.com** and see every Dover book in print. Dover publishes more than 500 books each year on science, elementary and advanced mathematics, biology, music, art, literary history, social sciences, and other areas.

CATALOG OF DOVER BOOKS

TENSOR CALCULUS, J.L. Synge and A. Schild. Widely used introductory text covers spaces and tensors, basic operations in Riemannian space, non-Riemannian spaces, etc. 324pp. 5⅜ x 8¼. 0-486-63612-7

ORDINARY DIFFERENTIAL EQUATIONS, Morris Tenenbaum and Harry Pollard. Exhaustive survey of ordinary differential equations for undergraduates in mathematics, engineering, science. Thorough analysis of theorems. Diagrams. Bibliography. Index. 818pp. 5⅜ x 8½. 0-486-64940-7

INTEGRAL EQUATIONS, F. G. Tricomi. Authoritative, well-written treatment of extremely useful mathematical tool with wide applications. Volterra Equations, Fredholm Equations, much more. Advanced undergraduate to graduate level. Exercises. Bibliography. 238pp. 5⅜ x 8½. 0-486-64828-1

FOURIER SERIES, Georgi P. Tolstov. Translated by Richard A. Silverman. A valuable addition to the literature on the subject, moving clearly from subject to subject and theorem to theorem. 107 problems, answers. 336pp. 5⅜ x 8½. 0-486-63317-9

INTRODUCTION TO MATHEMATICAL THINKING, Friedrich Waismann. Examinations of arithmetic, geometry, and theory of integers; rational and natural numbers; complete induction; limit and point of accumulation; remarkable curves; complex and hypercomplex numbers, more. 1959 ed. 27 figures. xii+260pp. 5⅜ x 8½.
0-486-63317-9

POPULAR LECTURES ON MATHEMATICAL LOGIC, Hao Wang. Noted logician's lucid treatment of historical developments, set theory, model theory, recursion theory and constructivism, proof theory, more. 3 appendixes. Bibliography. 1981 edition. ix + 283pp. 5⅜ x 8½. 0-486-67632-3

CALCULUS OF VARIATIONS, Robert Weinstock. Basic introduction covering isoperimetric problems, theory of elasticity, quantum mechanics, electrostatics, etc. Exercises throughout. 326pp. 5⅜ x 8½. 0-486-63069-2

THE CONTINUUM: A CRITICAL EXAMINATION OF THE FOUNDATION OF ANALYSIS, Hermann Weyl. Classic of 20th-century foundational research deals with the conceptual problem posed by the continuum. 156pp. 5⅜ x 8½.
0-486-67982-9

CHALLENGING MATHEMATICAL PROBLEMS WITH ELEMENTARY SOLUTIONS, A. M. Yaglom and I. M. Yaglom. Over 170 challenging problems on probability theory, combinatorial analysis, points and lines, topology, convex polygons, many other topics. Solutions. Total of 445pp. 5⅜ x 8½. Two-vol. set.
Vol. I: 0-486-65536-9   Vol. II: 0-486-65537-7

Paperbound unless otherwise indicated. Available at your book dealer, online at **www.doverpublications.com**, or by writing to Dept. GI, Dover Publications, Inc., 31 East 2nd Street, Mineola, NY 11501. For current price information or for free catalogues (please indicate field of interest), write to Dover Publications or log on to **www.doverpublications.com** and see every Dover book in print. Dover publishes more than 500 books each year on science, elementary and advanced mathematics, biology, music, art, literary history, social sciences, and other areas.